D1201898

ИЗДАТЕЛЬСТВО
МИР

V. V. Beloussov

Geotectonics

With 134 Figures

1980
Mir Publishers Moscow
Springer-Verlag Berlin Heidelberg New York

V.V. Beloussov
Corresponding Member of the Academy of Sciences, USSR

В. В. Белоусов
ГЕОТЕКТОНИКА

Издательство Московского Государственного Университета

Translated from the 1976 revised Russian edition
by H. Campbell Creighton, V. Agranat, V. Shiffer

First published in English in 1980

Except for socialist countries, sole distribution by
Springer-Verlag Berlin Heidelberg New York

ISBN 3-540-09173-4 Springer-Verlag Berlin Heidelberg New York
ISBN 0-387-09173-4 Springer-Verlag New York Heidelberg Berlin

На английском языке

Printed in the Union of Soviet Socialist Republics

Preface

Geotectonics has a special place among the geological disciplines. In addition to ideas based on firmly established facts that constitute lasting scientific values, geotectonics, as a generalizing branch of geology, embraces broad constructions that link the planet's deep interior with its surface and are largely of a hypothetical character. The interpretation of the most general matters of the structure and evolution of the globe varies not only from one generation of geologists to another, but even within one generation. The interpretation depends not only, and not so much, on the state of geological knowledge, as on the progress of the related sciences of geophysics and geochemistry. In trying to discover the deep-lying causes of tectonic processes, geotectonics has to unite the results of all the Earth sciences, converting itself to some extent from a purely geological science into a general physical geographic or geonomic science.

The fluidity of the general ideas and the need for joint consideration of the geological, geophysical, and geochemical data to substantiate these ideas are the main difficulties facing the author of a textbook on geotectonics. There is undoubtedly, however, a need for a manual of this kind, particularly now when the literature on the various problems of geotectonics has grown so great and so varied in content that it is very difficult for the experienced researcher, let alone the student, to find his way.

The author has endeavoured to follow the syllabus of the Soviet university course on geotectonics as closely as possible, and has tried to concentrate attention mainly on the processes in the forming of crustal structure that are suitable for direct study. It would have been odd, however, if he had limited the exposition simply to what can be observed. The achievements of the modern Earth sciences are such that one can go quite far towards generalizing the factual data without risk of becoming hopelessly lost in a maze of hypothetical conjectures. Such generalizations, containing the main patterns of the tectonic development of the Earth's crust, constitute an essential part of the book, but it also includes hypothetical ideas, since the latter, though of passing value, usually serve as a guide for lines of further research. The author is aware that this part of the book will call for revision in the next few years.

Only a few of the matters usually treated in courses on structural geology are repeated in this manual. It is presumed that the student is acquainted with structural geology and the principles of tectonophysics to at least the standard of the author's *Structural Geology* (Mir Publishers, Moscow, 1978, in French).

The list of main sources on various matters of geotectonics and related sciences, and of the cited literature, in the end of the book, will help the student to broaden his knowledge considerably, and will also be found useful to post-graduate students in their research.

Since the author is well aware that he has by no means overcome all the difficulties facing the compiler of a textbook on geotectonics, he suggests that his book be treated more as a study aid than as a course textbook.

Contents

Preface V

Introduction 1

Chapter 1. The Content of Geotectonics 1

Chapter 2. Types of Tectonic Process 5

PART I

Tectonics of Continents

SECTION A **General Tectonic Movements of the Earth's Crust** . . . 13

Chapter 3. Present-day Oscillatory Movements 13
Categories of Oscillatory Movements 13
Present-day Oscillatory Movements 14

Chapter 4. Recent Oscillatory Movements 20

Chapter 5. Ancient Oscillatory Movements of the Crust 28
The Tectonic and Geomorphological Expression of Oscillatory Movements 28

Chapter 6. Oscillatory Movements of the Earth's Crust and the Formation of Thicknesses and Facies of Sedimentary Series 43
The Formation of Thicknesses of Sedimentary Series 43
The Formation of Facies of Deposition 46

Chapter 7. Methods of Studying Ancient Oscillatory Movements 54
The Method of Thicknesses 54
The Method of Facies 57
The Volume Method 60

Chapter 8. General Properties of Oscillatory Movements of the Crust 62
The Theoretical and Practical Importance of Studying Ancient Oscillatory Movements 66

Chapter 9. General Disjunctive Tectonic Movements of the Crust 69
Deep Faults 69
Deep (All-crustal) Slashes 70
Deep (All-crustal) Transcurrent Faults 73
All-crustal Overthrusts 78
Deep (All-crustal) Normal Faults 80
Deep (All-crustal) Separations 81
Present-day Manifestations of Deep Faults 82

SECTION B **Tectonic Movements Within the Crust** 84

Chapter 10. Block Folding 84

Chapter 11. Injection Folding 90

Chapter 12. The Folding of General Crumpling 101
General Characteristics of Crumpling Folding . . 101
Methods of Reconstructing the History of Crumpling Folding 101
The Link Between Crumpling Folding and Oscillatory Movements 103
The Origin of Crumpling Folding 103

Chapter 13. Deep Folding and General Considerations About Folding Movements 115
Deep Folding 115
General Considerations About Folding Movements 119

Chapter 14. Intracrustal Faults 124
Classification of Fractures. Faults of Different Orders 124
Dynamic Conditions of Faulting 128
General Factors Affecting the Position of Faults . . 132
General Fracturing of Rocks (Joint Systems) . . . 135

SECTION C **Patterns of the Evolution of Continents** 136

Chapter 15. Endogenous Continental Regimes. General Features and Geosynclinal Regimes 136
A General Characteristic of Endogenous Continental Regimes 136
Classification of Endogenous Continental Regimes 137
Geosynclinal Regimes 138

Chapter 16. Platform and Orogenic Regimes 146
Platform Regimes 146
Orogenic Regimes 150
Some Terminological Problems 155

Chapter 17. Rift, Magmatic Activization, and Continental Margin Regimes 157

The Rift Regime 157
Regimes of Magmatic Activization of Platforms . . 163
Regimes of Continental Margins 165

Chapter 18. Principal Patterns of the Endogenous Development of Continents 179
General Evolution of Endogenous Regimes . . . 179
Endogenous Processes in the Stable Geosynclinal-Platform Stage 186
The Rhythm of Endogenous Processes 192

Chapter 19. Principles of the Zoning of Continents by Endogenous Regimes 201
Tectonic Zoning 202
Zoning by Types of Endogenous Regime 204

PART II

Tectonics of Oceans

Chapter 20. Relief, Sediments, and Rocks of the Ocean Floor . . 209
The Relief of the Ocean Floor 210
Sediments and Rocks of the Ocean Floor 218

Chapter 21. Tectonics of the Ocean Floor 225
Structures of the Ocean Floor 225
General Palaeogeographic Data 230
Some Conclusions About the Geological History of Oceans 235

PART III

The Earth's Internal Structure, Composition, and Deep Processes

Chapter 22. General Data on the Earth's Shape, Rotation, and Structure 241
The Size and Shape of the Earth 241
The Earth's Rotation 241
The Earth's Inner Structure from the Data of General Seismology. The Earth's Seismicity 243
Gravimetric Data 246
The Earth's Heat Regime 247
The Earth's Magnetic Field 250
The Composition of the Earth 254
The Age of the Earth 258

Chapter 23. The Crust and Upper Mantle (Tectonosphere) 259
The Structure and Composition of the Crust . . . 259
The Structure and Composition of the Upper Mantle 264

The Heat Regime of the Tectonosphere 272
Certain Conclusions About the Tectonosphere's Structure and Evolution 275

Chapter 24. The Main Stages in the Development of Geotectonics 276
Geotectonics Before the Eighteenth Century . . . 276
Geotectonics in the Eighteenth Century 277
Geotectonics in the Nineteenth Century 279
Geotectonics in the First Half of the Twentieth Century 282

Chapter 25. Recent Views on Geotectonics 290
"The New Global Tectonics" 291
Critique of the "New Global Tectonics" 297

Chapter 26. The Search for a Synthesis. 304
A General Scheme of Deep Processes 304
Conditions for the Development of Continental Endogenous Regimes 306
The Interrelation of Continental Endogenous Regimes 314
Conditions for Differentiation of the Tectonosphere into Continental and Oceanic 317
Differentiation and the General Evolution of the Earth 319

Literature. 326
Index. 328

Introduction

CHAPTER 1

The Content of Geotectonics

The word *geotectonics* is derived from two Greek words *gê*—earth, and *tectonicon*—to build. It consequently means literally the science of the Earth's structure, but such a definition is too wide since it embraces, in essence, almost, the whole of geology. We need to give more precision to just what is the true content of geotectonics.

Being a geological discipline, geotectonics is concerned with the structure, not of the whole Earth, but mainly of its crust. In truth, however, it is necessary, in order to understand the causes determining the observed structure of the crust and the patterns of the evolution of its structure, to investigate the globe's deeper lying envelopes as well, but we are then already outside the scope of geotectonics proper and are relying on the data of geophysics and geochemistry.

By the structure of the Earth's crust geotectonics means the distribution in it of bodies formed from various rocks. This distribution can be pictured as an aggregate of various geometrical figures laid down in the crust. Some rocks form stratified bundles, which may be horizontal, inclined, or bent into folds, while others are encountered in massifs of irregular form, and so on. The geometrical bodies formed by various rocks are known as the *forms of occurrence of rocks*, as *structural forms*, or simply as *structures*. Geotectonics studies structural forms, their history, and the processes leading to their formation.

It is not difficult to establish that structural forms are always formed and subsequently altered as the result of movements of some sort in the material of the Earth's crust, movements that may affect its whole thickness when it upwarps or downwarps as a whole, or may be manifested in deformations of certain volumes of rock within the crust.

Movements in the material of the Earth's crust leading to the formation and alteration of structural forms are known *as tectonic movements*[1]). Geo-

---

[1]) Apart from the term "tectonic movements" the terms "tectonic deformations" and "dislocations" are used as synonyms; they also refer, however, to the structural results of tectonic movements. These two terms, moreover, are used primarily to signify tectonic movements that occur in a limited volume of rocks and do not affect the whole thickness of the Earth's crnst.

tectonics thus has the task of studying not only structural forms but also the tectonic movements creating them. *This allows us to define geotectonics as the discipline dealing with mechanical processes in the Earth's crust and their results.* Hence geotectonics does not study the process of the formation of the material of rocks itself and makes use in such cases of the findings of other branches of geology namely, petrology and lithology. It is, however, concerned not only with the form of occurrence of a rock as a whole but also with the features of its internal structure and texture, which are also created as the result of tectonic movements. Such a feature can be, for instance, the oriented arrangement of rock grains imparted to them during deformation.

To make the content of geotectonics more precise we must exclude structural forms of an obvious surface origin from consideration, forms, for instance, like landslide structures, or dislocations brought about through the movement of glaciers over the Earth's surface, etc. Geotectonics is concerned with structures of deep-seated origin. Tectonic movements belong to the category of endogenous processes.

The dividing line between endogenous and exogenous processes, however, it must be said, is not always easy to draw. Many dislocations, both tectonic and exogenous, for example, are produced by the same cause, namely, the force of gravity. A criterion for distinguishing exogenous, i.e., non-tectonic processes, can be that they involve only the very surface of the crust. Tectonic movements extend to considerable depths within the crust.

All these considerations allow us to define geotectonics most exactly as *a geological discipline studying movements of the Earth's crust to considerable depths, and also the structures formed by such movements.*

This general task of geotectonics can be divided into a number of problems that follow one another in logical sequence.

The *first task* is to study the morphology of structural forms and classify them by morphological features. Folds, for example, can be classed as linear or dome-shaped (quaquaversal), upright or inclined, etc., and fractures as faults, overthrusts, shear faults, etc.

The *second task* is to study the mechanism of the development of structural forms. As already indicated, structural forms take shape and develop during movements in the Earth's crust. By comparing the present occurrence of rocks with their earlier position, e.g., a bed bent into a fold with a horizontally deposited one, we can get an idea of the direction and extent of the movements that have taken place. The aggregate of these movements of individual sectors and rock grains is also the mechanism of the development of a given structural form. In other words, the mechanism is the *kinematics* of the shaping of structural forms. Its study enables the morphological classification of structural forms to be supplemented by a kinematic one.

These first two problems form the division of geotectonics known as *structural geology*.

The *third problem* consists in studying the spatial distribution and history of structural forms, which is done by the methods and techniques of regional geological investigations. In this way structural complexes characterized by certain general structural features can be distinguished, for instance, platforms[2]), folded zones, and transitional zones with structural features

[2]) The Russian term "platform", which will be used throughout this book, is sometimes taken as synonymous with "kraton", the term "kraton" commonly used in Western textbooks, but is not so. Ancient and young platforms are distinguished, for example, and only the former are more or less synonymous with "kraton". (*Translator's note*).

intermediate between platforms and folded zones, etc. The age of structural complexes is also established. A result of solving the problem may be the geotectonic zoning of the area studied, a task that falls to *regional geotectonics*.

The *fourth problem* is to elucidate the patterns of development of tectonic movements by which zones characterized by a certain sequence and interrelation of various processes are distinguished. For instance, geosynclinal zones with a typical, definite cycle of tectonic processes may be distinguished. By solving this problem, zones of different *geotectonic regimes* can be established and the tectonic part of the crust divided historically into *geotectonic cycles or stages*.

Finally, the *fifth problem* is a generalizing one, consisting in developing a theory of geotectonic processes and in elucidating their causes and motive forces. The geotectonist, as we have already said, inevitably has to overstep the bounds of geotectonics proper when setting himself this last task, and to draw on the data of other branches of geology, above all of petrology, that sheds light on the formation conditions of igneous and metamorphic rocks. He must also go deeply into the field of the related Earth sciences of geophysics and geochemistry, since he can only get an idea of the situation and processes in the deep interior of the Earth with their help; and it is there, in the interior beneath the crust, that the causes of geotectonic processes are concentrated.

The last two problems belong to *general geotectonics*. The fifth one, we must note, still remains unsolved, and there is as yet no fully substantiated geotectonic theory that causally explains the evolution of crustal structures. Our ideas in the field amount at present to a number of assumptions and *geotectonic hypotheses*. Since they are based on far from complete knowledge of deep-seated conditions and processes, they always contain elements that are more or less controversial (often very contentious). As our knowledge of the Earth deepens, hypotheses fail the test in the light of new data, and others take their place. Their positive significance is that they generalize the facts accumulated and suggest the most promising directions for further investigations.

In addition to the term *geotectonics*, the shorter one *tectonics* is also used. The two are often synonymous, but *tectonics* is employed also as a synonym for *structure*; we speak of the tectonics (i.e., structure) of some area or other, but not of its geotectonics.

Geotectonics serves both practical purposes and to solve theoretical problems. Understanding of the mechanism of the formation and patterns of distribution of structural forms is necessary for any regional geological investigation: viz. geological mapping; the search for, and exploration of, mineral deposits; the solving of problems of engineering geology, hydrogeology, and other branches of geology having to do with any of the peculiarities of the structure of the Earth's crust. The geologist, for instance, does not, as a rule, see the whole section of the surface sectors of the crust. The section is interrupted by frequent gaps in exposures, gaps that have to be filled, for which he must have correct ideas of the patterns of the combinations of structural forms. These patterns, in turn, depend on the general tectonic behaviour of a given locality, and the geologist has to find signs enabling him to guess what regimes he is dealing with in a given case. It is only possible to evaluate a mineral deposit correctly when its structure is known, which can be found with the help of the methods of geotectonics.

Tectonic movements of the recent geological past that, judging from a number of indications, continue to this day, have to be studied when long-term harbour and hydraulic structures are being planned, and also for seismic zoning, i.e., to establish the strength of the earthquakes that can occur in a given seismic area.

Geotectonics is of very great theoretical importance. By bringing out the patterns in the development of tectonic movements it throws light on a very important aspect of the evolution of the Earth's crust as a whole. It also offers considerable possibilities for theoretical generalization at the level of geology as a whole, and even at the level of all the Earth sciences. The interest shown in geotectonic generalizations by representatives of the most varied geological disciplines and the related Earth sciences, geophysics and geochemistry, is therefore not surprising. Since the deep-seated processes developing within the Earth are linked with its whole history, and even with the conditions of its formation, geotectonics borders on the cosmogony of the solar system as regards its most general problems. The latest investigations of the Moon have shown that geotectonics contributes to a correct understanding of many of the structures of the lunar surface, but also that it itself can, on the other hand, draw information from observations on the Moon of use for understanding the structures of the Earth's crust.

CHAPTER 2

Types of Tectonic Process

The fundamentals of structural geology are not dealt with in this book The reader is assumed to be familiar with them to the extent of the ordinary textbooks. See, for example: V.V. Beloussov, *Structural Geology*. Translated by A. Gurevich, Mir Publishers, Moscow, 1968. Here we shall only touch on its problems in passing, our book being chiefly devoted to general geotectonics, which means that our attention will be mainly focused on the movements of the Earth's crust that shape tectonic forms rather than on the forms themselves. It must be remembered, however, that ideas about tectonic movements themselves (or diastrophism) come almost exclusively from study of the structural forms that are the end results of these movements. Only by studying the folding of crustal strata can we form an opinion of "folding" tectonic movements. Only by considering the distribution of the thicknesses of the Earth's crust and the lithological composition of rocks and the broad gentle upwarps and downwarps of strata do we obtain data on vertical (so-called oscillatory) movements of the crust. The history of the uplifts and subsidences of the crust that occurred in past geological periods is also reconstructed from the changes in the outlines of land and sea imprinted in the distribution of regions (a) of the accumulation of marine and continental sediments, and (b) where the sediments of a given age have been eroded away. Only in a few limited cases does it prove possible to study tectonic movements from direct recording of them (e.g. contemporary uplifts and subsidences of the crust directly observed by means of geodetic methods), rather than from their end results.

Geologists have been trying for a long time to work out the best classification of tectonic movements.

From observations of crustal structures and signs of past changes in the outlines of land and sea, M.V. Lomonosov had already, in the eighteenth century, distinguished two categories of tectonic process: on the one hand, rapid catastrophic breaking up of "stony wildernesses" that had been "displaced from their former position", and formed "fissures, slopes, rapids, precipices and chasms of various size and bizarre shapes"; and, on the other hand, "imperceptible and long-period subsidences and elevations" in the Earth's surface.

A similar division of tectonic movements took firm root in the middle of the nineteenth and early twentieth century, and is held to this day. At the end of the last century G.K. Gilbert suggested distinguishing *epeirogenic* and *orogenic* tectonic movements. In this century these terms were given more precise definition by G. Stille, and his understanding of them has remained unaltered for decades. According to Stille, epeirogenic movements (*epeirogenesis*) found expression in the slow uplift and subsidence of separate segments of the Earth's crust, while orogenic movements (*orogenesis*) were represented by the crumpling of strata into folds and their displacement along faults. In this conception, it should be noted, the original sense of the terms proved to be distorted: translated from the Greek "epeirogenesis" means "continent-building", and "orogenic"—"mountains building". But vertical movements of the crust produce seas (through subsidence) as well as land, while folding and movements along faults by no means always create mountains. Mountains usually develop through slow vertical elevations of the crust, i.e., as a result of these same epeirogenic movements.

A little before Gilbert, and concurrently with him, several Russian scientists (N. A. Golovkinsky, A. D. Ozersky, and A. P. Karpinsky) termed the movements called epeirogenic in the West "oscillatory" or "wave-form oscillations of the Earth's surface".

An extremely simple classification that is fairly common (and in use to this day) divides tectonic movements into *vertical* (*radial*) and *horizontal* (*tangential*). It classes not only the epeirogenic or oscillatory movements of the preceding classifications as vertical movements but also vertical displacements along the discontinuities of crustal blocks, or the formation of dome-shaped (quaquaversal) folds. Linear folds, resulting from horizontal compression of the crust, overthrusts, and shear faults are classed as horizontal movements. This classification has the drawback that both vertical and horizontal movements operate in many structure-forming processes, for instance, when folding produces both horizontal convergence of crustal segments and vertical displacement of strata.

M. M. Tetyaev called tectonic movements "forms of geotectogenesis" distinguishing oscillatory, folding, and magmatic forms, and also a form of macro-oscillations (meaning by this high amplitude vertical movements of crustal blocks along faults).

The present writer used to follow Tetyaev's classification in general, but distinguished only oscillatory, folding, and disjunctive tectonic movements, so excluding the "magmatic form of geotectogenesis" and "macro-oscillations", and introducing a separate group of tectonic ruptures or disjunctive dislocations.

This last classification proved convenient for everyday use but (like the preceding ones, incidentally) had the character of a simple listing of observed processes or of their structural results, and in no way reflected either their interrelations or their origin. Their origin, it is true, cannot be reflected in the classification until we know the authentic causes of tectonic movements, but the interrelations and mutual subordination of processes ought to be included to some extent.

A step in this direction was made by V. E. Khain, who tried to relate various tectonic movements to different depths. He distinguished the following movements: exotectonic, occurring at the very surface of the Earth and connected with exogenous factors (landslides, glacial dislocations, etc.); surface, whose causes lay in the sedimentary layer of the crust (e.g., diapiric salt domes); crustal, which develop in connection with processes in the consolidated part of the crust (e.g., linear folding); deep, linked with processes in the upper mantle; and superdeep, caused by processes in the lower mantle and core of the Earth.

In our view, exotectonic and superdeep movements should be dropped from this classification. The first represent an exogenous process and as such are not part of geotectonics. The second belong more to geophysics than to geotectonics. There remain surface, crustal, and deep-seated movements. Although this division has a certain significance in principle, the terms cannot, however, be regarded as apt, since all these movements take place in the crust.

In posing the problem of reflecting the co-ordination of the various categories of tectonic movement in a classification, one has to start from the fact that their most important characteristic is not so much the depth of their "location" as their involvement of a certain volume of crust. Tectonic movements can be divided very roughly into (a) *general* or *all-over crustal* movements and (b) *intracrustal* ones. The first involve the whole thickness of the crust, i.e., the whole crust moves in these cases within some sector or another. Intracrustal movements manifest themselves in a limited volume of the crust and find expression in the deformation of this volume without affecting the displacement of the whole crust. As an example of all-over crustal processes we may take slow uplifts and subsidences of the crust involving extensive areas. These "oscillatory" movements obviously involve the whole crust. Examples of intracrustal tectonic processes are diapiric salt domes and strongly compressed linear folding. Both these processes are always localized in a certain volume of rocks and do not involve the crust throughout its whole thickness. Several of these classification schemes of tectonic movements are given in Table 1.

We shall begin our exposition with *all-over crustal tectonic movements.* Among them we distinguish, first of all, a most important class of *oscillatory movements of the Earth's crust*, i.e., slow uplifts and subsidences.

It would be natural to counterpose horizontal crustal movements to this class, but the point cannot be considered clear at the present time. If one were simply to take everything now being published in the geological literature at its face value, one would have to include in this category such phenomena as continental drift, ocean floor spreading, transform faults or large-scale strike-slip displacements, large-scale crustal overthrusts, and large-scale faults and rifts. We shall be trying later, however, to show that continental drift and ocean floor spreading do not, judging from all the geological evidence, actually happen in nature. Large-scale crustal overthrusts possibly do now exist around the Pacific ("Benioff zones"), though the nature of these discontinuities is not clear. Over-all transform or transcurrent faulting of the crust does apparently exist but its scale and significance in crustal development is greatly exaggerated in the current literature. All-over crustal faults and rifts occupy an important place in the development of continental crust.

All-over transcurrent faults, overthrusts, vertical faults, and rifts, it should be noted, come under one heading in that they are associated with fractures that cut through the whole thickness of the Earth's crust and apparently penetrate into the upper layers of the mantle. These are known as *deep faults* and play an important role not only in connection with large-scale horizontal crustal movements but also with oscillatory ones, and in general can be thought largely to determine the structural evolution of the crust.

Because of the many vague points regarding the types of all-over crustal movement we shall adopt a simplified preliminary classification here in which these movements are divided into oscillatory and fault-generating, the latter leading to the origin of various kinds of deep fault. In the text that follows fault generating movements will be called, for brevity's sake, simply faulting.

Table 1. Selected classifications of types of tectonic movement

Gilbert, Stille, and others (end of the nineteenth century and beginning of the twentieth)	Tetyaev (1941) (forms of tectogenesis)	Beloussov (1954, 1962)
Epeirogenic (radial)	Oscillatory	Oscillatory (general and wave)
Orogenic (tangential)	Folding	Folding
	Macro-oscillatory	Disjunctive
	Magmatic	

Khain (1973)*			
Essentially vertical		Essentially horizontal	
Surface			
Folding (contraction folds)	Block movements (along vertical and reverse faults)	Folding (general contraction, gravity collapse, along overthrusts and shear faults)	Transcurrent faults, regional overthrusts, mass overthrusts (gravitational)
Crustal			
Folding (basement and investment folds)	Block movements (along transcurrent and reverse faults)	Folding	Transcurrent faults, regional overthrusts, mass overthrusts (non-gravitational)
Deep			
Undulatory	Block movements (along deep transcurrent and reverse faults)	Undulatory (megafolds)	Deep faults and overthrusts

Beloussov (1975)	
Intracrustal	
Folding (block, injection, general crumpling, and deep)	Disjunctive (transcurrent faults, reverse faults, slashes, overthrusts, rifts, mass overthrusts)
All crustal (deep)	
Oscillatory (all-over and block-wave)	Disjunctive (deep faults, slashes, shear faults, overthrusts, rifts)

* Exotectonic and superdeep movements have been omitted from the table.

All-over crustal movements lead to the formation of large-scale crustal structures since they take place in various directions or at varying rate in neighbouring sectors of the crust. When these sectors are linked together by smooth flexures of the crust, large-scale all-over crustal flexures or large-scale buckling and troughs develop during oscillatory movements, and with large-scale transcurrent faults large-scale horizontal flexures develop. When there are primary or secondary deep faults on the margins of sectors with different conditions, then block structures arise (large horsts and grabens with oscillatory movements and large-scale all-over transcurrent faults with horizontal movements).

We shall divide *intracrustal tectonic movements* below into cohesive and disjunctive ones, distinguishing more detailed separate types among them.

All the tectonic movements and the structures created by them will be considered first for continents. A special section will be devoted to the tectonics of oceans.

PART I

Tectonics of Continents

SECTION A

General Tectonic Movements of the Earth's Crust

CHAPTER 3

Present-Day Oscillatory Movements

Categories of Oscillatory Movements

Oscillatory movements of the Earth's crust, uplifts and subsidences, have been going on throughout the whole of the geological history of continents known to us and are still occurring. They have occurred everywhere within the continents and continue to go on. At the same time other forms of movement, both those affecting the whole crust and those within it, are always of a local nature and each of their manifestations is restricted in time. We are therefore justified in treating oscillatory movements of the crust as the most general form of tectonic movement.

Various methods are employed to study oscillatory movements: viz. geodetic, historical, geomorphological, and geological.

The geodetic measurements enable us to study the oscillatory movements that are taking place now. The historical method is based on using relicts of the past or evidence that indicates what happened sometime earlier and during a historical time, when the height of some sector of the land above sea level differed from the present one. The geomorphological method consists in using the shape of the present relief of the Earth's surface to reconstruct the history of the vertical movements of the crust that were involved in its formation. Since today's relief was formed predominantly in the Neogene and Quaternary periods, these methods are suitable for studying oscillatory movements belonging to these two very young geological periods.

Geological methods include study of sedimentary series, the composition and thickness of which reflect the subsidence and uplifting of the crust during their formation. They are used to study the oscillatory movements of every geological period and of the sedimentary strata remaining after them.

Although there is no doubt that oscillatory movements that have taken place in every geological epoch, from the present to those most remote from us, represent a single process, it is nevertheless convenient to divide them into three categories according to the time of their manifestation, namely *present-day*, *recent*, and *ancient*. This division is justified by the difference in the methods used to study movements occurring now or those that took place in historical times, belonging to the neotectonic period of the formation of today's relief and, finally, those that were manifested during earlier geological periods. The methods of studying recent and ancient oscillatory movements, however, it should be noted, differ only in part, and otherwise coincide. The difference is the geomorphological method that can only be used for recent movements. It is not applicable to ancient movements, since the relief from older periods has not, with rare exceptions, been preserved. The only applicable geological methods for ancient movements can also be used for recent ones.

Present-Day Oscillatory Movements

In 1620 the harbour of Torneo was built in deep water in the Gulf of Bothnia. In a hundred years it had become so shallow that it was no longer usable. Shoaling of the sea around the whole of Scandinavia continued through all this time, as the coastal population can remember it. Shallows were gradually turned into islands, and islands into peninsulas, bays and straits grew shallow, and the coastline receded further and further. From the historical evidence it has also been possible to determine the rate of the fall in sea level approximately in relation to the land. It proved to be close to 1 cm per year.

At first people tried to explain this phenomenon by a general fall of the level of the World ocean, but this explanation had to be dropped when a reverse phenomenon, i.e. a rise of sea level relative to the land had occurred at the same time in other coastal areas. The inhabitants of the coastal areas of Holland, for example, beginning from the tenth to the eleventh centuries, had had to build and continuously extend dykes to protect their land against the encroaching sea. These dykes have reached a height of 15 m in our day. At a depth of 18 m a layer of peat has been discovered in the coastal strip, the age of which has been determined by radiocarbon dating at 7200 years. Since the peat was not originally formed below sea level, the depth of its occurrence and its age point to subsidence of the coastal part of the land at an average rate of 0.25 cm per year. During recent centuries, however, rate of subsidence has been quicker, 0.5 to 0.7 cm per year.

Comparison of these two cases—the uplifting of the land relative to sea level in Scandinavia and its subsidence in Holland, indicate that the cause of these changes lies not in fluctuations in the level of the ocean, but in vertical movements of individual sectors of the Earth's solid crust, some of which (like Scandinavia) rise, while others subside.

Historical evidence of a rise or subsidence of land relative to sea level has also been discovered in many other places on seacoasts. In Sukhumi Bay, for instance, the submerged buildings of the old Greek colony Dioskurias have been discovered. Classical Greek vases, found in the sea at a depth of tens of metres off the Crimean Peninsula, also indicate subsidence of the coastline. In Northern Italy old pavements in the city of Ravenna, of the Adriatic coast, have been submerged to 45 cm in the first half of our century. Near Marseilles the remains of an old Roman city have been found in the sea at a depth of 6 m. Ancient towns that have sunk beneath the sea are known along the Dalmatian coast of the Adriatic. The Belgian city of Bruges, on the contrary, once a busy medieval sea-port is now separated from the sea by a strip of land 15 km wide.

Fig. 1. Changes in sea level at Ratan and Pitea (Sweden) (after Tsuboi, 1939). The figures on the curves denote the years of observation

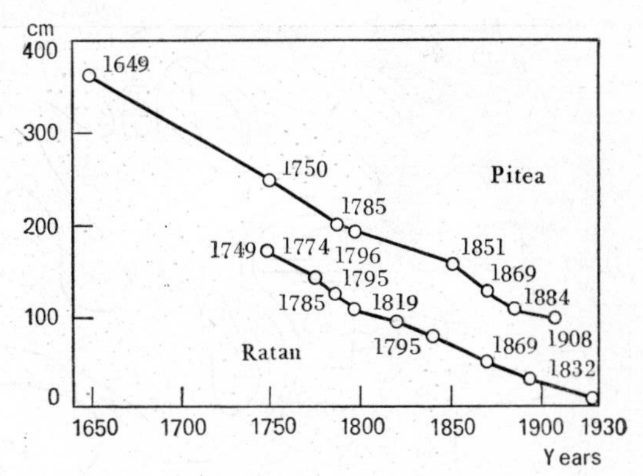

In some cases changes have been observed in the direction of land movements in historical times; sectors that had previously subsided have later risen, and vice versa. On the coast of Italy, not far from Naples, in the city of Pozzuoli, there is an old Roman building of the second century B.C. It is usually referred to as the Temple of Jupiter Serapis, but in fact represents the ruins of a market place. Only a few columns are left. Their pediments today lie 225 cm below sea level. However, traces of the stone-boring common piddock (*Pholas dactylus*) can be found on the columns up to a height of 6.3 m. There was consequently a time when the columns were submerged to this depth. According to historical evidence subsidence went on until the fifteenth century, i.e., for 17 centuries. Next, a rise began, which went on until the middle of the eighteenth century, i.e., for three centuries. Thereafter, subsidence resumed, which is now proceeding at a rate of 2 cm a year.

Archaeological excavations at Khersoness in the Crimea have shown that low-lying sectors of this classical Greek and later Byzantine colony were periodically drowned by the sea and drained again. Such flooding occurred in the sixth to the fifth centuries B.C.; from the first century B.C. to the first century A.D.; in the fourth and fifth centuries A.D., and in the tenth and eleventh centuries. In this case one cycle of the oscillatory movement lasted five or six centuries.

Present-day oscillatory movements have been measured precisely by means of geodetic instruments and methods, including water gauges (tide gauges), repeated levelling and tilt indicators.

Water gauges are mainly used on sea coasts. Hundreds have now been installed at various coastal points of all the continents. As early as 1649 a water gauge was set up on the coast of the Gulf of Bothnia in the town of Ratan. A hundred years later the same device was installed in Pitea. The results of records of these devices to the year 1930 are shown in Figure 1. The rate of the sinking of sea level relative to the land is quite constant and equals 1 cm a year.

The most universal method of measuring present-day oscillatory movements is that of repeated precision levelling. It can be employed in any area on land at any distance from the sea. Reliable results are obtained by repeating levelling traverses in ten or twenty years.

Such levelling has been carried out, for example, in many directions in western and south-western areas of the Russian plain. By comparing levellings made between 1913 and 1932 with those made between 1945 and 1950,

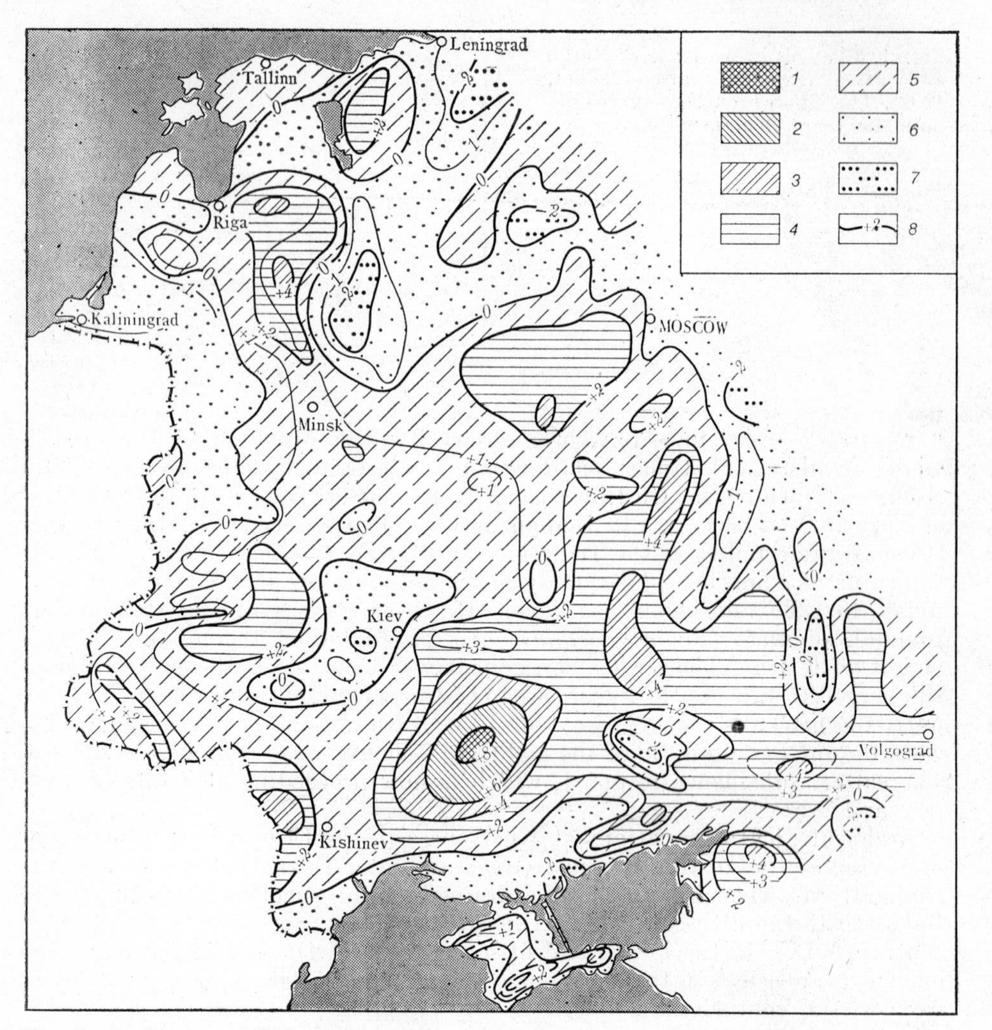

Fig. 2. Outline map of the rates of present-day vertical crustal movements on part of the East European platform (edited by Meshcheryakov, 1958):

1-5—uplifts of various intensity (mm/year) (*1*—from +8 to +10; *2*—from +6 to +8; *3*—from +4 to +6; *4*—from +2 to +4; *5*—from +0 to +2); *6-7*—subsidences of various intensity (mm/year) (*6*—from —2 to —0; *7*—from —4 to —2); *8*—isolines

Meshcheryakov and colleagues compiled a map of present-day movements in these areas of the European part of the USSR. It has since been improved. In 1973 a map of present-day vertical movements of the crust over the whole of Eastern Europe was published, a part of which is given in Figure 2.

As the isolines show, covering the West of the Russian plain, a modern north-south Esthonian-Carpathian uplift zone has been depicted in the west of the Russian plain. Another uplift zone, occupying a central position may be called the Central Russian zone. The Moscow basin is subsiding.

In the Great Caucasian Range geodetic measurements have discovered a present-day rise with a rate of 10 to 12 mm a year. A rise at approximately the same rate has been measured in the Carpathians. In the foothills of the Caucasus, in Poti, a subsidence is occurring at a rate of 6.2 mm a year.

Extensive levelling has been done in Japan, which has revealed a very complex mosaic of areas of contemporary uplift and subsidence, each one tens or a few hundreds of kilometres across. The rates usually vary at around 4 to 5 mm a year.

Measurements of present-day oscillatory movements of the Earth's crust have been conducted in many countries and are being continued at the present time under an international programme. The preparation of this programme and the organization of the work on an international scale owes much to the initiative of the Soviet scientist Yu. A. Meshcheryakov.

The material assembled shows that present-day oscillatory movements are taking place everywhere on the continents. Their rate usually varies within limits of a few millimetres per annum, but is as much as 1 cm a year in places or even higher. High rates of contemporary elevation (around 1 cm a year) are observed, in particular, in areas covered by the last glaciation (e.g. the Baltic and Canadian Shields). But, in general, a very little difference is observed in the rates of present-day movement between plains and mountains. The rates of uplift in the plain of the Donets Basin and mountainous Caucasus, for example, are the same.

An important standard of oscillatory movements in general and at the present time in particular, is their gradient, i.e., the change of the relative height of two points related to the distance between them and to a unit of time. To determine the maximum gradient it is necessary for the points between which it is measured to lie on a line perpendicular to the isolines of the rates of movement. The dimensionality of the gradient is expressed as follows:

$$\frac{[\text{distance}]}{[\text{time}]\times[\text{distance}]}=\frac{1}{[\text{time}]}$$

If we take a year as the unit of time, the dimensionality of the gradient will be $\frac{1}{[\text{year}]}$, or $[\text{year}]^{-1}$.

The gradients of present-day oscillatory movements usually lie between $2\cdot10^{-8}$ and $5\cdot10^{-8}$ year^{-1}, which corresponds to an increment of the difference in height between points located 100 km apart, i.e., 2 to 5 mm a year. These gradients increase at margins of local structures, that are either rising or subsiding intensively, and at faults.

It is interesting to compare the distribution of modern uplifts and subsidences of the Earth's crust with its structures that are the result of much longer previous vertical movements. In some cases the comparison reveals a coincidence of direction of present-day movements with that of the older ones, and in others a discrepancy. As a rule, mountain ranges formed as a result of elevation of the crust during the Neogene and Quaternary, are continuing to rise today, but the Apennines and Central Hungarian Hills, for instance, are now subsiding. Structural depressions are also subsiding now as a rule (e.g., the Moscow syneclise is sinking), but the Kuban depression is rising. For the East European platform Meshcheryakov found a coincidence of present movements with the character of older structures in 70% of cases; in the remaining 30% the present-day movements prove to be "superimposed", i.e., foreign to the older structure.

If present-day oscillatory movements were to continue to develop in the direction now observed and at the same rate for a long geological time, high mountains and deep depressions would be formed where they do not now exist.

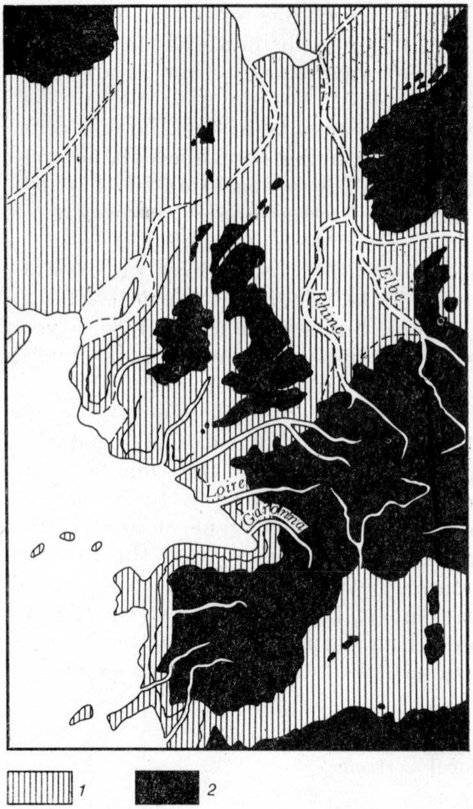

In the area of the East European platform, for instance, which is rising at a rate of 2 mm a year, an elevation of 2000 m would be formed in a million years. These rates are observed on the platform but there is no high relief.

This gives one to think that the rates of oscillatory movements observed in our time cannot continue long and that the direction of the movement alters quickly, e.g. that the movements are oscillatory in nature. We can represent oscillatory movements in general as a complex process consisting in oscillations of different orders and periods, superimposed upon one another. In studying present-day movements we record very small and very rapid oscillations with a period, probably, of several hundred years, while the formation of large-scale sagging and buckling of the crust, depressions and uplifts, is determined by much slower oscillations with a much longer period.

Fig. 3. Ancient river network of North-western Europe, now submerged by the sea (after Beloussov):
1—former land; 2—present-day land

The causes of oscillatory movements, including present-day ones, are not known, but certain processes specific for recent times affect them to some extent or another. We pointed out above, for example, that present-day uplifts are intensive in regions of Quaternary glaciation, where the unburdening of the crust operates after the melting of the ice cover with a lag due to the viscosity of the subcrustal material. The weight of the ice sheet pressed the crust into the substratum, displacing it. When the weight was removed, the substratum began to resume its place and the crust began to move upward. Geological history indicates, however, that the Baltic and Canadian Shields had already been rising for a long time, from the end of the Proterozoic, but the rise was very slow. We can, therefore, think that the "glacioisostatic factor" played only an additional, accelerating role, since the rise of the crust had been caused fundamentally by deep-seated processes.

Great difficulties are introduced into the study of present-day oscillatory movements by the oceans' own (*eustatic*) fluctuations of level. Their significance is clear from the fact that all geodetic measurements of heights are run from sea level, and if this level changes the errors inevitably creep into comparison of levellings made at various times. Eustatic fluctuations in the oceans' level may be due to various causes, but the main one, apparently, is a change in the amount of water in the ocean. Such a change happens, for example, with glaciation of the land, when a considerable amount of water becomes locked up in the form of ice on the land without reaching the ocean, or, on the contrary, with melting of the ice, when the previously locked-up water is liberated and flows into the ocean.

During the maximum of Quaternary glaciation, it has been established, the level of the oceans was approximately 100 mm lower than the present level and a considerable area of the marine shelves was exposed. Even now it is possible to find traces of former river beds, not buried under an accumulation of sediments in many places on the surface of the shelf beneath the sea (Fig. 3).

It is extremely difficult to establish and measure present-day eustatic fluctuations of the ocean level, because of the absence of datum marks. In connection with the continuing contraction of the area of glaciers it is presumed that the level of the ocean is rising at present at a rate of a millimetre a year.

CHAPTER 4

Recent Oscillatory Movements

By recent oscillatory movements are meant the vertical tectonic movements of the Earth's crust that are the cause of the formation of major features of the contemporary relief of continents—mountain ranges, depressions, and plains. The observable relief of the Earth's surface is geologically young. The beginning of the periods of present-day relief-formation dates to a rather different geological time in various places, but in most cases falls into the Late Miocene, and only rarely goes back to the Palaeogene. It has been established that on the site of such mountain ranges as the Caucasus, the Alps, the Tien Shan, and the Cordilleras, there was, at the end of the Palaeogene and beginning of the Miocene, if not quite a flat relief, at any rate, but slightly undulating one. Only later did the stage of "recent tectonic activation", also called the "neotectonic" stage begin, during which contemporary mountain ranges, including such grandiose ones as the Tien Shan and Pamirs, rose on the surface of the continents. The length of this stage, which as a rule embraced the Neogene and Quaternary, is not more than ten or fifteen million years.

To study the history of oscillatory movements during this stage both geomorphological and geological methods are used. The former are used mainly where uplifts predominate in the whole and elevated forms of relief take shape, while the latter are mainly used to study zones of downwarping and the accumulation of sediments.

Geomorphological methods consist in studying certain features of the relief.

A general, very rough idea of the orientation and intensity of recent oscillatory movements can be obtained simply from the height of major forms of relief: a high mountain range is always the result of a recent tectonic uplift, while the low-lying lands surrounding it belong to regions of recent subsidence or a lesser uplift. From the hypsometric map of the Soviet Union we see that its whole southern and eastern periphery is occupied by mountain ranges and massifs that enclose the country, as it were, in a huge amphitheatre. To the north lowlands and plains predominate. We would be right, if we interpreted this picture as a manifestation of a general recent tectonic uplifting of the southern and eastern margins of the country, as a result of which the surface as a whole was given a tilt towards the north.

Much greater possibilities of geomorphological methods are associated with the main property of the movements being considered, i.e., their oscillatory character. As we have already said, the process of the crust's oscillatory movements always breaks down into movements of different orders, superimposed on one another. The protracted uplifting of a mountain range, embracing the whole neotectonic stage, reflects the highest order of recent oscillatory movement. It determines the general orientation of the evolution of the relief. Against its background oscillations of the higher orders, with correspondingly shorter periods take place. The minor oscillations, being superimposed on the larger ones, are algebraically summated with them. When both minor and major oscillations tend in one direction, their rates built up and the resulting rate increases. On the other hand, when a minor oscillation is in the opposite direction to a major one, the rate of the first is subtracted from that of the second. It then depends on the ratio of the rates whether there is a slowing down of the basic movement or its stop as a result of the superimposition, or whether the movement will temporarily tend to be in the opposite direction. It is these slowing downs of oscillatory movements of the crust, their temporary stops and reversals that leave traces in elevated forms of land relief that can be used by the geologist to divide the process of recent oscillatory movements into stages and to determine the amplitudes and rates of the movements.

The point is that changes in the rate and direction of vertical crustal movements affect the rate of erosion and so the nature of erosion forms of relief.

Any flow of water, as we know, tends to a certain "profile equilibrium" which is the very gentle gradient at which water will still run off, but at which further cutting into the underlying basement ceases. While the slope of the channel is much steeper than the profile equilibrium, downcutting proceeds quickly and streams cut narrow gorges and canyons with steep walls. As the gradient of the channel comes closer to the profile of equilibrium, downcutting gets slower and lateral erosion plays an increasing role: the stream begins to meander and widens its valley, on the floor of which alluvium accumulates to an ever greater extent.

Variations in the rate and direction of the crust's vertical movements lead to the elevation of mountain ranges above the neighbouring depressions, and accordingly of land gradients, which now gaining speed and now slowing down, so that the already existing channels of streams become now steeper and now gentler. At the same time periods of rapid downcutting by streams are succeeded by periods of lateral erosion.

Such a change in erosional regimes leads to the formation of stepped terraces in valleys. Each terrace is the remains of a more or less broad valley floor formed by lateral erosion in a period of the slowing down or arrest of downcutting. When the latter resumes, a new channel is cut to a deeper level, leaving a terrace behind at a certain height above it.

In order to judge the tectonic history of a locality from terraces we have to know the age of the terraces (albeit relatively) and to be able to compare their present position with the initial profile of equilibrium on which they were cut.

In mountain ranges the relatively older terraces lie, as a rule, higher than the younger ones. This reflects the main trend of the process: namely, continuous elevation of the range, though interrupted by oscillations of a secondary order. Since the axial part of a range usually rises faster than its margins, terraces acquire in time an ever steeper gradient, the greater and older the terrace. They, therefore, usually fan out from the foothills of a range toward its crest, and the vertical distances between them increase. From the slope of terraces it is possible to judge the amplitude of the elevation of a range from the time of the formation of one terrace or another and how much the inclination of the slopes has been altered.

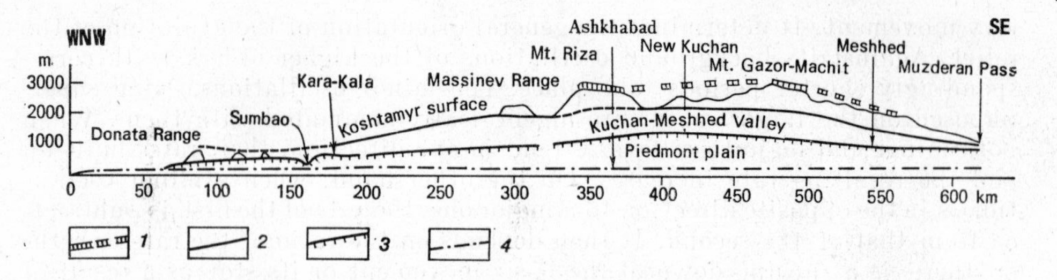

Fig. 4. Outline geomorphological profile along the strike of the Kopet Dagh Mountains (after Rezanov, 1959):
Denudation surfaces: *1*—Miocene; *2*—Pliocene; *3*—Early Quaternary; *4*—Late Quaternary piedmont plain

When the rise of a range is accompanied with the formation of steep flexures or faults on its slopes, these dislocations are imprinted in the occurrence of the terraces, and can be established by careful tracing of the terraces.

When elevation of a locality is succeeded by lengthy subsidence, the sequence of the terraces is broken and older ones prove to be buried beneath younger ones (so-called overdeepened valleys).

Stepped terraces are also formed on sea-coasts. There, too, they are due to the irregular development of oscillatory movements. Each marine terrace matches a time of marked retardation or halting in the uplift of the coast, when the tide has time to cut an abrasion platform in the coastal slope. If the uplift of the coast is temporarily succeeded by subsidence, littoral sediments accumulate on such a platform sunk beneath the sea. Resumed uplift will lead to the bench being raised above sea level and becoming the next terrace.

The method of studying recent oscillatory movements in regions where uplifts predominate is also based on observations of what are called denudation surfaces. These are similar to terraces in their formation conditions, but are planation surfaces that go far beyond the bounds of individual valleys and take in broad strips on the slopes and peaks of mountain ranges. A denudation surface is a peneplain or bordering plain formed where all streams, both permanent and temporary, have reached the profile of equilibrium and their valleys have almost merged. It is a gently sloping hilly surface that is formed during especially long halts in uplifts of the crust. Protracted arrests happen less often than the short-term ones that lead to the formation of terraces. The number of peneplanation periods is therefore always fewer than the number of terrace-forming ones. Nevertheless, several denudation surfaces are usually observed in mountainous regions, lying one above the other. The oldest surfaces are found high up at the crest of the range. Their formation preceded the beginning of the ridge's uplift. The younger surfaces are located lower down. The old surfaces are usually more or less broken up and their relicts are preserved on separate peaks only. The past existence of a very old and very high surface, now already completely broken up, can, however, often be guessed from the 'level of the peaks', which have been inscribed, as it were, in a gently convex arch (Fig. 4).

Surfaces of denudation, like terraces, are originally extremely gently dipping, almost horizontal. Therefore, when one now finds them dipping steeply or they are broken by faults, one can affirm that they underwent tectonic dislocations after their formation.

Now let us turn to zones of protracted recent downwarping of the crust. In recent troughs Neogene and Quaternary sediments accumulate and their history, as indicated, is studied by geological methods, based on consideration of the composition and thickness of the sediments. Geological methods are used much more extensively to study the history of ancient oscillatory movements, and they will be dealt with fully below in connection with them. Everything said about them is also applicable to the study of recent oscillatory movements in the areas of downwarping. Here we shall confine ourselves simply to a few brief observations.

Sediments accumulate gradually in regions of subsidence, layer upon layer, as downwarping develops. The thickness of the amassed sediments is usually close to the amplitude of the downwarping and can be used to measure the latter.

In the sections of areas of downwarping the older layers naturally lie at the bottom and the younger ones at the top. The character of detrital sediments thus reflects not only the physical-geographic conditions of the area of accumulation, but also events occurring in the area of uplift from which the detritus is carried. When uplifting in the adjacent region is intensified and the corrosive energy of streams increases, coarser clastic material is carried into the trough, but when uplifting slows down or stops altogether, the material carried into the depression is finer. The alternation of periods of intensification and weakening of erosion which is thus reflected on the slopes of uplifts by the formation of terraces is expressed in the section of troughs by an alternation of layers of more and less coarse detrital sediments. Since the older terraces on uplifts usually lie higher up than younger ones, while the older strata occur beneath the younger ones in troughs, there is thus a "crossing-over" on the boundary between zones of uplift and downwarping of the elements, simultaneously being formed and genetically connected, of the geological structure. The river terraces and surfaces of denudation not only descend lower and lower toward the foothills of a range, but also converge. On the line of transition to the zone of downwarping terraces and denudation surfaces of various age merge and continue in the trough zone in the form that accords with them by the time when the boundaries between layers developed, the thickness of the strata between boundaries increasing toward the axis of the trough.

Presently both recent and present-day oscillatory movements are being studied in many countries under the programme of a special international project. A map of recent oscillatory movements has been compiled for the Soviet Union (edited by N.I. Nikolaev and S.S. Schultz) on which the total amplitudes of both uplifts and subsidences are shown for the whole neotectonic stage. A simplified version of this map is given in Figure 5. Similar maps are now also being compiled for other countries.

Examination of all the available data on recent oscillatory movements indicates that, like present-day ones, they occur everywhere on the continents, where no regions are free of them.

The shape of terraces and denudation surfaces indicates that an arch-like flexure of the crust usually underlies mountain ranges. The character of the bedding of the sediments in areas of recent subsidence shows that the troughs had a U-shaped primordial form.

It has been noted that not only is a divergence of terraces in mountain ranges from the foothills to the axis observed that indicates quicker uplift of the axial zone, but also an increase in their number. There is a splitting of

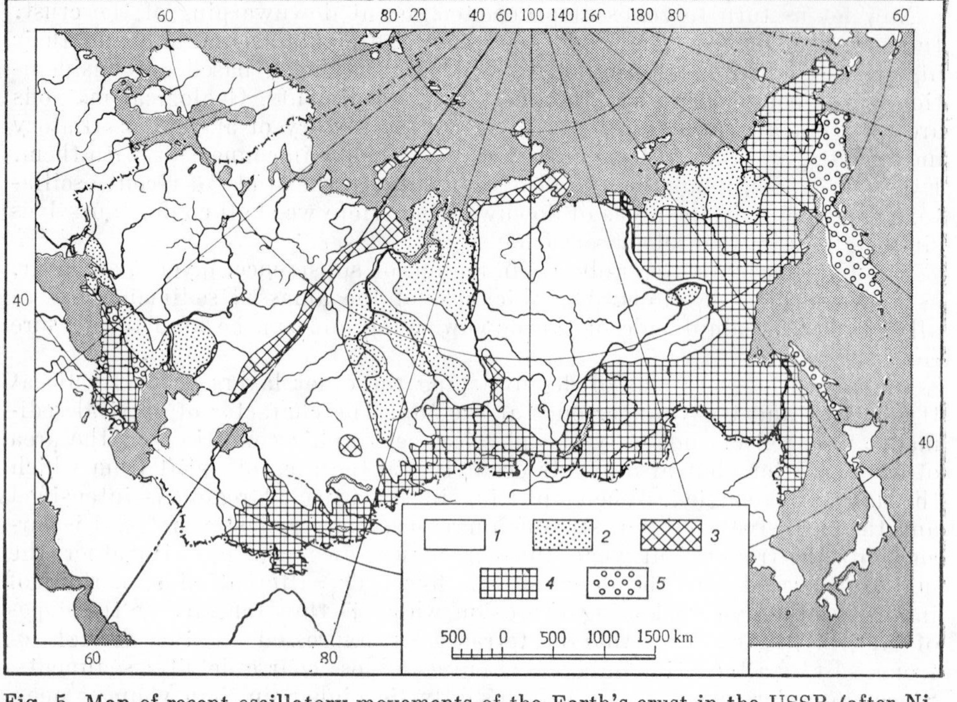

Fig. 5. Map of recent oscillatory movements of the Earth's crust in the USSR (after Nikolaev and Schultz, 1959, with simplifications):

1—regions of very weak manifestations of recent movements; *2*—regions of weak subsidence; *3*—regions of weak uplift; *4*—regions of intensive movement with a preponderance of uplift; *5*—regions of intensive movement with a preponderance of subsidence

terraces, i.e. two or three are formed from a single terrace, which can mean that a slower uplifting of the crust was interrupted by more seldom and more protracted halts, while there would be more, but shorter, halts with a relatively rapid rise.

Fewer terraces are accordingly formed in the same period of time on plains where recent oscillatory movements are characterized by low amplitudes and rates than in mountains, where the amplitude and rate of oscillatory movements are much higher (three or four terraces on plains as against 12 to 15 in mountains). This ratio is also observed in the case of surfaces of denudation (one or two on plains and up to seven in mountains).

The initial arches and U-shaped troughs are quite often disturbed during intensive recent oscillatory movements by sizable local dislocations, both plastic and fault. In the ranges of the Tien Shan, for example, there are surfaces of denudation with slopes up to 20°, whereas their original slope could not have exceeded 1° or 2°. In places the surfaces of denudation in the same mountainous country are broken by vertical faults with vertical displacements along them up to several kilometres long.

A combination of arches and vertical faults turns mountain ranges into arched-block structures.

An arched structure associated with recent oscillatory movements, though also very gentle, is Scandinavia. Tracing of the terraces of the Yoldian Sea which existed around 8000 years ago, made it possible to establish that

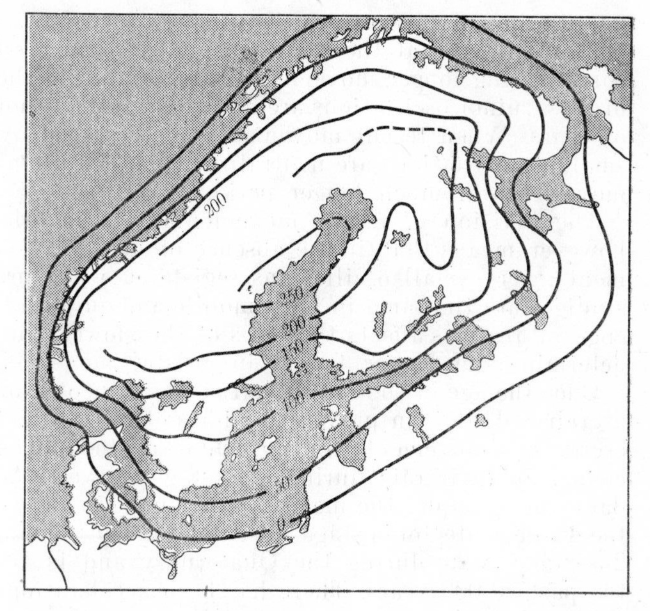

Fig. 6. Map of the isoanabases of Scandinavia (after Högbom, 1924):
lines of equal uplift of the bottom of the Yoldian Sea (in metres)

the central part of Scandinavia, located in the area of the Gulf of Bothnia, rose in post-glacial times by approximately 250 m; toward the periphery of this region the uplift gradually diminishes and further out gives way to sinking (Fig. 6).

Are recent oscillatory movements a continuation of similar ancient movements? Do they succeed them or are they "superimposed" on these old movements with a different pattern of distribution of uplifts and subsidences?

As in present-day oscillatory movements, both succession and the superimposition of recent movements are encountered in individual cases. The recent subsidence of the Caspian or Pechora depressions, for example, may be considered a continuation of downwarping that had already begun in the Palaeozoic. The recent slow uplifting of the Urals, Donets Basin, and Central Kazakhstan have an ancient heredity. Examples of "superimposition" are the ranges and depressions of the Tien Shan and Pamirs, where there was a weakly undulating plain before the neotectonic stage. As already noted, the relief of the site of the ranges formed from Alpine geosynclines at the beginning of neotectonic times was low and smooth. These last examples indicate that the neotectonic stage led to significant changes in the structure of the crust, though not everywhere.

The rate of recent oscillatory movement in mountains is up to 5 mm a year or more, but on plains it does not exceed half a millimetre a year and is one order of magnitude slower there than the rate of present-day oscillatory movements. Unlike present-day movements, whose rate is the same on average both in mountains and on plains, recent movements consequently prove to be differentiated; two regimes are manifested: namely, mobile (mountain) and quiescent (plain).

Both phenomena, i.e., the slower rate compared with present-day movements, and division into different regimes, may be due to the fact that the time interval for which rates are averaged is much longer for recent movements than for present-day ones. As already noted, the extremely short period of observation of present-day movements leads to our measuring the rate of very small, and at the same time very rapid oscillations, that are simply complicating longer movements, more stable in their direction. The periods of these minor oscillations are measured only in thousands, or even hundreds, of years. When recent movements are averaged over millions of years, the smallest oscillations are neutralized, and the mean rates reflect much slower movements of much longer period.

The division of recent movements into mobile and quiescent regimes, however, means that, in the absence of such a division for present-day movements, very small oscillations registered when present-day movements are studied, are the same in both mobile and quiescent regions, while the difference of regimes affects the rates of the slower, long-period oscillations that determine the longer trends in crustal movements.

Once the age of the individual stages of neotectonic development has been established, it then proves possible not only to determine the mean rate of oscillatory movements for the whole neotectonic stage, but also to note a change in their rate during this stage as well. According to Milanovsky's data, for example, the mean rate of uplift in the Great Caucasian Range for the whole neotectonic stage does not exceed 0.5 mm a year whereas it reached 2-3 mm a year during the Quaternary and is already 15 mm a year over the past 20 000 years. There has been acceleration by a factor of 30 during the neotectonic stage. On the other hand, the centre of the Scandinavian arch was found to rise at a rate of 3 cm a year in the period directly following the melting of the ice; this has slowed in our day and now does not exceed 1 cm a year.

The gradients of recent oscillatory movements in mountain regions average $1 \cdot 10^{-8}$ year^{-1}, i.e., are rather less than present-day movements; on plains, however, they are 10 to 30 times smaller. Consequently, the division of recent movements into mobile and quiescent applies also to gradients.

The amplitudes of neotectonic vertical movements reach many kilometres in mobile regions. In the Great Caucasus, for instance, the neotectonic uplift is as high as 4000-5000 m, whereas the amplitude of neotectonic subsidence of the Kura-Rioni intermontane trough is up to as much as 5000 to 7000 m. In the Tien Shan the total amplitude of neotectonic movements is around 12 km.

The study of present-day and recent oscillatory movements has great practical and theoretical importance. A considerable group of mineral deposits is associated with recent ones through erosion and the transport of sedimentary material over the Earth's surface and weathering. They include various placers and a large number of deposits of the zone of weathering. The laws of their distribution can only be understood, and so the most effective direction for prospecting them determined by studying the history and distribution of recent oscillatory movements. The regime of rivers depends on present-day and recent movements: meandering and swamping occur in areas of subsidence and gradually get stronger; in areas of uplift, on the contrary, the channels of rivers become straightened and ravines grow more intensively. A change of gradient, even if very slow, affects the direction and rapidity of run-off. Oscillatory movements affect the level of ground waters, which rises with a sinking of the crust, or lowering of its gradient, and sinks with a rise. On the sea coasts present-day oscillatory movements lead to coastal zones becoming either shallower or deeper.

It follows from this that study of present-day and recent oscillatory movements should also interest prospectors and mining engineers, hydrogeologists, hydrotechnicians, land reclaimers and improvers, and agronomists.

Seismic zoning is largely based on knowledge of present-day and recent oscillatory movements, i.e., the defining of zones of possible occurrence of earthquakes and their strength. Zones of heightened gradients of these movements are usually the most dangerous seismically.

The theoretical significance of present-day movements is that they present a very rare opportunity of studying tectonic processes in action; in most cases our opportunities are confined to examining their end results. Combined registration of present-day movements and observations of changes in the geophysical fields (gravitational, magnetic, thermal and electric) can furnish the material for judging their deep-lying causes.

CHAPTER 5

Ancient Oscillatory Movements of the Crust

The Tectonic and Geomorphological Expression of Oscillatory Movements

Ancient oscillatory movements have left their traces in the thicknesses and lithological composition (facies) of sedimentary rocks.

Let us consider the section of the coal measures of the Upper Viséan stage[3]), and Middle and Upper Carboniferous of the Donets Basin. These strata consist of alternating thin layers of sandstones, shales and schistous clays, limestones, and coals. Paralic coals form in swampy lagoons and their level of deposition is close to that of the sea; the other rocks of the same series carry a shallow-water marine fauna. They were all formed at a depth not exceeding a few tens of metres, although the limestones were formed at a rather greater depth than the clays, but the latter occur at a rather greater depth than the sandstones. The limiting depth can be taken at 50 m, for the limestones contain many remains of algae and colonial corals.

The sequence of the rocks making up this series consequently indicates that oscillations occurred there in the Earth's surface during the Viséan stage and the Middle and Late Carboniferous, now rising to water level and now sinking to a shallow depth. These surficial oscillations can be represented as a curve; the part of the curve relating to the C_2^5 age is shown in Figure 7.

One can draw another conclusion from consideration of this section. The coal measures of the Carboniferous in the Donets Basin are very thick, at least 6000 m. Throughout the whole of this thickness the coal-bearing series consist of a uniform complex of shallow-water sediments, and in this respect there are no differences between the bottom and the top of the series. Such an enormous series of uniformly interbanded, shallow-water sediments can clearly only be accumulated in a situation of gradual subsidence of the Earth's crust, occurring simultaneously with the accumulation of sediments. A formed depression did not exist in advance that was subsequently filled with sediments. The crust was sagging and emerging trough was continuously being filled with sediments, the surface of which remained the whole time near that of the water, maintaining shallow conditions in the basin.

A change in the facies of rocks from one stratum to another thus indicates oscillations of the Earth's surface, i.e., its rise and subsidence, while the thickness of the sediments is a sign of subsidence of the crust as a whole.

The Earth's surface is the level of the top of the sediments. Oscillations of the surface are governed by the ratio of the rates of accumulation of sediments and of subsidence of the crust.

---

[3]) In the Soviet stratigraphic scheme, the Viséan stage of the Lower Carboniferous corresponds to the Middle Dinantian or Upper Tournaisian stage of the scheme for the West European Carboniferous. (*Translator's note*).

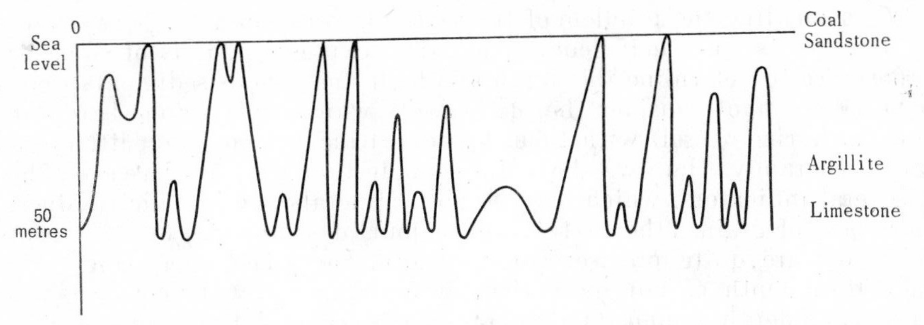

Fig. 7. Oscillations of the Earth's surface in the Donets Basin during the C_2^5 age (after Stepanov, 1932)

The following cases can occur:

1. the accumulation of sediments proceeds more slowly than subsidence of the crust;
2. accumulation proceeds at the same rate as subsidence;
3. accumulation proceeds more rapidly than subsidence.

In the first case the surface of the accumulating sediments sinks and the basin's depth increases, but more slowly than the subsidence of the crust. In the second case the basin's depth remains unchanged. In the third case, since accumulation outpaces subsidence, the sediments' surface rises and the basin becomes more shallow in spite of the subsidence of the crust. The surface of the sediments can continue to rise even after it is above sea level, further accumulation taking place in subaerial conditions.

Similar, but opposite cases can occur in regions of tectonic uplift of the crust. As soon as the Earth's surface rises above sea level denudation processes lowering the surface come into play. The rate of elevation of the surface will therefore always be slower than the rate of structural uplift of the crust. If the rate of denudation exceeds that of structural uplift, the Earth's surface will subside, in spite of the structural uplift.

Accumulation and denudation consequently distort the surficial result of tectonic oscillations, and this distortion tends in the opposite direction to the movement: owing to the accumulation of sediments the surface subsides more slowly than the tectonic subsidence of the crust; and owing to denudation, it rises more slowly than the structural uplift. In some cases the change in height of the surface may even take a direction contrary to that of the tectonic movement.

We therefore have to distinguish between the tectonic processes of uplift and subsidence of the Earth's crust, on the one hand, and their reflection at the surface on the other hand, where it is being added to the phenomena of the accumulation of sediments and denudation. This surface reflection of oscillatory movements of the crust can be termed their *geomorphological expression*.

It is necessary to differentiate between a tectonic process and its geomorphological expression, since misunderstandings may otherwise arise. The geologist, in viewing the upward succession of a stratigraphic section, from relatively deep-water sediments to ever shallower ones, and finally to continental deposits, is apt to speak of "uplift of the crust" when, in the vast majority of cases, the crust has not been raised but has subsided, and the raising of the surface has been caused by sediments accumulating faster than the tectonic subsidence.

When tackling the problem of the particular interrelations between tectonic movements and their geomorphological expression, it is of considerable importance to determine the depth at which the various sediments were laid down. We cannot consider the methods by which this is done here but one can familiarize oneself with them by consulting textbooks on lithology and palaeogeography. Here we shall simply note that they are based on the use of several indicators, which may be both minerals and organic residues, the coarseness of grains, the texture and colour of sediments, etc. Some of the indicators are quite precise: colonial corals or green algae, for example, indicate a depth of not more than 50 m; epigenetic glauconite is usually formed at a depth around 100 m; paralic coals accumulate in lagoons at water level; rocks of evaporite origin, like salt, gypsum, and anhydrite, are also formed practically at sea level; remains of land flora point to subaerial conditions of deposition; while a lack of benthic fauna is usually taken as a sign of hydrosulphuric pollution and a quite considerable depth, measured in hundreds of metres. Many other indicators are not so definite, but taken together help resolve the problem of the depth of sedimentation.

When we have the whole complex of the sediments comprising the coal-bearing series of the Donets Basin in mind, it does not vary along the section, as we noted earlier, and so the range of fluctuation of the surface of the sediments also does not vary. During the Middle and Late Carboniferous as a whole, there was full compensation of subsidence there through the accumulation of sediments. This can be shown on a graph; in Figure 8 time is indicated along the horizontal axis while both vertical movements of the Earth's crust (the lower curve) and changes in the height of the surface (the upper curve) are shown on the vertical axis. The surface fluctuations shown are arbitrary, but are given to scale. The fluctuations are so small that even with a thickness of sediments of several hundred metres they can be disregarded and the subsidence considered to be compensated quite well by the accumulation of sediments, while the thickness can be taken as a measure of the tectonic subsidence. This conclusion will be the more correct if the whole section of the coal measures is considered. The maximum possible deviation from a compensation equal to the maximum range of elevation and subsidence of the surface (50 m) is then less than 1% of the total thickness. This deviation is also very short-term, because the surface of the sediments soon returns during subsequent rapid oscillations to the position it occupied at the very start of the accumulation of coal measures, which means that the compensation was re-established.

It is clear, however, that if we were to take a very limited part of the section, the thickness and duration of formation of which is commensurate with the amplitude and period of the elevations and subsidences of the surface, the deviation from compensation can prove very great, while during a steep change in the basin's depth compensation may have been fully disturbed so that the thickness does not serve as a measure of subsidence.

Let us now turn to the section of the Cretaceous and Cenozoic in the northern slope of the central part of the Great Caucasian range (around Ordzhonikidze) (see Fig. 9).

Glauconite is the characteristic mineral of the Lower Cretaceous deposits of this area, represented by marly and argillaceous sandstones, sandy clays, and also marls and limestones. Judging from the conditions of its formation the depth of the basin may be taken to have been around 100 m. The thickness of the Lower Cretaceous is 1100 m. While the Lower Cretaceous depos-

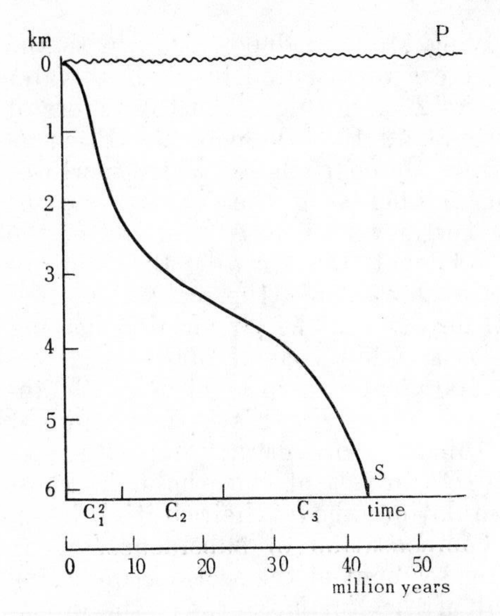

Fig. 8. Graph of oscillatory movements in the Donets Basin during the Middle and Late Carboniferous (after Beloussov):

P—the curve of the morphological expression of the oscillatory movements; *S*—the curve of tectonic movements of the crust

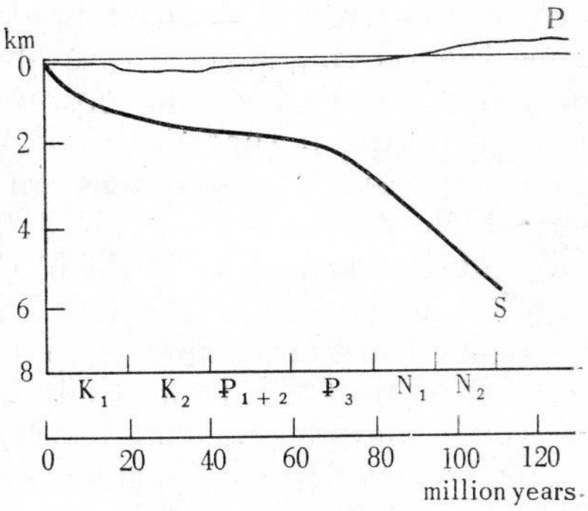

Fig. 9. Graph of oscillatory movements in the North Caucasus (around Ordzhonikidze) (after Beloussov). Symbols as in Fig. 8

its were being accumulated the depth varied little, and we can infer that there was good compensation of subsidence by accumulation during the whole of the Early Cretaceous.

The Upper Cretaceous limestones consist almost exclusively of foraminiferas and fragments of a prismatic layer of Inocerami. These rocks were deposited at a greater depth than the preceding ones, probably at not less than 200 m. The thickness of the Lower Cretaceous is 300 m. Again, since the depth did not vary essentially during this period, compensation was maintained.

With the passage from the Early to the Late Cretaceous, however, the basin's depth increased by 100 m. When the rocks of the Upper Cretaceous roof itself are taken into account, as well as the Lower Cretaceous, then, with a thickness of sediments around 300 m the true degree of subsidence from the end of the Early Cretaceous to the end of the Late Cretaceous will be 400 m. That means a deviation from compensation by 25%. This figure, however, will be higher if our survey covers only part of the total thickness of the Upper Cretaceous and not all of it. With a section thickness around 150 m, for instance, the true scale of subsidence from the end of the Early Cretaceous to the middle of the Late Cretaceous will be 250 m the deviation of compensation will consequently be 40%. On the other hand, if not only the whole of the Upper Cretaceous is taken into consideration, but also the whole of the Lower Cretaceous, then, adding together the whole period (with a total thickness of 1400 m), the deviation from compensation will not be more than 7%, taking the increase in depth during this time as 100 m.

In the foothills of the Central Caucasus the Palaeocene, Eocene, and Lower Oligocene are represented by marls around 170 m thick. Above them lies the Maikopian suite (Middle and Upper Oligocene and Lower Miocene), consisting of 850 m of clays. The overlying Miocene deposits down to the lower strata of the Upper Sarmatian stage are represented by shallow-water marine argillo-arenaceous sediments around 2000 m thick. The top layers of the Upper Sarmatian stage, Meotic stage, and the whole of the Pliocene (about 800 m) consist of continental sediments, the coarseness of which increases from the bottom upwards. The surface of sediments in the North Caucasus was consequently rising throughout the Tertiary period. At the end of the Miocene the Earth's surface rose above sea level. Determining the depth of formation of the Palaeogene rocks from the character of the rocks as around 200 m and taking it that the continental deposits of the Upper Miocene and Pliocene represent the sediments of deltas and low coastal plains, the surface of which was raised not more than 100 m above sea level, it can be inferred that the total amplitude of the uplift of the Earth's surface did not exceed 300 m. On the other hand, the thickness of Tertiary deposits is as much as 4000 m, which means that only 7% or 8% of the whole thickness of sediments resulted from their accumulation above their initial level, while the remainder is the result of compensation of subsidence.

Again, however, if we consider a shorter segment of the section, including in it the period of changes in the height of the Earth's surface, the deviations from compensation will be greater. If, for instance, we survey the epoch to which the 1000 m of the marine Miocene sediments and 400 m of the continental Neogene respond, then, with a thickness of 1400 m, the subsidence is only 1100 m and the deviation from compensation consequently 20%.

Let us now turn to the East European platform, where the sedimentary strata, unlike the Caucasian geosyncline, are comparatively thin.

Over the time from the Middle Devonian to the Kazanian stage of the Permian inclusive, roughly over 2000 m of sediments accumulated in the Central Volga region. The depth of the basin did not differ in Kazanian times from its depth at the beginning of the Devonian: in both cases it was extremely shallow, measurable in tens of metres. For the Middle and Upper Palaeozoic as a whole, consequently, there was good compensation of the subsidence by accumulation.

If we consider shorter intervals of time, however, as in the examples above, we arrive at rather different ratios.

Let us consider, for example, the Tournaisian and Viséan sections separately. The former consists of shallow-water limestones 100 m thick and the latter—of coal measures 50 metres thick. A graph of the processes taking place at that time is given in Figure 10. The depth of formation of the limestones is taken as 50 m, while the coals lay down at sea level. After that 100 m of limestones accumulated, their base having settled to a depth of 150 m. This was followed by silting and a rise of the surface of the sediments to sea level, and of their base to a depth of 100 m. After the accumulation of coal-bearing deposits the base of the limestones again lay at a depth of 150 metres. Consequently, with sediments 150 m thick, the tectonic trough measured altogether 100 m, i.e., there was 70% compensation, while 30% was linked with a change in the depth of the basin (with uncompensated accumulation of sediments).

After the coal-bearing suite of the Viséan stage, Viséan limestones 250 m thick accumulated. Since they are of reef origin, their depth of formation was around 50 m. There was consequently a deepening of the basin by 50 m and new subsidence, compensated by accumulation to a depth of 250 m.

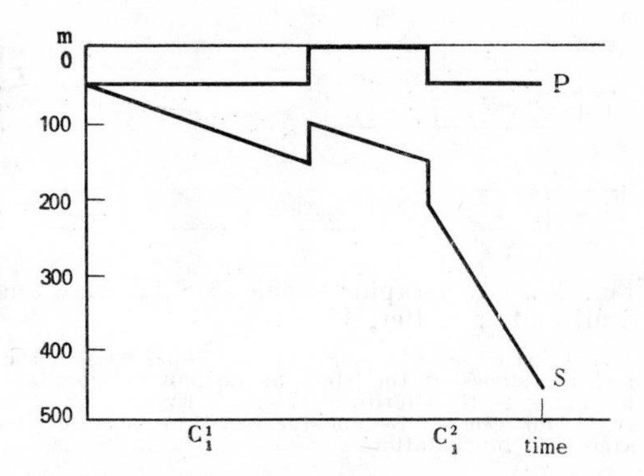

Fig. 10. Graph of oscillatory movements for the Tournaisian and Viséan stages on the East European platform (after Beloussov). Symbols as in Fig. 8

As a result of adding the Tournaisian and Viséan stages together we get full compensation of the subsidence: the scale of the subsidence and the thickness of the accumulated sediments both equalling 400 m.

A similar consideration of many other geological sections leads to the conclusion that there is a general trend in most cases in the interrelation of tectonic subsidence of the Earth's crust and the accumulation of sediments, consisting in compensation of the former by the latter. This trend is more clearly manifested the bigger the geological section considered. It may be disturbed when short segments of the section are considered, when the sequence of the rocks belonging to this segment reflects a change in bathymetric level of formation of the sediments, i.e., elevation or subsidence of the Earth's surface.

This survey indicates that there were considerable deviations from compensation at times and places linked with the formation of "empty" depressions, when the volume of incoming sedimentary material proved far from sufficient to fill a very rapidly developing tectonic trough in the crust.

For example, in the area of Ishimbai in the Cis-Urals massifs of reef limestones (called shikhans) of the Sakmarian and Artinskian stages are known around 1200 m thick. The same stages close to them on the east are represented by a series of clays, marls, and dolomites with a total thickness of 200 m. One facies passes into another quickly, but gradually, without a break in the continuity of the crust in this sector (see Fig. 11). The only possible interpretation is that the whole sector with both facies of deposition subsided by 1200 m during the Sakmarian and Artinskian stages. But whereas this subsidence was compensated within the "shikhan" zone by the accumulation of reef limestones, accumulation lagged far behind subsidence in the clay-marly facies zone: only 200 m of rocks of these facies accumulated so that, consequently, an "empty" depression 1000 m deep developed there.

As we shall see later, there is a stage in the development of geosynclines that is characterised by extremely rapid growth of intrageosynclinal troughs. At this time in them rocks are laid down that indicate a very deep marine basin, possibly as deep as several kilometres (radiolarites, jaspers, etc.). During the transition to this stage of siliceous sediments from the preceding stages of shallow-water deposition, the compensation of subsidence by accumulation was considerably disrupted.

These major deviations from the basic tendency, however, prove to be limited in both area and time. A deviation from compensation in the Ishimbai district is only observed in a small area. Even more important, during

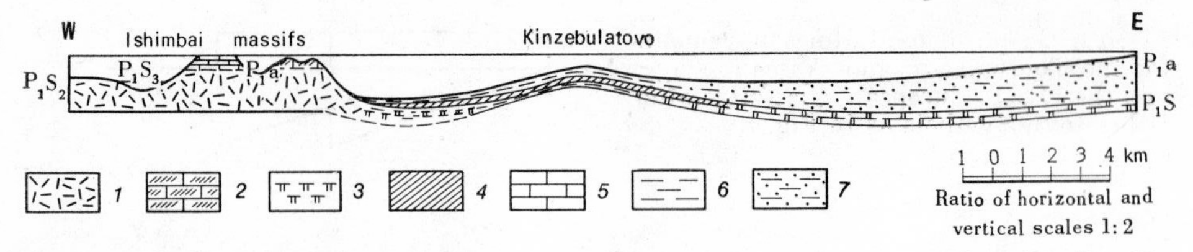

Fig. 11. Stratigraphic profile of Sakmarian and Artinskian deposits in the Ishimbai Cisuralia (after Keller, 1948):

Sakmarian deposits: *1*—massive reef limestones of the Tingskian and Tastubian horizons; *2*—massive reef limestones of the Sterlitamakskian horizon; *3*—dolomites, marls, and argillites; *4*—the same, belonging to the Sterlitamakskian horizon.
Artinskian deposits: *5*—massive reef limestones; *6*—marls and argillites; *7*—marls and argillites with sandstone intercalations

the Kungurian stage that followed the Artinskian, compensation was fully restored; the "empty" depression was filled with Kungurian lagoon deposits 1000 m thick. Since marked disturbances of compensation have not been observed for earlier stages of the Late and Middle Carboniferous anywhere in this district, the episode of disturbed compensation described was not only quite local but was also temporary, confined solely to the Sakmarian and Artinskian stages, which probably did not last more than five to seven million years. "Empty" troughs, probably, last longer in geosynclines for several tens of millions of years, but they, too, are compensated by accumulation in the preceding and subsequent stages of the geosynclines' development. One must, moreover, bear in mind that such "empty" troughs only develop where there is a eugeosynclinal regime (see Chap. 15), which always occupies a restricted area within geosynclinal zones.

The existence of the general tendency mentioned above is confirmed by a method that is interesting in that it eliminates the need to rely on determinations of the depth of formation of any particular sediment, thus avoiding the risk of errors in this always more or less contentious matter.

Let us try to compile the distribution of sedimentary thicknesses of various age and composition for a long interval of geological time and a considerable area, for instance, for the Palaeozoic and Mesozoic of the East European platform. This has been done in part in the attached series of schemes (Figs. 12, 13, 14, 15, 16).

By comparing these schemes we see that the plan of distribution of sedimentary thickness, i.e., the sites of relatively greater and smaller accumulation, remain very stable for hundreds of millions of years. In one and the same place either a relatively great or a relatively thin thickness of sediments, accumulated, although the composition of the sediments and physical-geographic conditions of their formation altered many times. From the Upper Palaeozoic to the end of the Mesozoic relatively greater thicknesses form a band from the Caspian Sea in the south to the lower reaches of the Pechora River in the north. A bay of relatively greater thicknesses reaches out westward from this band into the Moscow syneclise. In all directions away from these areas the sedimentary thicknesses decrease, but south of the east-west "promontory" of the Voronezh anteclise, the thicknesses again increase.

The distribution of thicknesses varied in many details from one system to another, but the general plan was maintained in a strikingly steady way, which completely contradicted the variability of the facies regime. During the time interval considered limestones of various type, marls, and chalk,

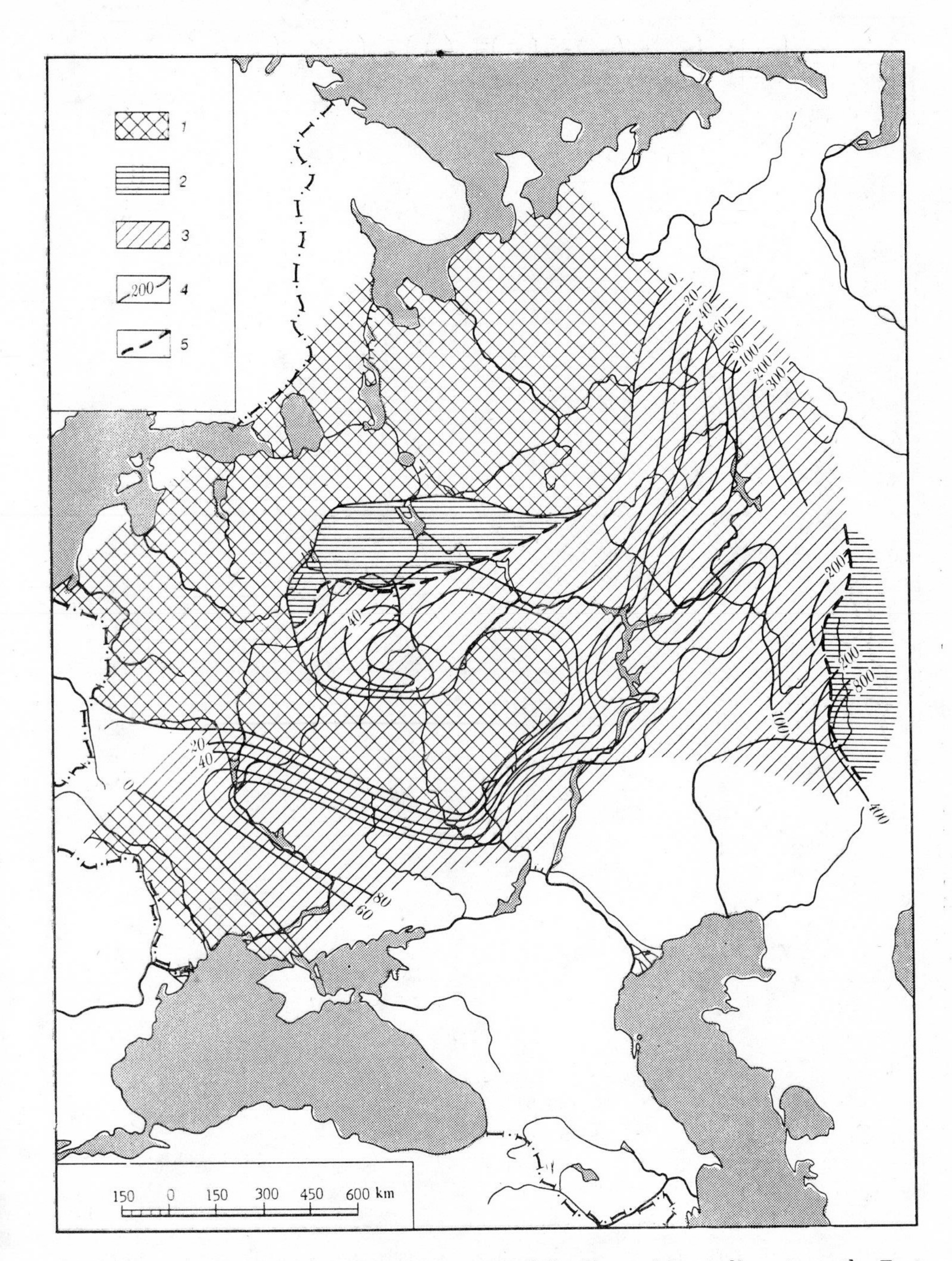

Fig. 12. Scheme of the facies and thicknesses of the Tournaisian sediments on the East European platform (after Ronov, 1949, and Vinogradov, 1967-9):

1—regions of erosion; *2*—sands and clays; *3*—limestones, dolomites, and clays; *4*—lines of equal thickness (in metres); *5*—boundaries of facies zones

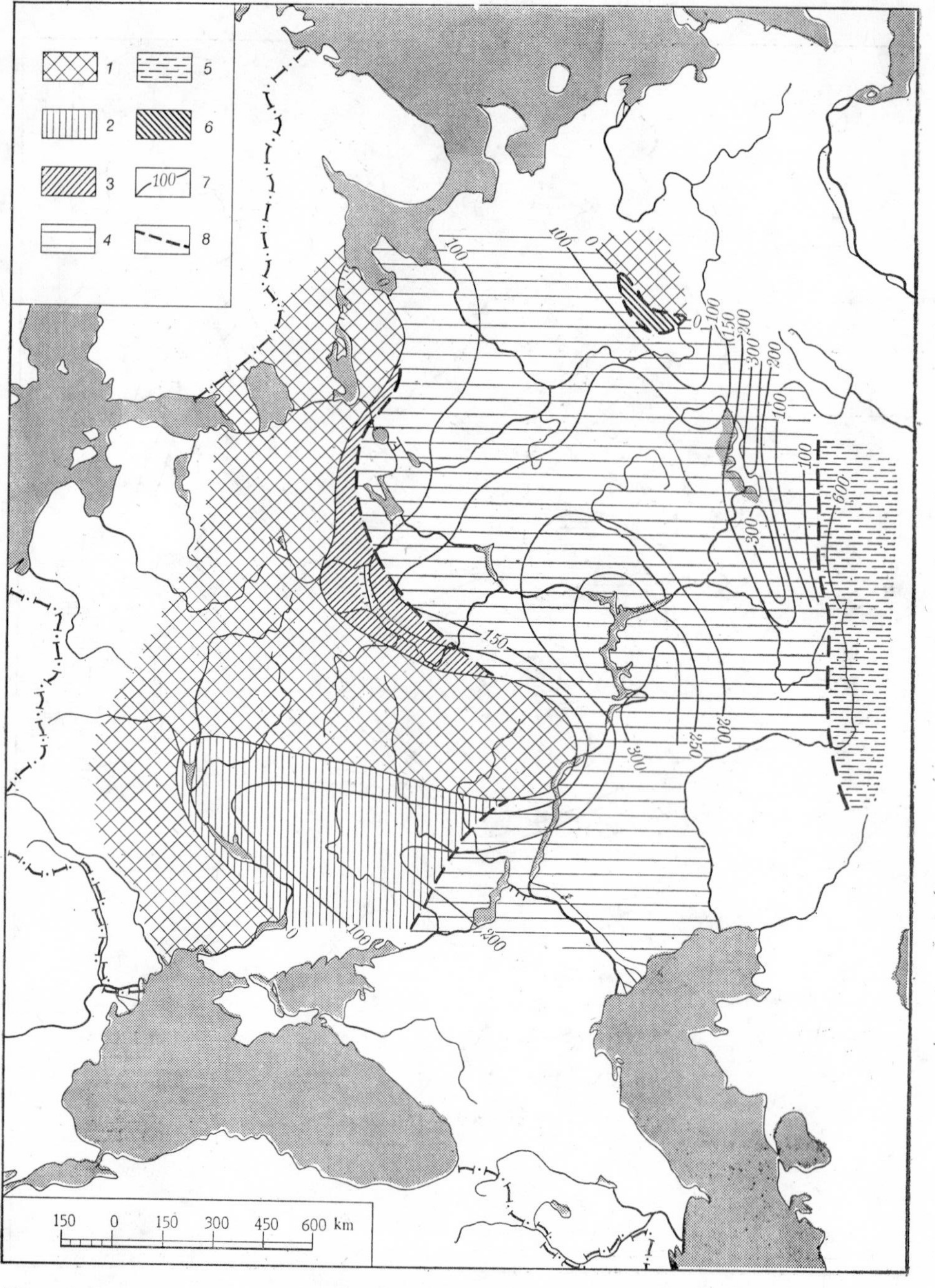

Fig. 13. Scheme of the facies and thicknesses of the Upper Carboniferous on the East European platform (after Ronov, 1949, and Vinogradov, 1967-9):

1—regions of erosion; *2*—clays, sandstones, coals, limestones; *3*—clays, limestones, and dolomites; *4*—limestones and dolomites; *5*—sandstones, clays, and conglomerates; *6*—lagoonal sediments; *7*—hickness isolines (in metres); *8*—boundaries of facies zones

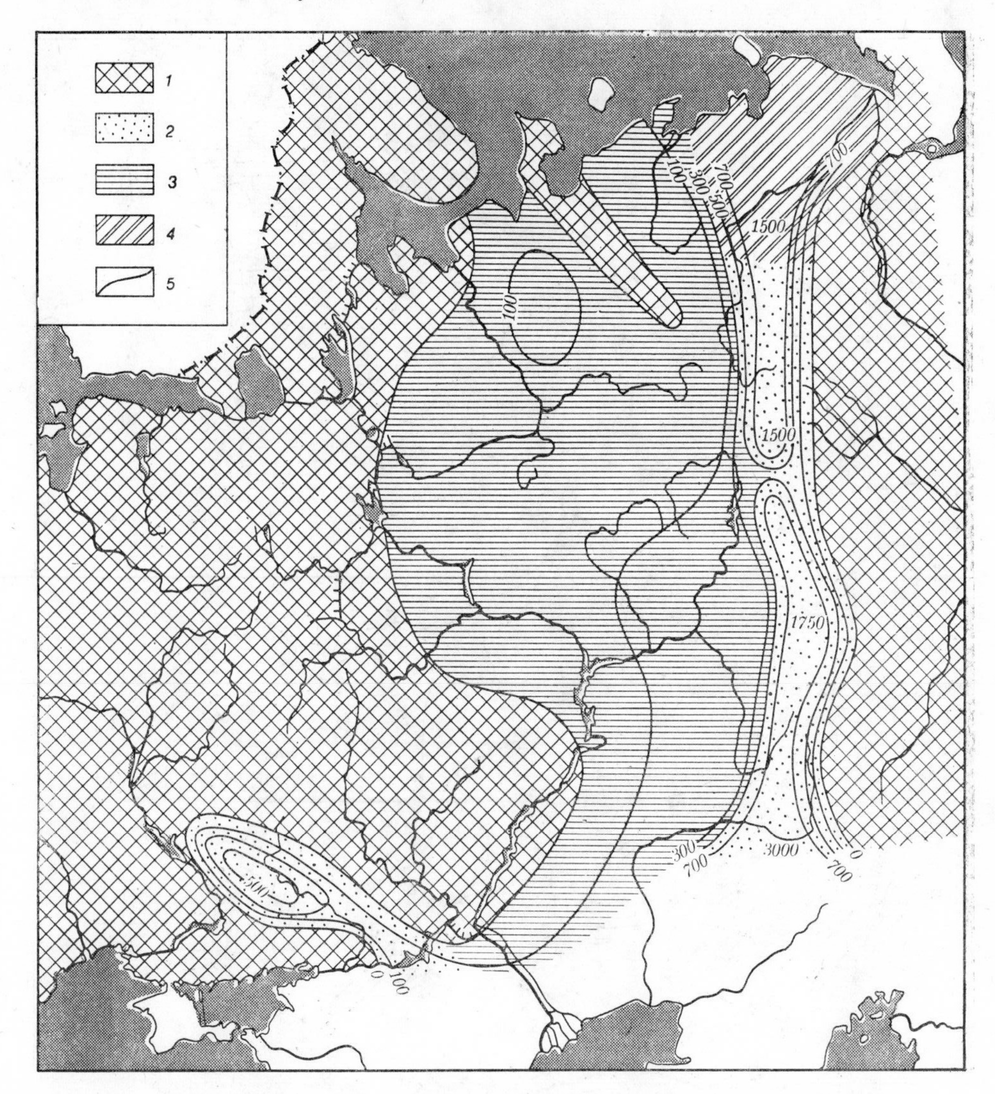

Fig. 14. Scheme of the facies and thicknesses of the Artinskian stage of the Lower Permian on the East European platform (after Ronov, 1949, and Vinogradov, 1967-9):

1—regions of erosion; 2—sands and clays; 3—limestones and dolomites; 4—coal-bearing deposits; 5—thickness isolines (in metres)

various sands and clays, lagoon gypsiferous and salt-bearing sediments, and continental laterites were laid down on the East European platform.

This example applied to platform conditions, but it turns out that we also observe lengthy stability of the plan of distribution of thicknesses in geosynclines. Figures 17 to 19 present schemes of the distribution of thicknesses for three stratigraphic subdivisions in the Caucasus from the Jurassic to the Palaeogene. It will readily be seen that during this whole space of time two zones of relatively thicker accumulation are distinguishable in the Great Caucasus, a southern (the axial part of the present-day range and its southern slope) and a nothern. Between them there stretches a narrow zone of

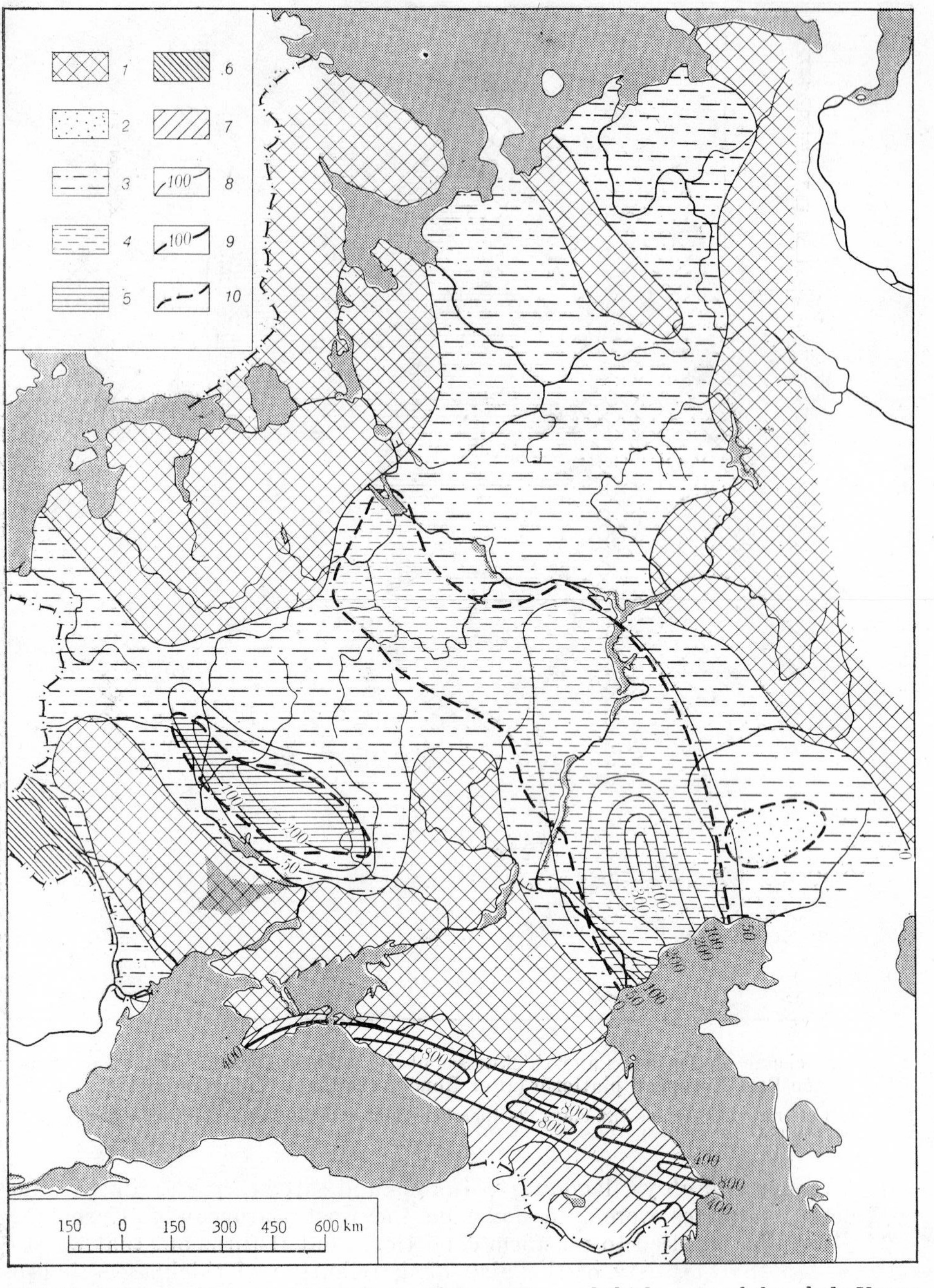

Fig. 15. Scheme of Kimmeridgian facies of deposition and thicknesses of the whole Upper Jurassic on the East European platform (after Ronov, 1949, and Vinogradov, 1967-9): *1*—regions of erosion; *2*—sands; *3*—sands and clays; *4*—clays; *5*—marls, clays, and sands; *6*—marls; *7*—limestones, marls, and clays; *8*—50- and 100-metre thickness isolines; *9*—400-metre isolines; *10*—boundaries of facies zones

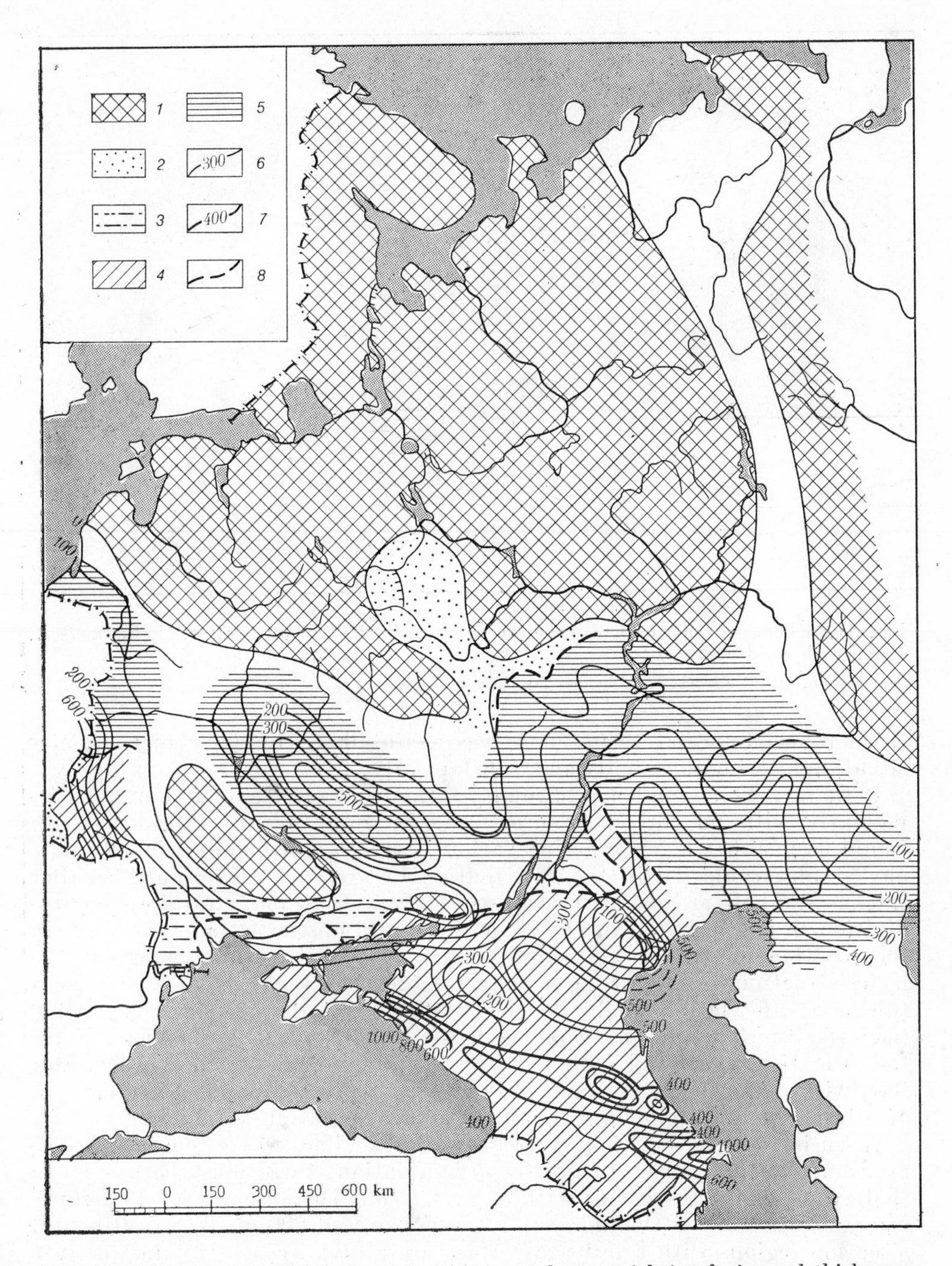

Fig. 16. Scheme of the Santonian, Campanian, and Maastrichtian facies and thicknesses of the whole Upper Cretaceous on the East European platform (after Ronov, 1949, and Vinogradov, 1967-9):

1—regions of erosion; *2*—sands; *3*—sands and clays; *4*—marls, limestones, and clays; *5*—chalk, marls, clays, and opokes; *6*—100-metre isolines of thickness; *7*—200-metre isolines; *8*—boundaries of facies zones

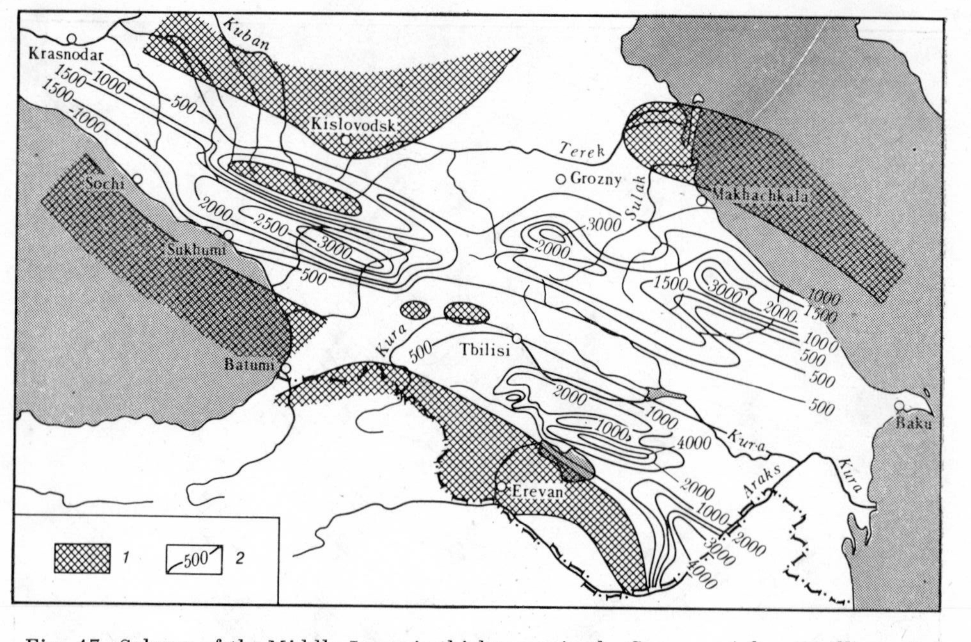

Fig. 17. Scheme of the Middle Jurassic thicknesses in the Caucasus (after Kirillova *et al.*, 1960, with modifications):

1—region of zero thickness; *2*—thickness isolines

thinner accumulation. This plan was maintained, in spite of the facies situation altering repeatedly and substantially.

If there were no tendency to compensation and sediments accumulated irregularly with no correspondence with the amplitude of subsidence, we would be unable when comparing various stratigraphic sub-divisions to note any stable regularity in the distribution of areas of greater and smaller thickness. Their areal distribution would vary from one stratigraphic subdivision to another, and in any event we would observe a radical change in the location of greater and smaller thicknesses with every change in the facies situation. In fact, if the scale of subsidence of the crust did not govern thickness, the latter would depend wholly upon the physical-geographic (facies) situation, on physical-geographic conditions more favourable for the formation of a given sediment in one place and less so in another. But then with a transition from one sedimentary facies to another, the location of sites favourable and unfavourable for accumulation would vary.

When in today's situation, we want to find regimes where there was general compensation of subsidence by accumulation, these must be sought in shallow-water marine basins with a levelled-out bottom such as shelf zones where depths vary little and usually do not exceed 200 m. In a continental situation regions with a levelled surface are alluvial coastal plains (mainly deltas) that dip gently toward the sea.

The geophysical and drilling data indicate that the thickness of sediments on shelves and alluvial plains differs widely in different places. The good levelling of their surface indicates good compensation of irregular subsidence of the Earth's crust by accumulation. Study of the conditions of formation of the various sedimentary rocks in geological sections shows that

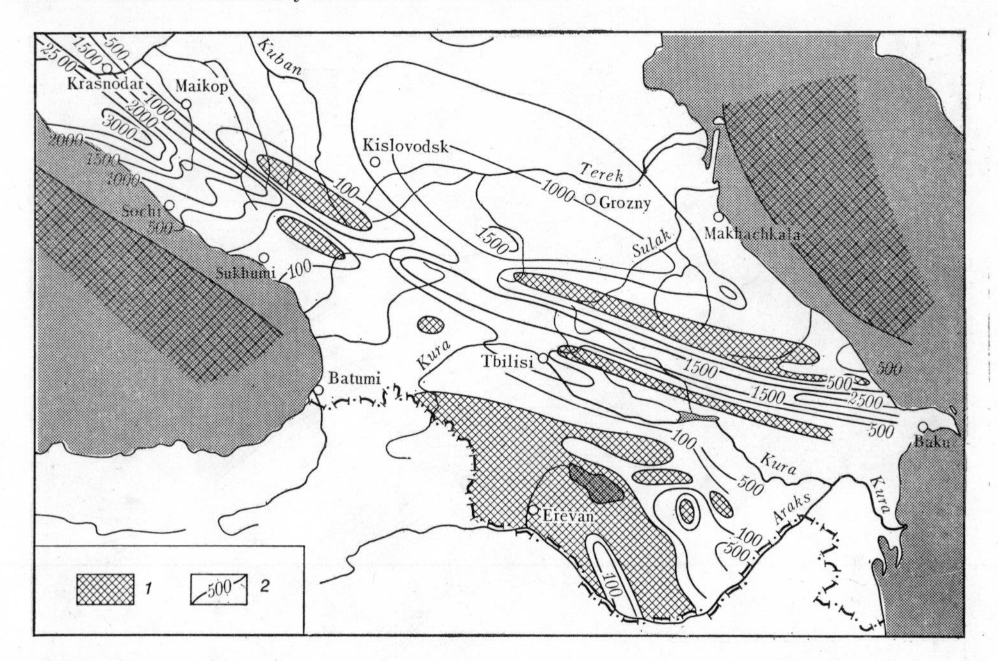

Fig. 18. Scheme of Lower Cretaceous thicknesses in the Caucasus (after Kirillova *et al.*, 1960, with modifications):
1—region of zero thickness; *2*—thickness isolines

it is these two situations, shelves and alluvial plains, that we also predominantly encounter in the geological sections of continents. Compensation is well known to be absent in deep oceanic troughs, but of these we agreed to speak separately.

Is there a similar tendency toward compensation of tectonic movements by surface processes in regions of uplift of the Earth's crust? Are tectonic uplifts levelled down by denudation processes in the same way as the troughs are levelled by accumulation of sediments?

At first glance the answer to those questions should be negative: a large number of mountain ranges now on the globe would seem to indicate an absence of compensation phenomena in this case.

The present-day relief, however, as we have seen, is geologically very young and has existed for a mere ten or fifteen million years. The older relief has not survived. Nowhere in geological sections do we find buried mountain ranges, comparable in size to the Caucasus, Tien Shan, or Andes. In some areas buried landforms are known, but they are only shallow valleys, karst holes, and low hilly areas.

There is no doubt, however, that high mountains also existed on the Earth's surface in past geological periods. On the site of the Urals, for instance, there was a mountain range in the Late Palaeozoic much higher than the present one. Highlands also existed then on the site of the present West Siberian lowland. In the Proterozoic there were also mountains on the site of the present Russian plain.

The existence of ancient mountains is ascertained in all these cases not from buried landforms, but from the products of their denudation, e.g., the coarse

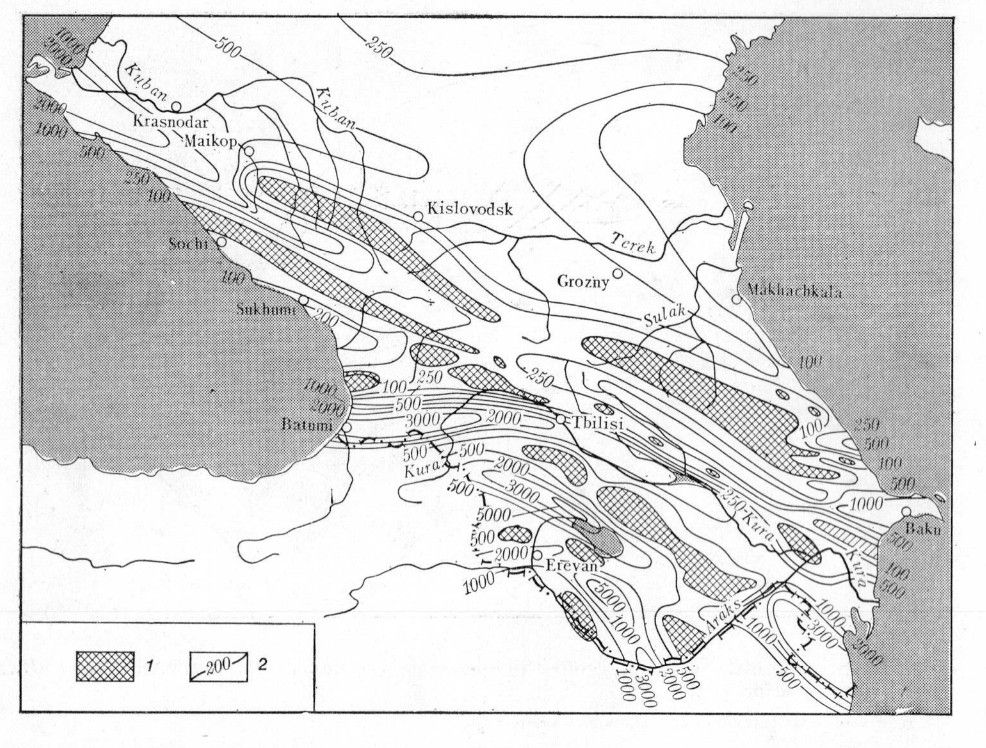

Fig. 19. Scheme of Palaeocene and Eocene thicknesses in the Caucasus (after Kirillova *et al.*, 1960, with modifications):
1—region of zero thickness; 2—thickness isolines.

sediments of a molasse formation. Mountains are fully worn down by denudation in a short time, and subsequent sediments are laid down already on a levelled surface where once mountains were.

Thus, in the perspective of geological time, not only troughs but also uplifts are compensated by denudation and the existence of the present-day alpine relief of mountainous regions, is a local, and mainly temporary deviation from compensation, similar to the local and temporary, but opposite in sign, deviation, observed in regions of subsidence.

We may say, consequently, that within the range of a geological section there is a transition to a levelled surface; depressions prove to be filled with sediments, while uplift is cut down by denudation. Ultimately, the relief is eliminated and only its traces remain in the geological section.

The geologist should not, however, forget that this is only the most general pattern and that here or there he may encounter deviations from it one way or the other in the form of uncompensated changes in the depth of a marine basin, or small ancient landforms buried in sedimentary series.

CHAPTER 6

Oscillatory Movements of the Earth's Crust and the Formation of Thicknesses and Facies of Sedimentary Series

The Formation of Thicknesses of Sedimentary Series

The general tendency towards compensation of subsidence by the accumulation of sediments needs explaining. A first hypothesis that can be put forward is that the Earth's crust sags beneath the weight of the sediments; where they accumulate more the subsidence is deeper, and where there is less sedimentation the amplitude of the subsidence is smaller.

It is not difficult to demonstrate, however, that an attempt to link vertical movements of the crust with a change in the weight of its surface layers leads to results that contradict what is observed in natural geological sections. The volumetric weight of sedimentary rocks is much lower than that of the subcrustal substratum that has to be squeezed out from below the sagging crust. The density of sedimentary rocks is about 2.5 g/cm^3, while that of the subcrustal material is 3.3 g/cm^3. Consequently, even if the crust's resistance to bending is disregarded, it will never sag beneath the weight of sediments with the amplitude equal to their thickness. As the cited densities indicate it would not sag by more than 75 m for every 100 m of sediments, this means that if accumulation began at a certain depth a marine basin would become shallower and would finally be completely filled with sediments. A change in the character of sediments can thus only be regressive. If the initial depth was shallow, there would not be the conditions for a thick series of marine sediments to accumulate; if the initial depth was 200 m, for example, the basin would prove to be filled when the thickness of the sediments was 800 m. Only with a very deep original depression can a thick sedimentary section accumulate in it. That, however, we repeat, must be regressive and, moreover, the development of the initial depression remains unexplained. The transformation of a region of accumulation into one of erosion is also not explained. This theoretical picture is obviously quite out of keeping with what is observed in natural sections with their intricate alternate transgressive and regressive series of sediments, continuous changes in the basin's depth, and signs of uplifts on the site of the previous subsidences and the formation of new troughs where there used to be uplifts.

For the same reasons uplifts of the crust cannot be linked with a diminution of its weight as a result of erosion of surface layers. The light rocks eroded will be compensated by seepage into the depths of the denser material. Because of that the bulging of the crust will be eroded at the surface faster than uplift proceeds and the relief will become lower. The whole process will cease when a peneplain develops on the surface. This picture, too, has nothing in common with the real, complex history of the relief of the terrestrial surface.

A change in the weight of the crust's surface layers thus cannot be the source of oscillatory movements of the crust. These must have a deep cause, independent of surface processes of accumulation and erosion, which, on the contrary, are determined by the oscillatory movements, uplifts and subsidences of the crust.

This conclusion is confirmed by the stability of the distribution of thicker and thinner sedimentary thicknesses, that we spoke of previously, from the examples of the East European platform and the Caucasus. In fact, if subsidence had been caused by the weight of the sediments, its rate would have depended on the rate of accumulation, but the distribution of sectors of faster and slower accumulation would inevitably have to vary with a change in the facies of sedimentation. Every sediment accumulates faster where the physical-geographic conditions are favourable to its accumulation; with a transition from the sediments of one facies to another, the rate of accumulation of which is determined by other factors, the distribution of places of faster or slower accumulation would change. If subsidence of the crust therefore depended on the rate of deposition of sediments, then, with each change of facies state, the distribution of thicker and thinner thicknesses would have to alter radically with each variation. This is not the case, however. The stability of the distribution of thick and thin thicknesses, which does not depend on the facies situation, indicates that there is a more general cause determining the thickness of sediments which can only be oscillatory movements of the Earth's crust.

To throw light on the character of the link between oscillatory movements of the crust and the formation of sedimentary series and their thickness and composition, let us consider what happens to the clastic material washed down from eroded land into an adjacent marine basin. On reaching the basin the material comes under the effect of movements of the water medium, among which the commonnest everywhere are the oscillations of water particles arising from waves. When waves are formed (and in the open sea they practically always are, although they may vary in height), the water particles move in a certain closed orbit, which is circulated far from the shore, but a more or less flattened ellipse near the shore, in shallow water. In the open sea water mass moves upward and forwards in the direction of the wind and then returns back along the base or lower branch of the orbit. The closeness of the coast forces waves to deviate from the direction of the wind; the crest of the ellipsoid orbit then moves at a wide angle towards the shore, while along the base of the orbit the particles recoil away from the shore.

The wave motion of water is weaker with depth, and at a certain depth dies away. At a depth equal to the wave length, the oscillations become quite insignificant, their amplitude being only 0.002 of the waves on the surface, i.e., a surface wave 5 m high becomes an oscillation of 1 cm. With a strong steady wind wave lengths in the open sea may reach 150 to 200 m. In consequence, these figures characterize the maximum depth of penetration of waves, or the depth of the "wave base". On approaching the coast, and in shallow water, wave lengths decrease, while the depth of the wave base also becomes less. This means that stable accumulation of sediments is only possible below it. Above the wave base sediments are subjected to continuous rewashing and transport and over quite a long time will be ultimately dumped from sectors of the bottom that are higher than the wave base to deeper places. The wave motion of the water thus prevents raising of the bottom (i.e., surface of the sediments) above a certain level. Wave motions grind the bottom, as it were, levelling it at a certain depth.

When the whole area on the shelf below the wave base has been filled with sediments, newly arriving material will be carried out beyond the limit of the shelf to the continental slope. But as soon as the crust subsides in the shelf zone a new "area of possible accumulation" beneath the waves action base level is opened up, in which a certain portion of the sedimentary material can be retained and accumulated. The thickness of these sediments will obviously correspond to the amplitude of the subsidence, provided that the influx of clastic material is sufficient to fill the opened-up area of possible accumulation. If certain sectors of the shelf are raised and others subside, then material will be eroded from the former as soon as their surface rises above the wave base and their thickness will diminish, while the area of the latter sediments will continue to accumulate and will increase in thickness.

The natural mechanism thus consists in a tendency to constantly maintain conformity between the scale of subsidence and the thickness of the accumulated deposits. Exact correspondence is obviously only possible given the following conditions:

a) the depth of the wave base remains unaltered;

b) the sedimentary material is enough to fill the area of possible accumulation;

c) the wave motion of the water distributes the material over the area of possible accumulation at the same rate as it is brought in.

These conditions are not maintained exactly in nature but the existence of a general tendency towards compensation shows that break of these conditions leads in most cases to only insignificant, temporary deviations from the compensation of subsidence by accumulation.

The depth of wave action varies with changes in the strength of the wind and configuration of the coastline or of the whole basin, but the changes in depth from this cause are not great [condition (a)].

The second condition (b) is met best in the zone of the continental shelf, to which clastic material is carried directly from the adjacent land. This zone is also most favourable for the life of organisms, a circumstance that ensures enough material of organic origin to fill an area of possible accumulation, even if the influx of material from the land is insufficient.

The third condition (c) largely repeats the first. The intensity of the rewashing of sediments on the bottom depends on the size of the waves. In landlocked bays or lagoons it is less intensive than in the open sea. However, even with very strong waves the redistribution of sediments over the bottom may lag behind their influx if the rate of delivery is very high. In that case the surface of the sediments will rise. The rewashing action of waves gains rapidly in strength with diminishing depth. For that reason, the higher the surface of the bottom is raised, the more it is subjected to strong wave action and the more difficult further accumulation of sediments becomes.

For all that, accumulation of sediments may lead, as study of geological sections shows, to a certain sector of the sea becoming completely filled with sedimentary material, and with continuation of the same conditions, their surface will rise above sea level. Further accumulation will then take place in subaerial conditions in which underwater wave action gives way to agents of subaerial denudation, which on the whole possess a greater energy than waves, and therefore hamper elevation of the surface to a greater extent.

All this argument allows us to infer that a quite strong mechanism exists in nature, that ensures a thickness of sediments corresponding to the amplitude of the subsidence of the crust. It operates, however, only in shallow-water environment within the zone of the continental shelf, or on low alluvial plains. Since, as said earlier, in the geological sections on present-day continents sediments formed in just such conditions definitely predominate, this mechanism is of great geological significance.

It is not out of place at this point to touch on the comparison of the ancient sediments that form geological sections with the conditions of the accumulation of sediments today. It will be readily noted that there is a very marked dependence for the sediments being formed today, of their rate of deposition on the place and mode of their formation. Near the mouths of rivers masses of clastic material accumulate quickly. Corals build their structures so quickly, that coral limestones accumulate much faster than shell limestones, or limestones of chemical origin, etc. For that reason, the thickness of the youngest sediments, formed on the sea floor during the Quaternary period, varies widely depending on the way the sediments are formed and on the physical-geographic conditions in which sedimentation takes place.

At the same time one established an inverse peculiarity for the ancient geological deposits: the independence of thickness from physical-geographic state and the submittance of thickness to another factor, i.e., the amplitude of the crust subsidence.

This difference becomes understandable in the light of the conditions of protracted sedimentation. Sediments that have only just been deposited on the sea bottom or on the coastal alluvial plain have not yet been subjected to rewashing and redistribution. They still lie where they were initially brought to and their distribution over the bottom is not uniform, but in some places the waves and in other places the subaerial denudation cause their resorting, which ultimately leads, viewed from the perspective of geological time, to a distribution of strata, conforming to the amplitude of the final subsidence of the crust.

This comparison of present-day processes of sedimentation with their end result consolidated in geological sections, provides a good example of the statistical character of geological patterns. An average picture of the distribution of sediments passes into the geological section and the factors determining it are not the same as those that controlled the distribution of the sedimentary material at the beginning of its deposition. Over a short interval the slowly operating factor, oscillatory movements of the crust, remains imperceptible and other factors occupy the foreground, changeable, inconstant, but violently operating factors, linked with transient physical-geographic conditions. As time passes, however, and the sedimentary material is redistributed, the influence of these changeable conditions diminishes, as they mutually neutralize each other and the role of the slow, but constant factor, the degree of subsidence of the crust, comes to the fore.

The Formation of Facies of Deposition

The oscillatory movements of the Earth's crust affect the lithological composition of deposits as well as the distribution of their thicknesses. We shall employ the term *facies* from now on for lithological composition, giving it, however, a narrower meaning than is usually done. For us *facies* here is above all the lithofacies and in particular those of its qualities that help us judge the relative distance of the site of their formation from the area of erosion, i.e., from the source of the clastic material. We adopt a simplified list of "facies" in this sense that includes for marine deposits, shingles (conglomerates), sands (sandstones), clays, and the group of biogenetic rocks, called limestones. As regards calcarenites, they are considered together with terrigenous rocks (shingles and sands). In addition, summarily distinguished is a general continental facies that includes coals and evaporites.

When there are no continental deposits the boundary between regions of erosion and deposition runs along the seacoast and the series of sedimentary facies begins at the coast with the coarsest marine deposits. When deposits of a continental facies accumulate, the boundary between the areas of erosion and deposition runs on the land and a continental facies develops first in the series of sedimentary facies along it. Further away from the boundary of erosion the continental facies gives way to marine ones.

The distribution of sedimentary facies over an area depends on the following processes: the sorting of the clastic material by particle size; the change of separation of clastic material, and of migration of the facies zones in time and space.

With subaerial deposition the grading is caused by the gradual diminution of the transporting energy of streams, as they flow down an increasingly gentle slope to the base level of erosion. As a result of the reduction in energy first the heaviest particles and then ever lighter and finer ones settle to the bottom. Thus, a zoning by coarseness arises in the distribution of continental sediments with distance from the boundary of erosion.

A meandering of streams from side to side over low alluvial plains is superimposed on this process which leads to a very broad area of distribution of ancient deltaic and other alluvial sediments.

On the sea bottom the grading of material by particle coarseness happens under the action of the same waves that also influence the distribution of the thicknesses of sediments. The grading action of waves is linked with the fact that water particles move towards the shore in the coastal zone with greater energy than their backwash. The loss of energy is associated with its partial expenditure on overcoming the internal friction of water, on striking coastal rocks and stones, and also with the part of the water thrown by wave onto the shore that seeps through the sand and shingle of the beach and completes its return flow inside the detritus. As a result the more vigorous motion of a wave to the shore throws onto it all the material that can be shifted by waves of a given force, while the less vigorous backwash does not carry back the whole of the material from the shore into the deeper water of the basin but only the lighter fractions, leaving the largest and heaviest fragments on the shore. The same thing also happens further out within the zone where the surface of the sediments is accessible to waves oscillations. Facies zones are thus gradually formed on the bottom of a basin, differing in the coarseness of the sediments, which are distributed with distance from the region of erosion area in the sequence listed above. Beyond the boundary of distribution of clastic material we are already in the zone of deposition of nonterrigenous rocks, predominantly of biogenetic origin (in our simplified list of facies—limestones).

It will be seen from the foregoing that the grading of material and formation of facies zones, like the distribution of thicknesses, takes time, which explains the observed differences in the areal distribution of old and present-day facies of sediments. Any map of the bottom sediments of contemporary seas shows an extremely complex, patchy distribution of the various types of sediment.

At the same time, when we are reconstructing the distribution of sediments for ancient geological periods we find much greater order; the sediments of one facies completely cover vast, very uniform areas, and the various facies are arranged in bands that succeed one another in accordance with the coarseness of particles. This orderliness, and that of the distribution of thicknesses, is the consequence of long rewashing and grading of sediments. This process has not yet been completed in today's sediments since, while rewashing of sediments goes on, the configuration and position of the coastline, depth of the sea, sites of the influx of coarse material from land, and other local physical-geographic conditions change recurrently, these conditions become averaged over geological time and the patchiness in the distribution of sediments gives way to orderliness, which is another example of the statistical character of geological patterns.

Knowledge of the conditions determining the breadth of the grading of terrigenous material on the bottom of a basin is very important so as to understand the processes of the forming of facies zones. Palaeogeographic schemes indicate that with the passage from one geological epoch to another the breadth of the distribution of clastic material alters: in some cases it is concentrated in a narrow zone around the region of erosion, giving way further out to biogenetic sediments, and in other cases it covers the bottom of the basin in a very broad band.

As a result of redistribution over the bottom by waves all the clastic material, as we have seen, is ultimately included in the area of possible accumulation below the wave base. The breadth of the final distribution of this material is determined by the relation between the rate of subsidence of the crust in the region of accumulation and that of the delivery of clastic material from the region of erosion. With a constant rate of subsidence in the area of accumulation, accelerated influx of clastic material leads to a broadening of its distribution zone, while a slowing down of influx leads to contraction of the same zone. On the other hand, with a slowing of subsidence in the region of accumulation, material will be distributed over a wider area with uniform delivery, but will occupy a narrower zone in the area of possible accumulation given accelerated subsidence.

The rate of delivery of material depends, first of all, on the rate of uplift of the region of erosion, because greater uplift leads to an intensification of the erosive action of streams. One may say, therefore, that the width of spread of clastic material depends on the relation between the rate of uplift in the region of erosion and the rate of subsidence in the area of accumulation. An increase in the rate of uplift of the region of erosion leads to a widening of the distribution zone, and the result of a slowing down of subsidence in the region of accumulation is the same. Slowing down of the uplift of region of erosion and acceleration of subsidence in the region of accumulation lead to an opposite result, namely, narrowing of the distribution zone. At the same time, the width of the zones occupied by detrital sediments of various fractions is altered. A change in the breadth of their distribution from one stratigraphic unit to another causes an alternation of rocks of different facies in the geological section (see Fig. 20). A layered alternation of different sediments in the section can be regarded as a reflection of changes in the relation between the rate of uplift in the region of erosion and the rate of subsidence in the region of accumulation.

During oscillatory movements of the crust there is a redistribution of land and sea, and shifting of coastlines. If the relation between the two rates remains unchanged and only the boundary between these regions is shifted, the width of facies zones is maintained but they are shifted in area in consequence

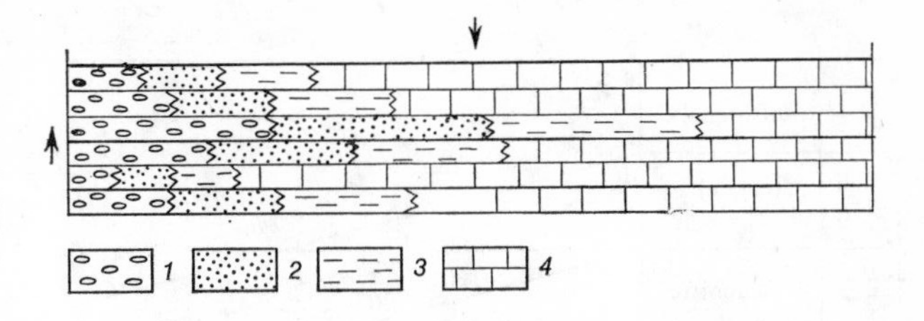

Fig. 20. Interstratification of sediments with a change in the width of facies zones depending on the relation between the rate of uplift in the region of erosion and the rate of subsidence in the area of accumulation (after Beloussov):

1—shingles; *2*—sands; *3*—clays; *4*—limestones

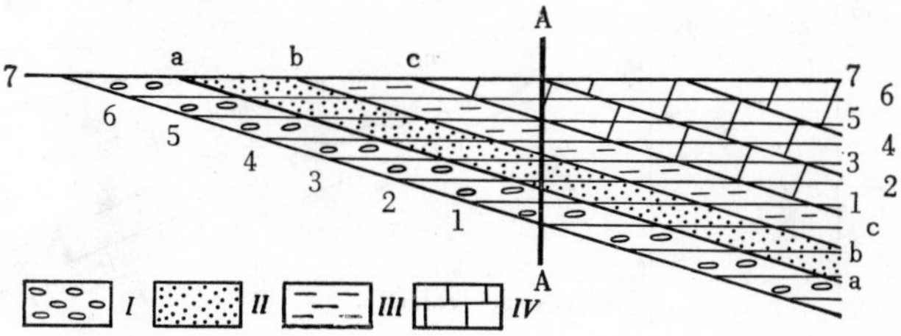

Fig. 21. Shift of facies zones with a transgressive change of the coastline (after Beloussov):

I—shingle; *II*—sands; *III*—clays; *IV*—limestones; *a*, *b*, *c*—facies boundaries: *1-7*—temporary boundaries

of a shift of the coastline. Such a shifting leads to the sediments of one facies, previously deposited, being overlain by those of another facies.

A scheme of the shifting of facies zones in a situation of transgressive shifting of the coastline is given in Figure 21. The digit *1* marks a certain position of sea level that we take as the initial one. Shingle, clays, and limestones accumulate on the sea bottom and are arranged in zones in that order with distance from the shore. The Earth's crust subsides and coastline moves to point *2*. All the zones, while retaining their breadth, are shifted and where shingle was previously deposited sands now settle, while clay accumulates above the earlier sands and limestones above clay. Further shifting of the coastline to points *3*, *4*, etc., leads to a continuation of this process.

As a result, a sequence will be observed from bottom to top in the vertical section along any line, e.g., *A-A*, of a replacement of shingle by sand, of sand by clay and, finally, by limestone. This transgressive series of sediments is evidence in this case of a recession of the coastline in the course of time from the spot where the section studied is located. As will be seen from this diagram sediments of the same facies have a different age in different spots: the sediments belonging to the same facies prove to be younger along the line of movement of the coastline. The stratigraphic boundaries (*1-1*, *2-2*, *3-3*, etc.) do not coincide with the facies boundaries (*a-a*, *b-b*, *c-c*, etc.). Facies migration proceeds in time, as well as in space.

Sections of Mesozoic sediments in South-east England and North-east Ireland are shown in Figure 22. The transgression begins in South-east England in the Aptian stage with basal conglomerates and sands, and reaches Ireland only in the Cenomanian stage. Deposition of the successive rocks of the transgression series accordingly lags behind in Ireland; the pure white chalk already being deposited in the Turonian stage in England, only starts to be deposited in Ireland in the Senonian.

Let us consider a case of the regression of a coastline (see Fig. 23). In the figure the line *1-1* corresponds to the initial position of the surface of the sea on the bottom of which sediments are deposited in the normal sequence. As

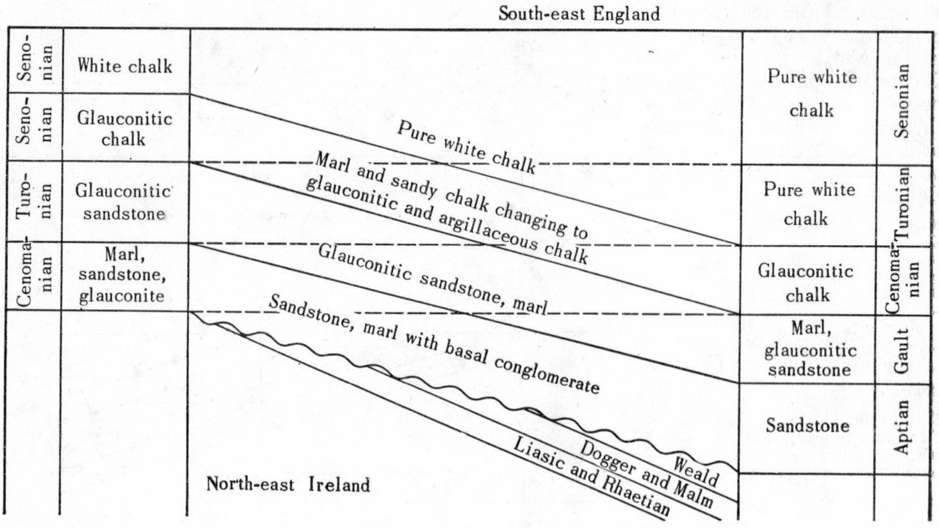

Fig. 22. Migration of facies between the South-east England and North-east Ireland after Grabau, 1960)

the coastline moves to points *2*, *3*, *4*, etc., all the facies are shifted to the right. In this case we observe a regressive series of sediments in the vertical section *A-A* starting at the bottom with limestones and ending at the top with shingle. Such a series is evidence of gradual approach of the coastline to the point at which the section considered is located. Once again, a time in the formation of sediments of the same facies in different places is established. It should be noted, however, that, in spite of the regressive sequence of rocks, their accumulation proceeds the whole time in a situation of subsidence of the crust.

An example of a regressive series is the section of Palaeozoic and Neogene sediments of the northern slope of the Caucasus described above. The regressive series of Permian sediments on the western slope of the Urals is quite typical; there Carboniferous limestones give way to Lower Permian sandy-argillaceous sediments and Upper Permian continental sediments. Here and there sediments accumulated during subsidence of the crust, and their regressive sequence, which reflected gradual raising of the Earth's surface, was due to sediments accumulating faster than the subsidence. The rapid accumulation was the consequence of an intensive uplift and erosion of the adjacent mountain ranges, the Caucasian in the first case and the Urals in the other.

An alternation of transgressive or regressive series with serrated introduction of different facies into one another occurs when the coastline shifts transgressively and regressively. Hence it will be clear that an alternation of strata of different composition in a section may be explained not only by a change in the width of distribution of detritus but also by migration of the coastline.

In the example of facies migration in England and Ireland cited above, the shifting of the coast occurred quite slowly, so that its history can be traced by ordinary stratigraphic methods (e.g., by the change in fossil fauna). But the shifting of coastlines can happen so quickly that there is no time for facies migration to leave its "fingerprints" in a change in the picture of the fossil organisms. Even in these cases, however, there are indirect indications of such a migration. Quite often, for example, an extremely broad distribution of shingle is found in the basement of some regressive series or other, the palaeontological data indicating that the shingle did not extend beyond a single stratigraphic unit. In such cases there seems to be a discrepancy between the area of distribution of the basal shingle and the width of the band in which shingle is usually deposited along a coast. This unconformity may be explicable simply by the basal bed of shingle covering so wide an area not having developed all at once, but gradually, during the shifting of the coastline. It developed "like a carpet" following the rapidly shifting coastline, and so represents, in essence, a trace of the shifting coast. At any single moment it was being formed in a narrow coastal zone. The same sediments of other facies, covering wide areas, can also be regarded as separate traces of shifting of a coast.

The two phenomena discussed here, namely, change in the width of the distribution of clastic material and shifting of the coastline, are inseparable in nature. They always occur together. That being so the following situations can arise:

1. With advance of the sea the relation between uplift and subsidence shifts toward a predominance of subsidence. The facies zones shift, following the coastline, and the zones of clastic facies gradually become narrower and narrower. In any vertical section a transgressive series of sediments comes into evidence (see Fig. 24*A*).

2. With advance of the sea the relation between the uplift and subsidence shifts toward a predominance of uplift. The zones of clastic facies widen and some at least of the boundaries of the facies zones (those relatively further away from the shore) may therefore shift in a direction opposite to the movement of the coastline. In that case a regressive series of sediments may be observed in some sections, in spite of the advance of the sea and the fact that the coastline is moving away from the point of observation (see Fig. 24*B*)

Since this case seems unusual, let us take an example from a natural situation. On the northern fringe of a projection of Palaeozoic rocks in the Donets Basin the Upper Cretaceous is represented by chalk, marls, and limestones, and only at its very base is a thin bench of sandy-argillaceous rocks observed. On the northern edge of the Donets Basin the Palaeogene already consists wholly of sandy-argillaceous sediments, which not only occur far to the north of the Donbass, but also transgressively overlap its Palaeozoic rocks, extending to the southern boundary of the Cretaceous deposits much further south. Thus, in this case, the marine basin widened from the Cretaceous to the Palaeogene, but the sequence of facies in the vertical section has a regressive character.

3. With retreat of the sea the relation between uplift and subsidence shifts toward a predominance of uplift. The facies zones are shifted in the direction of the movement of the coast and the zones of terrigenous sediments expand. In all sections a regressive series is observable (see Fig. 24*C*).

4. With retreat of the sea the relation between uplift and subsidence shifts toward a predominance of subsidence. The zones of clastic sediments narrow and some of them (those relatively further from the coast) may be shifted in a direction opposite to that of the coastline. In several sections a transgressive series can be observed, in spite of the regression of the sea (see Fig. 24*D*). The East European platform on the boundary of the Early and Late Creta-

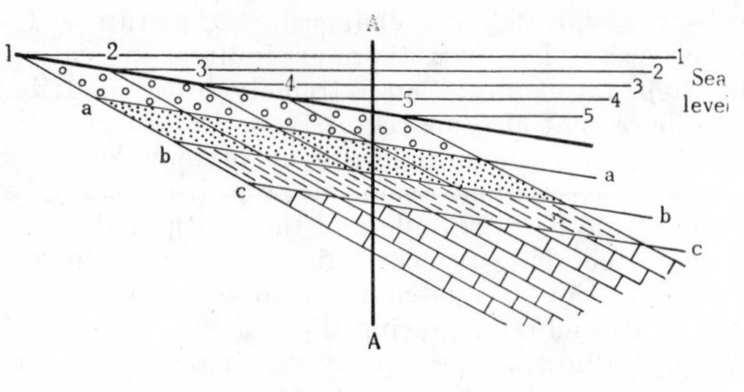

Fig. 23. Shifting of facies zones with regressive shifting of the coastline (after Beloussov). Conventional symbols as in Fig. 21

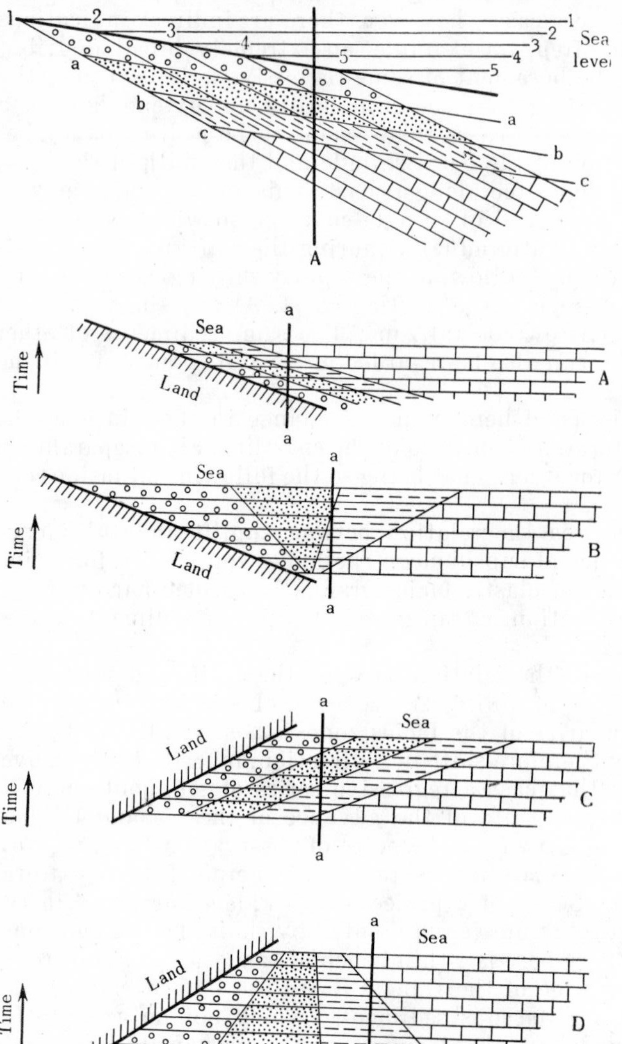

Fig. 24. Cases of a combination of change in the width of the distribution of clastic material with transgressive and regressive shifting of the coastline (after Beloussov): *A, B, C, D*—different cases described for section *a-a* in the text (after Beloussov). Conventional symbols as in Fig. 21

ceous can serve as an example from nature. The Late Cretaceous sea occupies a smaller area on the platform than the Early Cretaceous, but in spite of the regression of the sea, the change of rocks from Lower Cretaceous sandy-argillaceous sediments to Upper Cretaceous limestones is transgressive in character.

When a geological section is being studied it is usually possible to establish that it is divisible into major segments, each of which covers a considerable stratigraphic interval, and which are distinguished by a characteristic "set" of sedimentary rocks. Such a segment may, for example, be made up of alternating sandstones and clays, another of limestones and clays, a third almost wholly of limestones, while with a fourth a rhythmic alternation of sandstones, clays, marls, limestones, etc., may be observed. Such "sets" of sedimentary rocks represented by definite facies are called *sedimentary for-*

mations. The transition from one formation to another may be either abrupt or gradual.

Everything said above about the conditions of the forming of facies of deposition enables us to link each formation with a definite regime of oscillatory movements of the crust, stably maintained during the whole time of the development of the given formation. Such a regime is characterised (a) by a certain average relation between uplifts and subsidences, by certain mean deviations from this relation to one side or the other, which occur with a certain rhythm; (b) by certain medium-sized migrations of the boundary between the areas of erosion and deposition; (c) by an average degree of compensation of subsidence by accumulation; and (d) by an average degree of deviation from this state. The transition from one formation to another means a change in some or other of these characteristics of the regime of oscillatory movements.

What has been said should not create the impression that the picture of sedimentary series as a whole is determined by the tectonic factor. Climatic conditions, the temperature and salinity of the water, the degree of development of the organic kingdom, the composition of the rocks undergoing destruction on the surface of regions of erosion, are all decisive in determining the nature of the material entering sedimentary series. The tectonic factor is mainly responsible for the distribution of this material over the area and in the geological section. Such are the most general features of the division of the "spheres of influence" of the various types of factor on the sedimentary series in the Earth's crust.

CHAPTER 7

Methods of Studying Ancient Oscillatory Movements

In order to study the history of ancient oscillatory movements we use their various forms of influencing the distribution of the thicknesses and facies of sediments. In the language of physics an inverse problem is posed, viz., to infer the history of the development of oscillatory movements in ancient geological periods from the change in thicknesses and facies of sediments both areally and in section. Two main methods are at our disposal for tackling this problem: (a) the method of thicknesses, and (b) the method of facies.

We must note at once the rather limited character of the data on the history of ancient oscillatory movements as compared with the information available from study of recent ones. For the latter there is the possibility in principle of reconstructing the history of vertical movements of the crust, both upwards and downwards. Downward movements are studied from the Neogene and Quaternary sediments accumulated, and upward movements from the surface forms of relief. But the landforms that existed in older periods—in the Palaeogene and earlier times—have almost vanished. There is therefore no possibility of reconstructing the history of ancient rising vertical movements of the crust as fully as for recent ones. In reconstructing fully the history of ancient downward movements from the thicknesses and facies of accumulated sediments, we obtain only indirect data on the locations and intensity of uplifts.

The Method of Thicknesses

This method is not usually employed in its pure form, but by taking the facies of the sediments studied into account.

When the final geomorphological expression of the oscillatory movements (i.e., change in the level of the surface of the deposition of sediments) is small at some point in the geological time interval being studied, the thickness accumulated during that period corresponds quite accurately to the amplitude of the subsidence of the crust. It is possible, consequently, to employ the thicknesses of sediments as a measure of subsidence, provided that comparison of the facies of the sediments deposited at the beginning and the end of the studied interval points to the maintenance of an unchanged depth (in the case of marine sediments) or of the height of the sedimentation surface above sea level (for continental deposits).

It is necessary, when estimating thicknesses, of course, to avoid possible errors due to secondary changes of thickness through the deformation of strata, for example, through their buckling into folds. Where plastic deformations of strata are associated with considerable distortions of the original

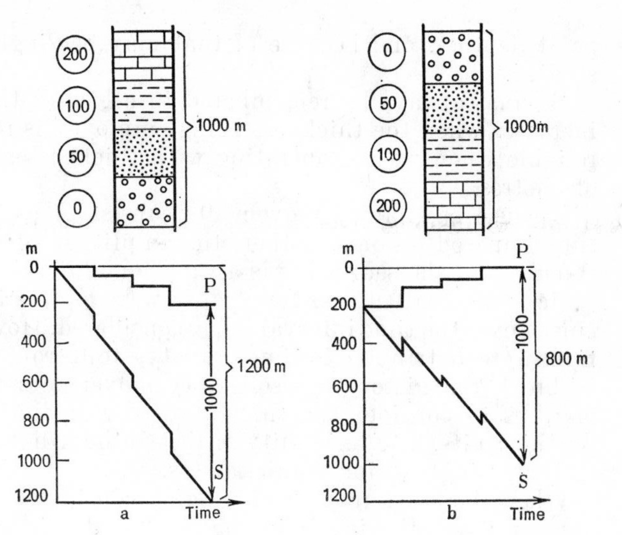

Fig. 25. Determinations of the amplitude of subsidence of the crust (*a*) in a situation of deepening of a basin; and (*b*) with shoaling (after Beloussov):

P—curves of the geomorphological expression of oscillatory movements; *S*—curves of tectonic movements; 1200 m and 800 m—the amplitudes of total subsidence; 1000 m—thickness; 0-200 m—depth of sedimentation

thickness of the series, not amenable to assessment, as happens for instance in metamorphic rocks, the thickness method is unsuitable.

While comparison of the facies of sediments occurring on the one hand at the bottom of the section being studied, with its top on the other, indicates a change in the level of the Earth's surface, one has to correct the thickness for the amplitude of the subsidence obtained, a correction that will be equal to the amplitude of the geomorphological expression of the oscillatory movements, i.e., the amplitude of the rise or lowering of the surface. With a raising of the surface the amplitude of the rise is subtracted from the thickness, and with a lowering of the surface, the amplitude of the change is added to the thickness. In the first case the thickness was not simply formed during the compensation of subsidence but partially also as a result of the accumulation of "excess" sediments above their initial level. In the second case, the accumulation of sediments lagged behind the subsidence, which remained partially uncompensated (see Fig. 25).

Experience of using the thicknesses method on many examples shows that, when a total thickness of at least several hundred metres is employed, the thickness of the sediments may be taken in the overwhelming majority of cases as a good approximation of the extent of the subsidence of the crust, even without any correction for a change in the level of the surface. This is associated with the fact that most of the sediments now observed in geological sections on continents were laid down in shallow marine conditions on the shelf, i.e., at a depth less than 200 m or were formed on coastal alluvial plains, where sediments of the continental facies were formed at a height not more than 100 m above sea level. In most cases, consequently, the level of sedimentation varied little. With very rare exceptions (see the example above of the Artinskian sediments in the Ishimbai district, p. 34), this holds for the whole sedimentary cover of platforms, and for parageosynclines and miogeosynclines. As for eugeosynclines, the exceptions for them are epochs of extremely intensive subsidence when "empty" troughs are formed (see p. 137) that are characterised by the formation of certain specific sedimentary facies (mainly of various siliceous rocks with signs of having been formed at great depths), and because of that may be singled out in the section being studied.

It must always be remembered in all cases that the possibility of errors increases when the thickness being employed is reduced. The method is most reliable when we are operating with thicknesses of hundreds and thousands of metres.

Because oscillatory movements consist of oscillations of different orders superimposed on one another, the amplitude of subsidence measured by the thickness of the sediments is always an additive one, representing the algebraic sum of all upward and downward oscillations of various order that occurred over the time interval being considered. How precisely the minor oscillations were distributed in time cannot be told from the total thickness. In order to break the history of oscillatory movements down into shorter intervals one has to consider the thicknesses corresponding to ever smaller stratigraphic units. The reliability of the method, however, diminishes as we pass to increasingly thinner thicknesses.

When the stratigraphic unit being considered is overlain by another unit not directly following it in the normal stratigraphic sequence, and there is a break in the sedimentation, the observed thickness is the total of all the oscillations that occurred from the beginning of deposition of the studied unit to the beginning of deposition of the series overlying it. If, for instance, the Lower Cretaceous is overlain by the Palaeogene, the thickness of the Lower Cretaceous will indicate the total of all the oscillations that occurred from the beginning of the Early Cretaceous to the start of the Palaeogene, which means that one cannot say from the thickness alone whether there was a break in sedimentation in the Late Cretaceous, or whether Upper Cretaceous sediments accumulated and were then eroded away before the Palaeogene. If the studied sediments outcrop on the surface or are overlain only by Quaternary deposits linked with the present-day relief and subjected to Quaternary erosion, the thickness preserved can be considered only the lower limit of the possible total thickness. Since a thickness is the sum total of all oscillations during accumulation, the correction for change in level of the surface of the unit being studied must be an aggregate one and based on comparison of the facies only of the sediments that begin and end the action of this unit.

The results of analyzing thicknesses can be expressed in a different form. It is possible to indicate the development of subsidence of the crust at a given point during some interval or other of geological time. To do so, geological time is plotted along the X-axis of a graph while the total thickness accumulated from the initial moment to each successive moment for which thickness is being calculated is plotted along the Y-axis. A change in the slope of the curve gives an idea of the change in rate of subsidence. If it is desirable to introduce corrections for a change in the surface level, a curve of geomorphological oscillations is also plotted, and the thickness plotted along it and not along the horizontal zero line. Examples of such diagrams were given above (see Figs. 8, 9, and 10).

A graph employing the change in the rate of subsidence, rather than its amplitude can also be plotted (see Fig. 26).

In order to get a picture of the distribution of the different amplitudes of crustal subsidence in a certain area over a definite period, maps of the distribution of thicknesses are compiled, based on the contour principle. Thickness contours are also known as isopachs or isopachous lines. As many points as possible, at which the thickness of the stratigraphic unit we are concerned

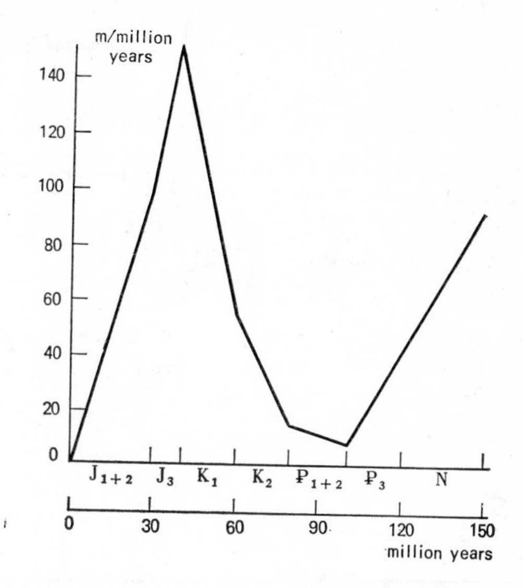

Fig. 26. Graph of a change in the rate of subsidence of the crust in the North Caucasus (metres/million years) (after Ronov, 1949) with corrections of the time scale

with have been determined, are plotted on the map and isolines are drawn through them for the section that is most convenient in the given case.

Thickness contours picture the relief of the bottom of the studied series as it would be if the top of the series were horizontal. If there is no need to introduce a correction for a change in the surface level, the picture obtained will reflect the relief of the bottom of the given series formed by the end of its accumulation. If the change in the facies of the sediments occurring at the top of the series makes it possible to suppose that the level of the surface of sedimentation was essentially different at various places in the studied area at the end of accumulation of the series, appropriate corrections are introduced so as to obtain the true relief of the bottom.

Sectors in which a given series is missing from the section are contoured by zero isolines.

Isopachous maps are compiled for the sequence of stratigraphic units developed in a given area. Comparison of them in historical succession then makes it possible to judge how the form and disposition of sectors or zones with a different degree of subsidence altered with time (see Figs. 12 to 19).

Isopachous maps can be accompanied by "palaeotectonic profiles" indicating the distribution of thicknesses along certain lines. If corrections for change in the level of the surface are not being considered when a profile is being compiled, the top of the suite being studied is plotted as a horizontal line, and thicknesses plotted downward from it. Composite palaeotectonic profiles can also be compiled on which the distribution of thicknesses of several units is shown at the same time (see Fig. 27).

The Method of Facies

So far we have used the facies of sediments only as additional data enabling us to introduce correction into thicknesses, so as to determine the amplitude of crustal subsidence, but certain independent tectonic problems can also be solved with their help.

When the change in facies is plotted on the map of a certain area of some stratigraphic unit, the direction from which detritus is brought in can be

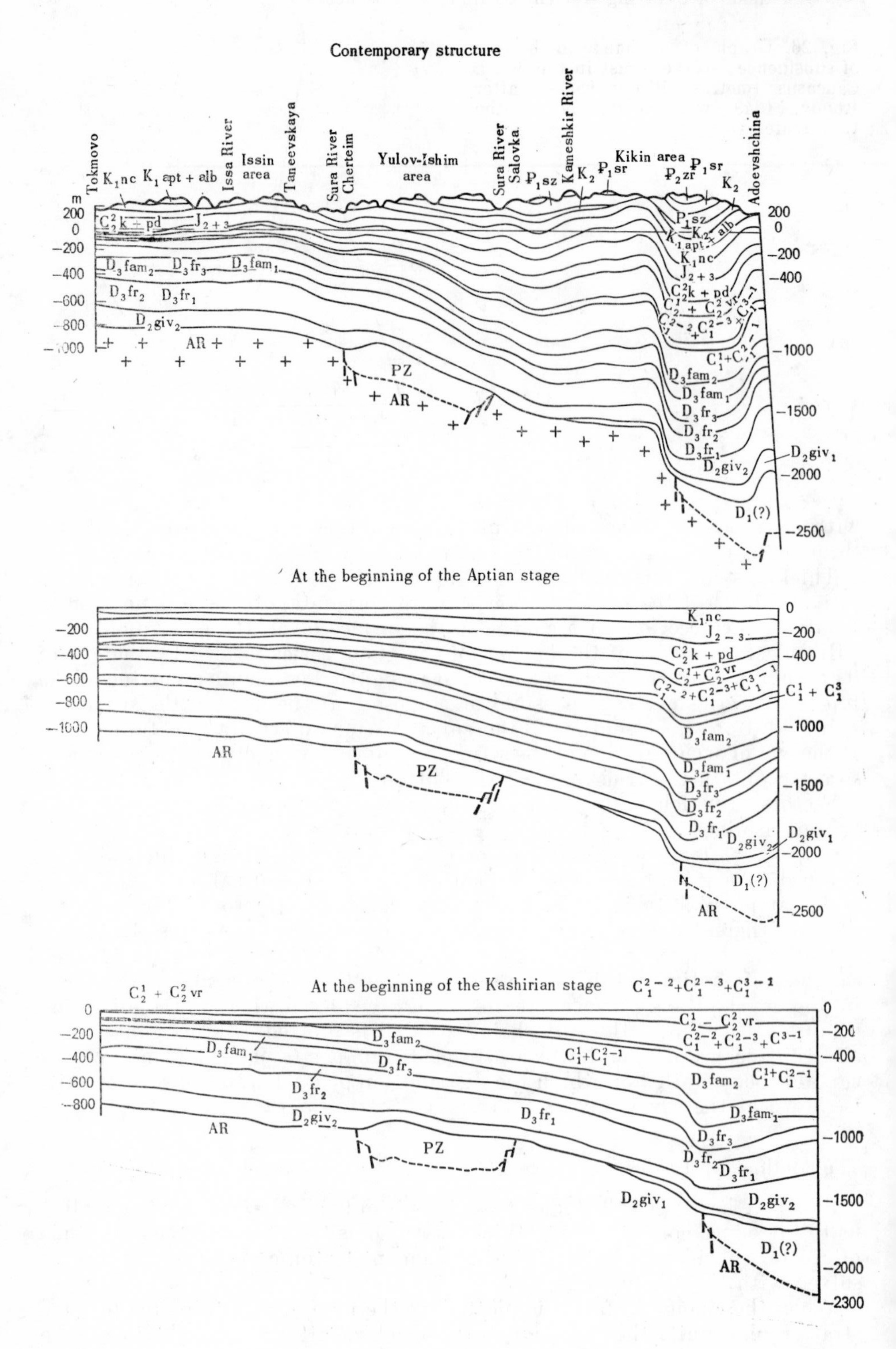

Contemporary structure
Tokmovo
Issa River
Issin area
Taneevskaya
Sura River Cherteim
Yulov-Ishim area
Sura River Salovka
Kameshkir River
Kikin area
Adoevshchina
AR
PZ
D1(?)
At the beginning of the Aptian stage
At the beginning of the Kashirian stage

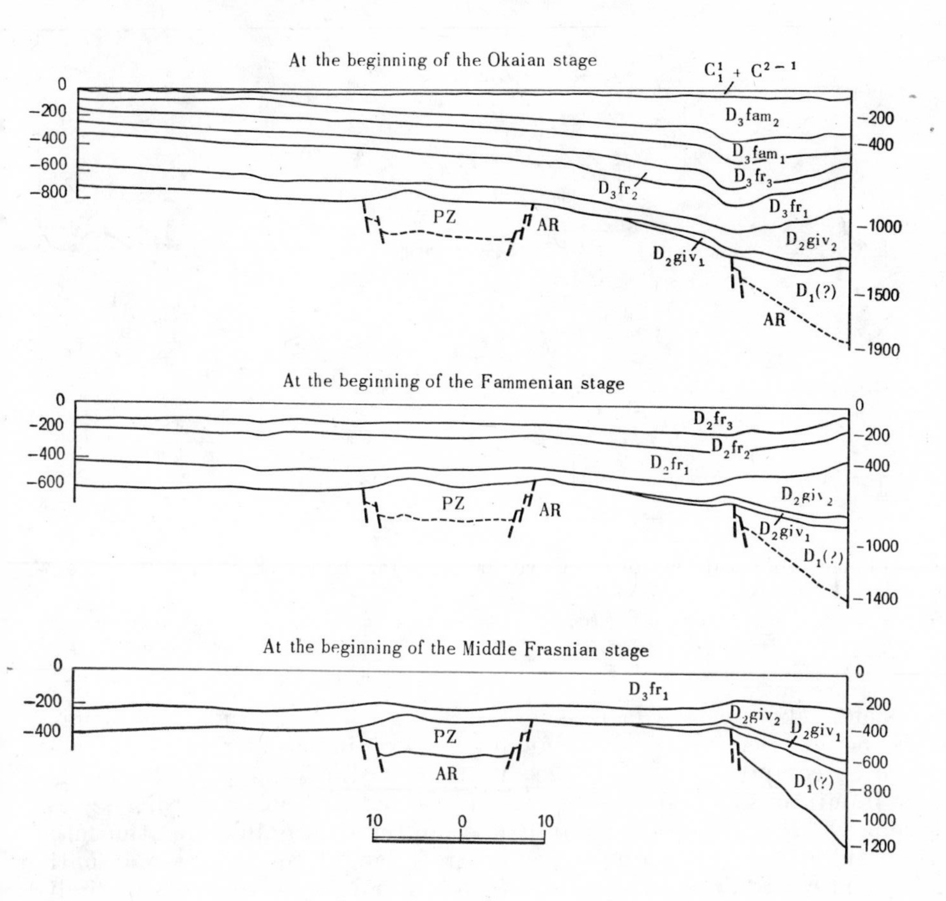

Fig. 27. A series of palaeotectonic profiles and the present-day structure along the line Tokmovo to Adoevshchina in the Transvolga area (after Khokhlov)

determined from the distribution of terrigenous sediments of greater or finer coarseness. If there are enough data we can outline regions of uplift of the crust; the nature of the zone of zero thickness can only be established from the facies. When sediments become coarse the nearer they are on all sides to this zone, we can say that the absence of sediments coincides with an area where uplift was occurring during the time interval being studied. When there is no such change in the nature of the sediments, it is possible that any sediments deposited in the area of zero thickness were eroded away some time later.

A change in the coarseness of sediments and of their areal distribution in time enables us to estimate the change in the rate of uplift in a region of erosion or of subsidence in an area of accumulation.

We spoke above of sedimentary formations as regular "sets" of facies. These "sets" can be used instead of individual facies to obtain a general idea of the distribution of regions of erosion and deposition, and the mean ratio between uplifts and subsidences.

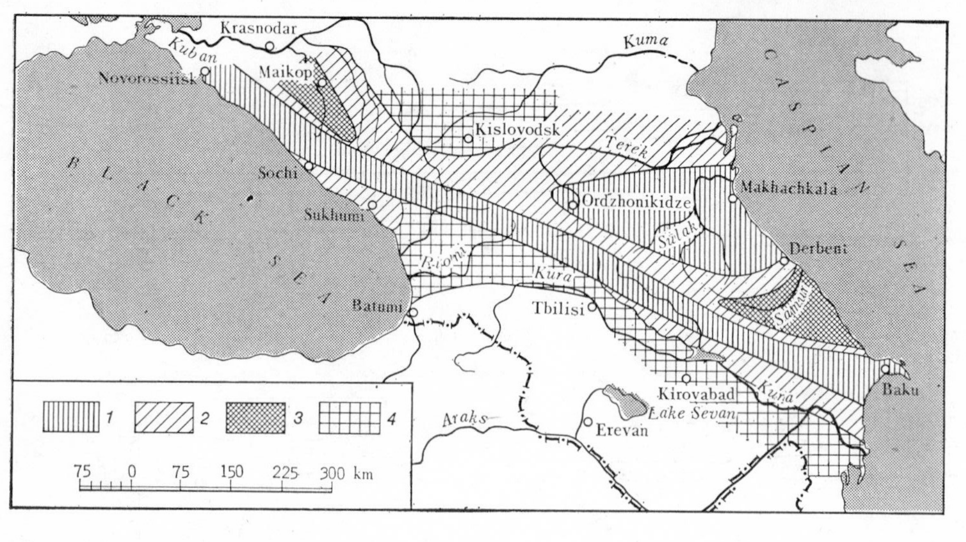

Fig. 28. Palaeogeological map of the Great Caucasus for the period of Precallovian erosion (after Beloussov):

1—regions of continuous Callovian occurrences; *2*—regions of transgressive Callovian occurrences in the Middle Jurassic; *3*—regions of transgressive Callovian occurrences in the Lower Jurassic; *4*—land in the Callovian era

Maps of thicknesses and facies can be combined in one single map. Facies can also be included in palaeotectonic profiles, but independent facies profiles can also be compiled on which the duration of stratigraphic units, for instance, in millions of years, can be shown rather than their real thicknesses.

In some cases it is useful to compile maps of the stratigraphic relationships between suites, distinguishing places where accumulation proceeded continuously and where there was a break; it is also possible in that case to indicate the size of the break (see Fig. 28). Such maps are, as it were, geological maps of the surfaces underlying a stratigraphic break and are therefore called "palaeogeological maps".

The Volume Method

The volume method was proposed by A. B. Ronov. Essentially it consists in the following.

Any quite large area of the Earth's surface is always divisible into sectors of uplift and subsidence of the crust, the rates of which differ from one another. One can speak of the volume rate of oscillatory movements, meaning by that the rate of increase of the total volume of all the tectonic rises and troughs found in a given area.

The volume of troughs can be taken as approximately equal to the total volume of the sediments accumulated in a given time interval. Since an uplift leads to erosion, the volume of uplifts in terms of geological time can be taken as approximately equal to the volume of detrital sediments carried away from these uplifts.

Consequently, if we estimate the volume of all the sediments accumulated during a certain time and, separately, the volume of detrital sediments in them, we will get the total volumes of all the troughs and all the uplifts formed

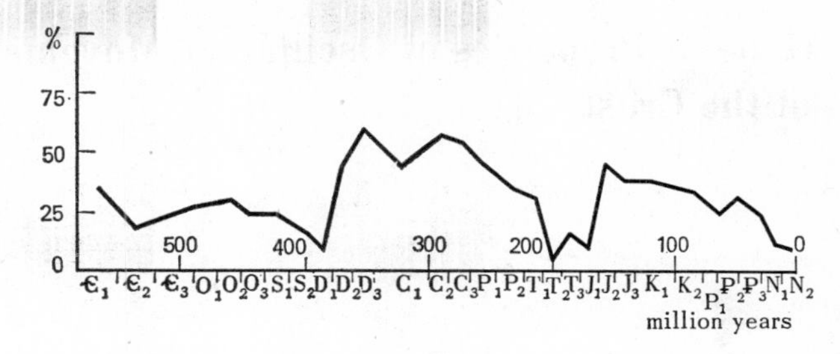

Fig. 29. Changes in the area of sedimentation as a percentage of the total area of the East European platform during the Phanerozoic eon (after Ronov, 1949)

during this time within the area being studied. We can also determine the ratio between the volumes of subsidences and uplifts, and this will be equal to the ratio of their volume rates. By making such calculations for the sequence of stratigraphic units, allowing for their absolute duration, we get the change in the volume rates of subsidence and uplift and their ratio in the course of geological time. The appropriate calculations are made from thickness and facies maps, and from detailed lithological sections that enable the role of detritus in the series being studied to be estimated.

Since we relate the volume of detrital sediments to the total volume of all sediments, the ratio obtained cannot exceed unity. Cases in which the volume rate of uplifts exceeds the volume rate of subsidences cannot, consequently, be established in this way. To discover them it is necessary to introduce a correction for the change of sedimentation level. If the rate of uplifts exceeds the rate of subsidences in total, more detritus is brought in than the troughs can accommodate, and the surface of the sediments in zones of accumulation will rise. This rise is also a measure of the preponderance of uplifts over subsidences.

The volume method is used to throw light on the development of oscillatory movements of the crust over wide areas, since it has to take in all possible sources of observable detrital accumulation. It yields a mean characteristic of the ratio between uplifts and subsidences for major intervals of time, i.e., for whole formations. A simplified version of the method consists in measuring the ratio between regions of sedimentation and of erosion, and the change in this ratio over time (see Fig. 29). The volume method can be used to bring out the general patterns of oscillatory crustal movements.

CHAPTER 8

General Properties of Oscillatory Movements of the Crust

Study of the history of oscillatory movements of the crust indicates that two principal types can be distinguished: viz.

1. general oscillations, and
2. undulatory or block-undulatory oscillatory movements.

The first type is expressed in concurrent uplifts or subsidences of extensive areas embracing whole continents or substantial parts of them.

The second type is manifested in long development of rounded or elongated uplifts and troughs tens to hundreds (in some cases thousands) of kilometres wide. The disposition of these uplifts and troughs can continue, as regards their general features, for several geological periods, although the boundaries between them always vary around a certain median position. But these same uplifts and troughs can shift slowly ("roll") along the continental surface, expand and contract considerably, and vanish and re-emerge.

An example of the manifestation of undulatory oscillatory movements is the case of the Moscow syneclise, which sagged with slight changes of its contours from the Middle Palaeozoic to the end of the Mesozoic. Another example is the Uraltau uplift, which was rising for almost the whole of the Palaeozoic.

Examples of general oscillations are the subsidence of the whole East European platform in the Middle Devonian when, after a long continental state, a broad marine transgression penetrated it, and the equally extensive uplift (as regards area) of the whole platform, with its complete drainage, at the end of the Palaeozoic and beginning of the Mesozoic.

General oscillations lead to transgressions and regressions of seas and to changes in areas exposed to the processes of accumulation and erosion, changes that affect the nature and distribution of facies of deposition. They are consequently discovered by the change in facies and breaks in sedimentation.

Undulatory oscillatory movements are expressed, primarily, in the distribution of thicknesses: they determine why greater thicknesses of sediments accumulate over a long period in some zones and thinner ones, or none at all, in other zones. But they also influence the distribution of facies, since undulatory movements result in the formation of erosion zones from which eroded material is borne to adjacent troughs.

Since both general and undulatory oscillations occur simultaneously in one and the same area, they are cumulative. Because general oscillations embrace much larger areas than the individual manifestations of undulatory movements, the latter can be held to occur against the background of the former. Thus undulatory movements of the same and opposite directions are superimposed on general oscillations of some one direction. Let us suppose

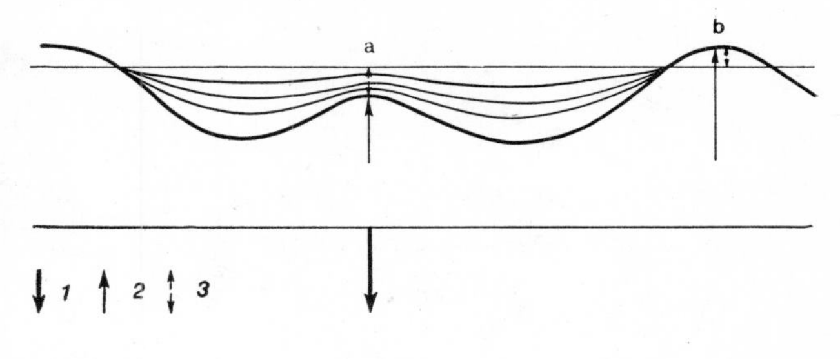

Fig. 30. Diagram of the superimposition of undulatory oscillations on downward general oscillations (after Beloussov). The length of the arrows corresponds to the rate of movement. At point *a* the rate of undulatory uplifting is less than the rate of general subsidence, and at point *b* it is higher.

1—rate of general oscillations; *2*—rate of undulatory oscillations; *3*—difference between the two rates

that a rising undulatory movement of more limited area is superimposed on a general downward movement (see Fig. 30). Then if the undulatory uplift develops faster than the general subsidence, there will be an absolute raising of the Earth's crust on the surface, and a zone of erosion will emerge. If, however, the rate of the undulatory uplift is slower than the general subsidence, the Earth's surface will subside at its crest but more slowly than the rate of general subsidence, and even more slowly than the subsidence of neighbouring troughs, where the downward undulatory movement is added to the general downward movement. In that case we can say that the uplifting in the zone of undulatory uplift is not absolute but relative. This point explains the succession of absolute uplifts in time by relative ones, and vice versa. With absolute uplift a region of erosion develops on its site while, with relative uplift, the same region becomes a site of accumulation of sediments, but the thickness of these is less than that in neighbouring zones.

A palaeotectonic profile of the Lower Cretaceous sediments in the area of the Belaya and Pshekha Rivers on the northern slope of the Great Caucasus is shown in Figure 31. During the Early Cretaceous this district subsided as a whole, evidence of which is the accumulation everywhere of sediments. We can consider this subsidence as a manifestation of general oscillatory movements (general subsidence); within this zone, however, two zones differing in the thickness of the accumulated sediments have been distinguished, a zone of relatively great thicknesses in the west and a zone of relatively thin ones in the east. This was the result of block-undulatory oscillatory movements superimposed on a general subsidence. In the west they tended downwards, like the general oscillations, and in the east they tended upwards. The block-undulatory uplift, however, was slower on average than the general subsidence, so that the undulatory uplift was expressed relatively rather than absolutely, i.e., in the form of relatively thin thicknesses of sediments. The ratio between the rates of general subsidence and of undulatory uplift, however, varied at times in such a way that the uplift in the east was faster than the general subsidence. In those periods the uplift was expressed absolutely; the surface sediments were above sea level and underwent erosion. The erosion preceded Valanginian deposition; the top of the Valanginian stage also suffered erosion before the beginning of Hauterivian depositions. The Albian stage on the uplift, moreover, is eroded away and the Cenomanian stage lies directly on the Aptian.

There were no breaks in the west and there, consequently, absolute subsidence always occurred. In this western zone great thicknesses of clays predominate in the Lower Cretaceous section as a whole, while sands play a

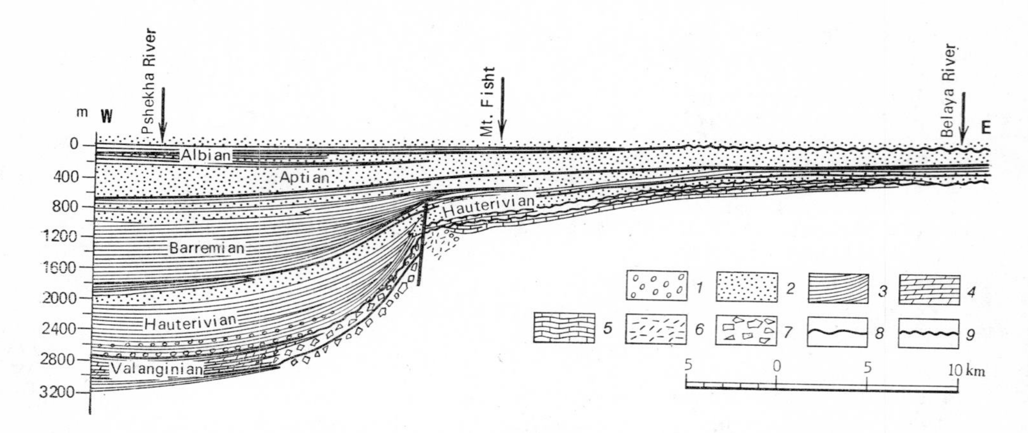

Fig. 31. Palaeotectonic profile of Lower Cretaceous sediments in the mesopotamia of the Belaya and Pshekha Rivers in the Caucasus (after Khain, Burlin, and Lomize, 1961):

1—conglomerates; *2*—sandstones and aleurites; *3*—clays; *4*—marls; *5*—limestones (oölitic, bioclastic, and pelitomorphic); *6*—reef limestones; *7*—limestone and conglomerate brecchia, cemented gravel, and sandstones; *8*—concordant stratigraphic contacts; *9*—erosion surfaces

much greater role in the thin thicknesses of the eastern zone. There, to the east, the sedimentary material underwent long erosion, which was encouraged by frequent general rising oscillations of lower orders that left no mark in the form of distinct interruptions of sedimentation or traces of erosion, but resulted in sum total in a slowing down of the general subsidence and a reduction of the final thicknesses of sediments. The finer argillaceous material was carried west through rewashing to the zone of deeper downwarping where the environment for accumulation was quieter, and only coarse material remained in the east. Sandy material, however, also spread westward from time to time forming benches among the clays there. This means that the area of distribution of coarse material varies, which points in turn to an alteration of the relation between the relative uplift in the east and the subsidence in the west.

A breaking down of general oscillations into movements of many orders superimposed on one another is characteristic of them. We have already mentioned this complexity of the crust's oscillatory movements in connection with present-day and recent ones. Attempts have been made to determine the periods of the oscillations of different orders, but they cannot be considered altogether successful. It has, however, been established that there have been oscillations with a very long period around 200 million years, on which oscillations of many orders with ever decreasing periods have been superimposed. The information on them, however, is rather uncertain; some periods have a length roughly equal to that of a geological period (30 to 40 million years). Others have a period of ten to fifteen million years, and cycles of four to six million years have been demonstrated. Very small oscillations with a period measured in hundreds of thousands of years determine the formation of separate strata.

The column of Palaeozoic sediments in the central areas of the East European platform from the Givetian stage of the Middle Devonian, i.e., from the beginning of the great Middle Palaeozoic transgression to the Tartarian stage of the Upper Permian and the Lower Triassic, preceding the draining of the whole platform at the beginning of the Mesozoic, is depicted in Figure 32. When we consider the change in the composition of the sediments in this column on a very general scale we see that the lower part has a transgressive character on the whole; the arenaceous-argillaceous deposits of the Givetian stage and of the lower horizons of the Frasnian are mainly succeeded by Upper Devonian and Carboniferous limestones. Conditions altered in Permian time in an opposite direction; the Carboniferous limestones give way to Lower Permian dolomites and anhydrites, at the top of which the red beds of continental clastic sediments of the Ufimian stage appear. Continental rocks decidedly predominate in the Upper Permian; their series also embrace the Lower Triassic, after which there was a long break in sedimentation on the East European platform.

On the whole the transgressive sequence at the beginning of the interval being considered and the regressive sequence (which was ended by uplift into a zone of erosion) at its close depict a single large cycle that should be taken as one of the first order lasting around 170 million years. This cycle is determined by the change in the relation of the total volume rates of uplifts and subsidences on the East European platform. The relative role of the uplifts that supplied zones of subsidence and accumulation with detritus diminished at first, but later increased again. Since changes in the relation between uplifts and subsidences is caused by general oscillations, the cycle singled out is one of general oscillations of the first order.

The transgressive series of the lower half of the column, however, is interrupted in the Upper Tournaisian and Lower Visean stages by signs of several uplifts accompanied by erosion and the development of arenaceous-argillaceous coal-bearing sediments. Another interruption of the same sort is noted at the beginning of the Middle Carboniferous, when uplifts again occurred and a certain erosion of earlier deposited sediments and accumulation of sands. There were two epochs of a break in sedimentation and erosion in Permian time; one occurs at the beginning of the Ufa stage and the other in the Tartarian stage. These disturbances of the directed change in composition of the rocks and breaks in sedimentation mark cycles of general oscillations of the second order, with a period of 40 to 60 million years.

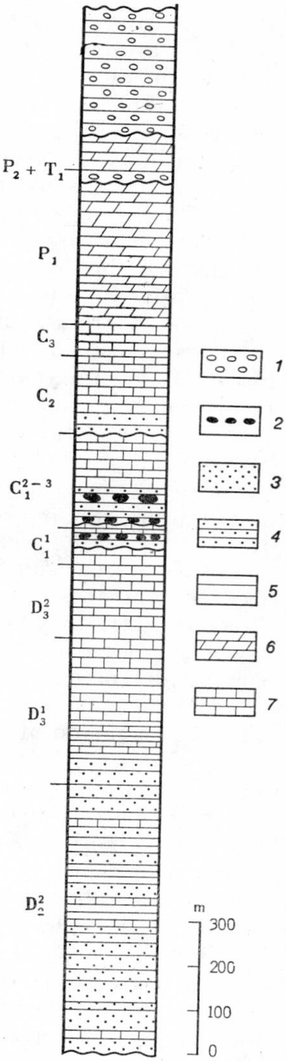

Fig. 32. Generalized section of Palaeozoic sediments in central areas of the East European platform (after Beloussov):

1—continental red beds; *2*—coals; *3*—sands; *4*—sands and clays; *5*—clays; *6*—dolomites, anhydrites, and gypsum; *7*—limestones

When, however, we study the sequence of rocks in this column in greater detail, we can distinguish cycles with ever shorter periods. In the lower, mainly terrigenous part of the column, for example, clays and sands alternate with intercalations of limestones among them. Intercalations of clays and marls are common in the middle, mainly Carboniferous part of the column, while an alternation of evaporites and sands and clays is observed in Permian time. All these alternations reflect minor cycles whose duration cannot be determined.

Undulatory oscillatory movements have a different rate in different places, and various amplitudes and gradients (contrast). Those with similar characteristics are concentrated in certain more or less clearly delimited zones. In this connection we can speak of different regimes of such movements, a problem that will be dealt with later; here, however, let us note that the highest rate, amplitude, and contrast are associated with geosynclinal regimes, while the lowest characterise the regimes of ancient platforms.

Under geosynclinal regimes the rate of ancient undulatory movements is close to 0.05 mm/year on average, but reaches 0.4 mm/year in places. The amplitude may be as great as 20 km, while the mean gradient is $0.7 \cdot 10^{-9}$ $year^{-1}$ (i.e. the difference in heights increases over a million years by 70 cm between two points 1 km apart).

In the regime of ancient platforms the same movements have a mean rate of 0.01 mm/year and an amplitude of several kilometres. The average gradient is $0.04 \cdot 10^{-9}$ $year^{-1}$.

It will be seen from these figures that the rate of ancient undulatory oscillatory movements is approximately one order of magnitude lower than that of recent oscillatory movements and two orders lower than present-day movements. These differences are due in the main to the fact that the averaging of rates for ancient oscillatory movements covers a much longer period than for recent movements, and even longer for present-day ones.

The conjugation of neighbouring uplifts and troughs may be either smooth or disjunctive. Smooth conjugations are observed with small gradients of the movements, although a seemingly smooth conjugation may in many cases be broken in fact by a great many discontinuities, each of low amplitude. Disjunctive conjugations are common in zones with high gradients of undulatory movements. In them discontinuities of high amplitude of displacement between the uplifts and troughs can occur. It was just such cases that justified V. E. Khain in calling this type of movement block-wave rather than wave.

Block-wave oscillations, by bending strata or other structural planes, e.g., the top of the crystalline basement, and displacing them up and down along deep faults, lead to the formation of corresponding structural forms in the Earth's crust, and the structures created by them may differ in height or depth and steepness of the articulation of uplifts and troughs, depending on their scale and contrast.

The characteristic morphological features of the structures created by undulatory oscillatory movements, and their nomenclature, will be pointed out below when we consider the patterns of development of various tectonic zones.

The Theoretical and Practical Importance of Studying Ancient Oscillatory Movements

The theoretical importance of studying the history of ancient oscillatory movements is very great. Since they are the principal type of tectonic movement, reconstruction of their history and revealing of the patterns underlying their development helps throw light on a most important aspect of the crust's history. Oscillatory movements are the background everywhere for other tectonic movements, e.g., folding and faulting, the mechanism and origin of which cannot be understood without establishing their connection with the oscillatory movements. Study of oscillatory movements in conjunction with the history of magmatism makes it possible to get an insight into several aspects of the deep-lying processes that underlie all endogenous geological processes. Many problems of the genesis of sedimentary rocks and, in particular, of sedimentary rock complexes (sedimentary formations) cannot be answered without drawing on findings about the regime of the oscillatory movements in whose conditions these rocks and complexes were formed.

Study of the history of ancient oscillatory movements is no less practically important.

By noting the changes in facies and thickness of sediments from one point of observation to another, and then joining these observations up in the form of a map of facies zones and distribution of thicknesses, the geologist can forecast, by interpolating and extrapolating, the probable composition and thickness of sediments in sectors beyond his points of observation, which he may have to do in order to compile geological sections and profiles, and in order to plan boreholes and interpret geophysical data.

It is especially important to know the distribution of the facies and thicknesses of sediments when they are associated with minerals. Minerals like coal, oil, natural gas, bauxites, phosphorites, and building materials are contained in sedimentary series, and their distribution cannot be forecast, and their reserves cannot be initially estimated without compiling facies and thickness maps, and without clarifying the history of the oscillatory movements of the crust by means of these maps.

By the way of example let us touch on certain of the relations between oscillatory movements and the accumulation of coals.

Paralic coals are formed on flat seacoasts and in swampy lagoons. Definite climatic conditions are needed for their formation, but we are not so much interested in those at the moment as in the regimes of oscillatory movements outside which, in spite of the most favourable conditions, accumulation of coal is impossible.

The bathymetric level of formation of coals coincides with sea level. Epochs favouring the formation of coal should therefore be those in which subsidence is exactly compensated by the accumulation of sediments and the surface of sedimentation is maintained at sea level. Such conditions arise when, during gradual "recompensation of a trough", i.e., when the accumulation of detritus is faster than the rate of subsidence, the state of sedimentation changes from marine to continental, or when development proceeds in the opposite direction, from continental to marine conditions. In both cases the intermediate stages are very favourable for the formation of coal where the level of sedimentation passes across sea level. Observations indicate that the first of these two possibilities is most favourable. Massive formation of coals is found in those zones and those epochs of geological time where and when the general trend of geological development consists in slow, gradual, but steady recompensation of subsidence by accumulation. Such a situation arises, for instance, in the marginal deeps of geosynclines during a certain stage in their development (see Chap. 15). An example is the Donets Basin, where the coal measures occur between typically marine Tournaisian and Lower Visean sediments, consisting predominantly of limestones, and typically continental Permian sediments. We have already mentioned that these coal measures of

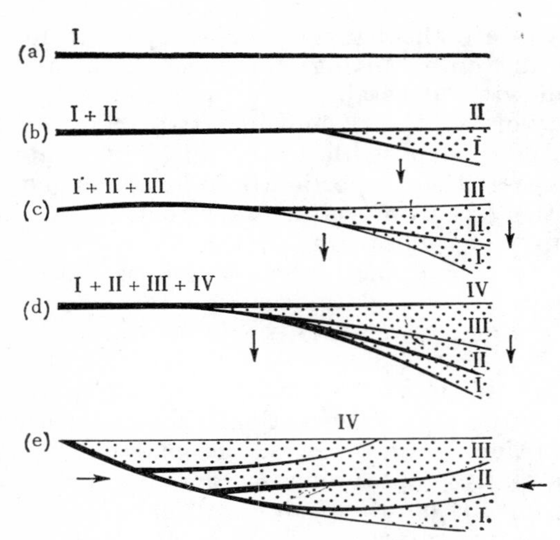

Fig. 33. Development of a coal field with a disintegrated coal seam (after Krashennikov, 1945):

a, b, c, d—successive development stages; *e*—present-day structure

the Donets Basin had a regime of protracted, on the whole very exact compensation of subsidence.

In this way a certain prognostic attribute of coals is elucidated that can help us predict in what part of a section the chances of finding coal are best, when the general character of the change in conditions of sedimentation is known in one zone or another.

For thick coal seams to accumulate there must be a favourable rate of subsidence of the crust corresponding to the rate of the process of coal formation itself. With a slower rate the coal seam cannot attain significant thickness and, moreover, with additional opposite upward oscillations it will be easily broken up and destroyed as it comes into a zone of erosion. Should the rate of subsidence exceed the rate of coal formation and fall to its rate only when subsidence slows down, a series of thin coal seams divided by benches of marine sediments will develop instead of a single thick seam. When regimes with different rates of subsidence lie in the vicinity and there is a gradual transition between them, a characteristic phenomenon of the splitting up of seams may be observed in the transition zone (see Fig. 33). Hence changes in the thickness of coal-bearing suites that reflect a change in the rate of subsidence from place to place can be a forecasting criterion, making it possible to predict the probable thickness and number of coal seams.

Study of the history of oscillatory movements helps us understand the direction and time of migration in oil and gas strata. These mobile minerals migrate upwards along the dip of strata and accumulate as a rule in the highest parts of structures. The present-day relief of the strata along which they migrated may, however, differ from that which existed when migration occurred. Reconstruction of the old dip of the strata by means of a palaeotectonic reconstruction, can lead to the discovery of oil and gas pools where they might seem unlikely judging by the present-day structure.

We shall limit ourselves to the examples given above, as they are sufficient to demonstrate the practical significance of studying the history of ancient oscillatory movements of the Earth's crust.

CHAPTER 9

General Disjunctive Tectonic Movements of the Crust

Deep Faults

The originator of the geotectonic hypothesis of contraction of the globe, Élie de Beaumont, was trying in the middle of the nineteenth century to find a certain regular pattern of trends in the strike of mountain ranges. He suggested that this pattern would correspond to the position of the edges of the polyhedron whose shape the Earth would assume after contraction.

At the beginning of this century W.H. Hobbs pointed out many examples of a geometrical "structuralization" of landforms in which rectilinear trends predominated. In the 1930s R. Sonder suggested the existence of primary faults in the Earth's crust that manifested themselves in the form of "lineaments", i.e., rectilinear structures and landforms. H. Cloos and R. Staub believed that the structure of Western Europe could best be understood if it were supposed that the crust was divided into blocks by deep faults, each block moving as a single whole.

This point received substantial development after A.V. Peive and several other Soviet tectonists advanced their views on "deep faults" in the 1940s. By deep faults are meant, in their views, tectonic ruptures of great extent, measured in hundreds of kilometres on the surface, that have a considerable effect on the character and localization of endogenous processes over a long period. Deep faults are usually expressed in a whole zone of parallel or distributed echelon ruptures, fractured, crushed, and sheared rocks, rather than in a single discontinuity of unbroken rocks. The width of such a zone is usually measured in thousands of metres, and may be more than ten kilometres. These properties of deep faults permit us to view them as ruptures that cut through the whole crust and go down into the upper layers of the mantle. This conception of deep faults is confirmed by outpourings of magma being confined to them, and by the geophysical data.

Types of deep fault, similar to those that are generally delineated along tectonic discontinuous displacements of any scale, are distinguished depending on the direction of the relative displacement of the sectors of the crust separated by them. Some deep faults cut the crust up vertically, and the dislocations along them also have a vertical throw. There is no special term in the geological literature to denote these dislocations, and they are called sometimes normal faults and sometimes reverse faults. By "normal faults", however, it is more correct to understand only faults with a fault plane sloping towards the downthrown side, while "reverse faults" slope, though steeply, towards the upthrown side. Here we propose to call vertical faults with vertical displacement the *slashes.* The first type of deep fault therefore is *deep slashes.*

The next variety is vertical, or nearly vertical, faults with a predominantly horizontal displacement. These are *deep* or *all-crustal shear* or *transcurrent faults.*

Then *deep* or *all-crustal reverse faults* and *overthrusts* are distinguished; *deep (all-crustal) normal faults* with the hanging wall on the downthrown side are also known. Their formation is associated with tension in the crust. The *deep* or *all-crustal separations* expressed in a build-up of gaping fissures in the crust are also evidence of similar tension.

Let us begin our survey of deep faults with deep slashes.

Deep (All-Crustal) Slashes

Deep slashes often serve as the boundaries between zones of uplift and subsidence caused by block-undulatory oscillatory movements. The movements of the planes of these faults are vertical, develop over a long time, and as is usual with oscillatory movement, may reverse their direction. Since these deep faults separate relative uplifts and subsidences, they form boundaries on which the facies and thicknesses of sediments alter abruptly.

Deep slashes develop everywhere. They are particularly distinct in regions where block-undulatory movements are broad in amplitude and high in rate. Examples are the deep faults of Central Asia, the Carpathians, and the Caucasus.

One of them, the Karakul deep fault, runs between the Alai Mountains and the Pamirs along the valley of the Muksu River. Vertical movements of opposite sign occurred on both sides of it over a long period. In Silurian and Devonian times, for instance, the zone of the present-day Alai Mountains sagged 6000 to 8000 m; at the same time the Northern Pamirs were rising. In the Early Carboniferous the Northern Pamirs, on the contrary, sagged by 3000 to 4000 m, while the Southern Alai rose relatively. In Middle Carboniferous time and later, to the end of the Palaeozoic, both the Alai and the Northern Pamirs were rising, and in the site of the Karakul fault there was a trough (see Fig. 34).

The Tyrnyauz deep fault stretches for over 300 km along the northern slope of the Great Caucasus. In Palaeozoic time it separated the sagging Hercynian geosyncline, lying to the north, from the zone of uplift to the south. The amplitude of the vertical displacement along the fault could have been as much as 10 km at that time. In the Mesozoic the same fracture lay on the boundary between the uplift of the Epihercynian platform in the north and a sagging Alpine geosyncline in the south. Later, when the Great Caucasus as a whole were rising, from the beginning of the Miocene, the fault lost its tectonic significance.

From the example of the Tyrnyauz deep fault we can get an idea of the complexity of this type of tectonic structure. In plan the fault is a band 2 to 10 km wide, divided by longitudinal ruptures into several slices or zones, with a section of different character and a different history of development. Five such slices or zones have been distinguished. Their geological sections are represented schematically in Figure 35. It is clear from comparison of their sections that their history begins with subsidence of the southern subzone in the Early and Middle Palaeozoic. In Early Carboniferous time subsidence developed in the northern subzone. In Late Carboniferous time this subzone was rising, while the southern and Tuzluk-Girkhozhan zones were again involved in subsidence. During Permian time only the southern zone subsided, all the others rising. In Triassic time and at the beginning of the Jurassic the whole band of deep faults was rising; subsequently the separate zones

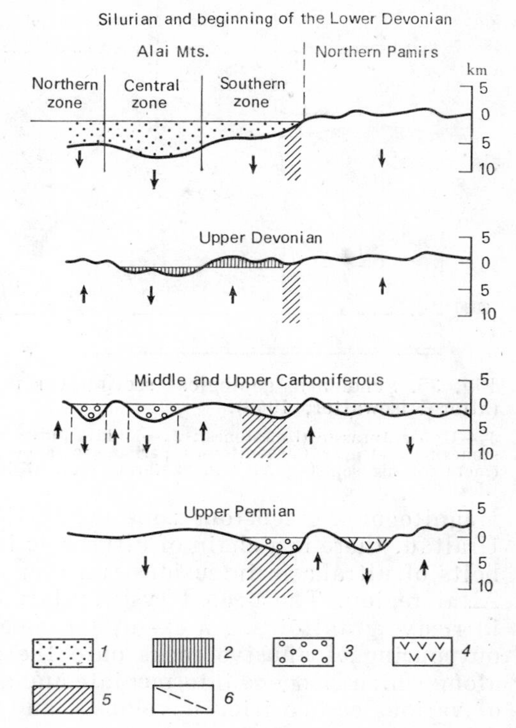

Fig. 34. Scheme of the Palaeozoic development of the Alai Mountains and Northern Pamirs and of the deep Karakul fault that separates them (after Marushkin, 1965):

1—shales, limestones, and sandstones; *2*—limestones and shales; *3*—limestones, argillites, and conglomerates; *4*—conglomerates, limestones, and effusive rocks; *5*—zone of the Karakul deep fault; *6*—partial fractures

began to subside, some from the Middle Lias, others from the Upper Lias or Upper Jurassic.

From the slope of the individual faults that complicate the band of the Tyrnyauz deep fault, it may be supposed that they all converge into a single fault at a depth of several kilometres.

Troughs, called peri-fault troughs by B.A. Petrushevsky, often develop along deep slashes. A big one is the East Ferghana trough adjacent to the Talas-Ferghana fault. The subsidence of the crust and accumulation of sediments in this trough continued throughout the Jurassic. During the Cretaceous there was an uplift in the same place. The amplitude of both the Jurassic subsidence and the Cretaceous uplift increased as they approach the fault which cut them both off abruptly.

A special form of the effect of deep slashes on the structure of the crust is the formation of what are called "crush zones". Examples are the Irtysh crush zone in the Altai Mountains and the Uspensky zone in Kazakhstan. The Palaeozoic rocks composing the locality have been crushed here and there into strongly compressed isoclinal folds pierced by a flow cleavage. These deformations are accompanied by metamorphism; the clays have been converted into phyllites or green schists. Such a zone, with a breadth of tens of kilometres, represents the surface outcrop of a deep fault expressed by a band of strongly deformed rocks heated by hot solutions rising along a dense network of fissures.

The deep slashes have an essential effect on magmatic manifestations. Outcrops of both intrusive and effusive rocks are associated with them. Along the Main Urals deep fault, for example, which divided the submerged Tagil-

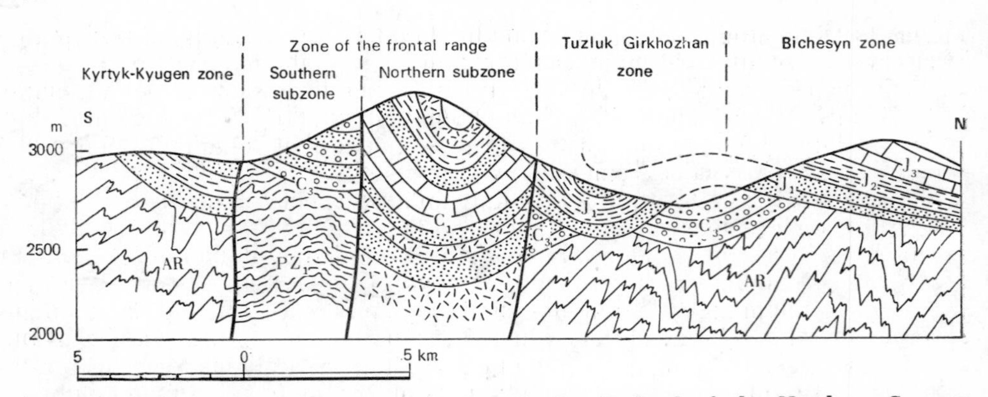

Fig. 35. Schematic geological section through the foreland of the Northern Caucasus (after Kisewalter, 1948):

J_3—Upper Jurassic (limestones); J_2—Middle Jurassic (argillaceous shales); J_1—Lower Jurassic (sandstones); C_3—Upper Carboniferous period (volcanogenic sedimentary strata); PZ_1—Lower Palaeozoic (metamorphic schists); AR—Precambrian (crystalline schists)

Magnitogorsk greenstone zone in the Palaeozoic from the uplift zone of the Uraltau, there is a chain of ultrabasic intrusions. Proterozoic and Palaeozoic belts of ultrabasic intrusions run along the many deep faults of the Sayan-Altai region. The deep Irtysh fault mentioned above is accompanied with intrusive granitoids. An example of deep faults that formed channels for the outpouring of effusive rocks onto the surface is the Transcarpathian fault along which Neogene intermediate and acidic lavas and hypabyssal intrusions of various composition developed, and the Chukotka-Okhotsk fracture with thick effusions of Cretaceous andesites and andesite-basalts.

Deep slashes are of practical interest because of the fact that they often control the location of mineralisation either directly or through the igneous rocks that accompany them.

They are common not only in regions with intensive vertical movements of the crust but also on platforms. Many have been established by geophysical methods.

From the available geological and geophysical data on the whole it can be suggested that the Earth's entire crust, plus the upper layers of the mantle, is split up by faults into separate blocks of various size and shape.

The history of deep faults is complex. They have arisen at different stages of geological history, some in the ancient epochs of the Precambrian, others during various periods of the Phanerozoic, while some are very young. Many ancient faults "closed" long ago and have not manifested themselves for long geological ages. Cases are known, however, of the "revival" of ancient faults and even of several "revivals". The concrete manifestations of one and the same fault at different stages in its "revival" can differ. The Mongol-Okhotsk deep fault, for instance, which has existed from Riphean time to the present, divided a rising zone of Precambrian massifs to the north in Palaeozoic time from a sinking geosyncline to the south. In the Middle Palaeozoic it was a channel for an outflow of basic effusives, and in the Late Palaeozoic it controlled the distribution of massifs of gabbro and gabbrodiorites. At the very end of Palaeozoic time and the beginning of the Mesozoic a narrow peri-fault trough was formed in which marine sediments of the Late Permian and Triassic accumulated. In Jurassic and Early Cretaceous time the one large peri-fault trough gave way to a chain of small, isolated troughs with continental deposits. Here, too, medium and acidic lavas were discharged and minor hypabyssal intrusions of various composition developed.

What can be said about the conditions for the formation of deep slashes? Can it be held that they are a consequence of oscillatory movements, that they are formed where, because of the great contrast of these movements, the crust cannot support the stresses arising in it and breaks?

It is probably exactly in this way that we must conceive the initial mechanism of the development of the faults that are directly observed on the Earth's surface. They arise on the boundaries between adjacent zones of intensive uplift and subsidence. That answer, however, can scarcely be considered the final one to the problem, when attention is drawn to the fact that the arrangement itself of block-undulatory zones of uplift and subsidence on continents often has a quite striking geometrical regularity. For instance, the vast number of uplifts and subsidences of various age on the latitude between Lake Baikal in the east and Western Europe in the west have the same north-western strike. This is manifested in the Eastern Sayan Mountains, the Enisei Ridge, the Altai Mountains, Central Kazakhstan, the western spurs of the Tien Shan, the Kopet Dagh Mountains, the Caucasus, the Donets Basin, the Eastern Carpathians, the Dinaric Alps, and the Apennines, and on the Epihercynian platform of Central and Western Europe, where the Mesocenozoic troughs and uplifts have a predominantly north-western strike. Timan and Pay-Hoy (plus their extension on the South Island of Novaya Zemlya) also have a north-western strike. In many places this strike is combined with a transverse north-eastern strike, and where the two meet the change takes place in an abrupt break at right angles, or they are united by a "butt" joint. A sharp break occurs where the Baikal Mountains meet the Eastern Sayan. The Western and Eastern Sayan have a butt joint at right angles. The Rudnyi Altai change their north-western strike to the north-eastern strike of the Tom-Kolyvan Mountains, while the latter, as the geophysical data indicate, change strike in just the same way ot a north-western one north of Novosibirsk beneath the platform cover of the West Siberian lowland. A butt joint is observed between the Kuramin Mountains and the Northern Tien Shan. The Polar Urals join Pay-Hoy with a 90° break.

Such geometrical regularity in the distribution of tectonic zones is seemingly dictated by just as regular a network of deep faults separating zones with different regimes of both oscillatory movements of the crust and other endogenous processes. Deep faults play their role in this as primary tectonic elements. Their regular geometric pattern is an extremely important feature of the structure and development of the Earth's crust.

Deep (All-Crustal) Transcurrent Faults

A group of deep faults is distinguished in whose history horizontal rather than vertical displacements predominate to a considerable extent. According to this feature these faults should be classed as shear or transcurrent faults. Their length, which may be up to a hundred kilometres, forces us to regard them as general all-crustal faults penetrating the whole thickness of the crust.

Horizontal displacements of sectors of the crust along faults, even when of great amplitude, are much less confidently established, it must be noted, than vertical ones of much smaller amplitude. To assess relatively vertical displacements there are always reference points available in the form of stratigraphic horizons, the thicknesses and facies of sediments, and terraces and

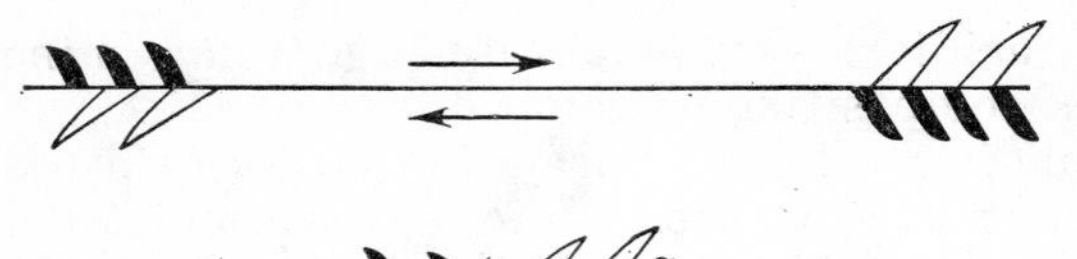

Fig. 36. Scheme of the zones of compression and tension along a transcurrent fault (after Beloussov):

1—separation faults in tension zones; *2*—folds in compression zones

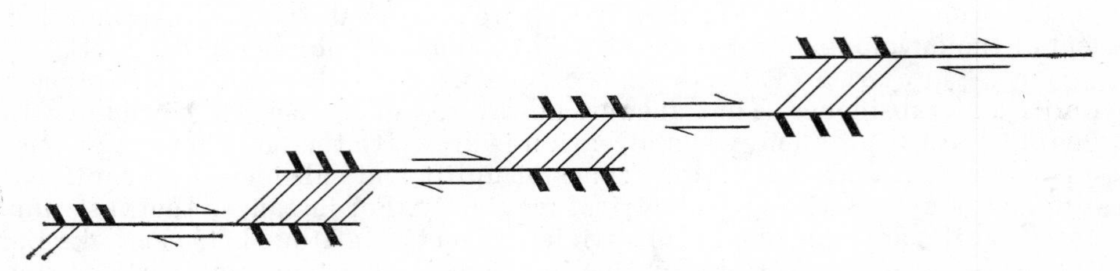

Fig. 37. Scheme of compression and tension zones in a system of imbricate transcurrent faults (after Beloussov):

short thick lines—separation fractures in tension zones; thin double lines—folds in compression zones

surfaces of denudation, all of which make it readily possible to establish both the fact of the displacement and its amplitude. To discover horizontal off-sets reference points of another kind are needed.

These reference points can be purely structural. Since every shear fault ends somewhere along its strike, a horizontal displacement must be "absorbed" over the length of the fault or by corresponding deformations at its ends. On one side there must be compression of the rocks and on the other side stretching (see Fig. 36). The former can be absorbed through the buckling of strata into folds and the second by various stretched structures (through boudinage, normal faults, or movings apart). The alternation of zones of compression and stretching for a system of imbricate transcurrent faults is shown in Figure 37.

All these signs have the drawback that it is impossible in practice to establish the amplitude of the horizontal offset from them: they indicate solely the direction of the movement.

To determine amplitudes we need structures that intersect the fault, cut by it, and are displaced along with it. Such structures can be a fold cut by a shear fault and offset, a cut-up and displaced igneous body, displaced facies zones, etc. These structures, however, are far from always possible to find, and it is also far from always possible to prove that similar, disunited, and relatively displaced structures are in fact parts of a once single structure.

A discussion of long-standing still goes on in the literature around major transcurrent faults in connection with these methodological difficulties. Attitudes to them vary depending on the general moods and speculations prevailing at any time in geotectonics. Sometimes these faults almost vanish from geotectonics, at other times every large fault is elevated to the rank of a transcurrent fault.

The classic example of a deep transcurrent fault is the San Andreas fault in California that runs almost parallel to the Pacific Coast and can be traced on land for a distance of 800 km. Its northern end stops at the coast north of San Francisco, while the southern end passes under the waters of the Gulf of California (see Fig. 38). The fault intersects the major structures of Southern California at an acute angle (the uplift of the Coast Range, the trough of the

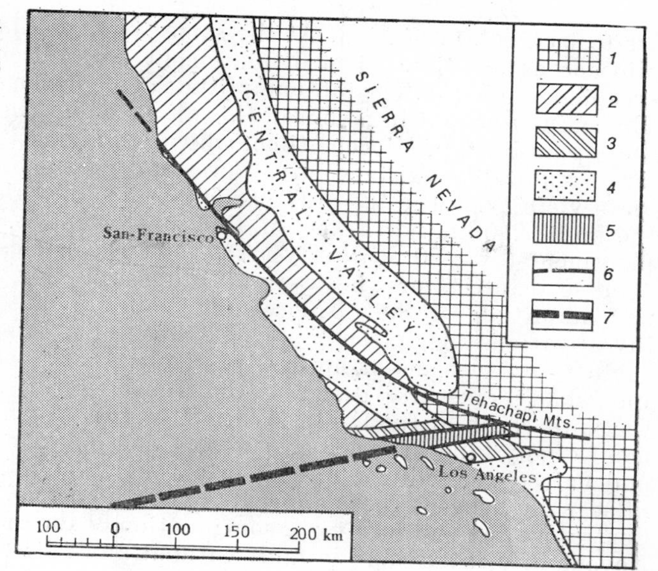

Fig. 38. Geological sketch map of California (compiled by the author after various sources):

1—Metamorphic Mesozoic rocks and granites of the Cimmerian (Nevadan) eugeosyncline; *2*—predominantly Franciscan complex of the eugeosyncline of Alpine age (Upper Jurassic and Lower Cretaceous); *3*—predominantly Cretaceous and Palaeogene rocks of the Transverse Ranges; *4*—Cenozoic deposits; *5*—predominantly Neogene deposits in the graben of the Transverse Ranges; *6*—San Andreas fault; *7*—submarine continuation of the Transverse Ranges—the Murray fracture

Central Valley, and in part the median massif of the Cordilleras). It is expressed by a belt of crushed rocks more than a kilometre wide in places.

The San Andreas fault attracted attention after the catastrophic San Francisco earthquake of 1906 that destroyed the city. The earthquake was caused by a sudden dextral horizontal offset along the fault. The displacement had a different amplitude in different places, with a maximum of 7 m. Comparison of the data obtained from several earlier geodetic surveys indicated that slow elastic deformation began in the zone of the future rupture several decades before the earthquake. For a long time it was not accompanied with displacement along the fault itself. Only when the accumulated stresses exceeded the friction along the fault did the catastrophic displacement occur that caused the earthquake.

It has been established that horizontal displacements along the same fault also occurred earlier, in the nineteenth century, and that the dislocation in all of them was dextral.

The topographical map of the area indicates that all the valleys intersected by the fault are cut up by it and that their segments are displaced relative to each other by a dextral amplitude of around a metre. The present-day relief is Quaternary, and if the length of the Quaternary age is taken as a million years, the mean rate of displacement will be 1 mm/year.

For this time interval of a million years the reference points are quite reliable and the transcurrent displacement along the fault can be confidently established. Geologists, however, have focused attention on the facies of the older, Palaeogene and Mesozoic sediments found on both sides of the fault, and have discovered features in their distribution that could be interpreted as signs of much older displacements along the same fault. A hypothesis was put forward that the fault was formed at the end of Jurassic time and that since then the total amplitude of the dextral horizontal displacement has reached 560 km.

After lengthy discussions such an ancient period for formation of the fault and such a large amplitude of displacement have been found by most researchers to lack substantiation. At the same time, the data pointing to the Late Miocene as the time when displacement began along the fault, have been accepted as reliable. The total amplitude of displacement is now put at 240 km. The main argument in favour of this offset is that the coarse conglomerates of the top horizons of Miocene time, which were deposited in an area east of the fault, require the existence of a source of corresponding detritus in the immediate vicinity, while the only feasible source now lies on the western side of the fault, 240 km north-west of the conglomerates. The mean rate of displacement (over 20 million years) again proves close to 1 mm/year.

Some investigators, however, also question this deduction about the age and amplitude of the displacement, pointing out that detailed geological investigations along the fault find structures intersecting it that witness against any such major horizontal displacements. One of these structures is the broad Transverse Range of the Tehachapi Mts., which stretches across the fault south of the Central Valley, and in which no substantial displacement of any kind has been observed. This range consists of Upper Cretaceous and Palaeogene sediments, bent into a massive anticlinal fold with a graben filled with Neogene rocks along its axis. The belt of granites surrounding the Central Valley and forming a continuation of the granite massif of the Sierra Nevada also intersects the fault without any noticeable horizontal displacement (see Fig. 38).

Those who doubt large horizontal displacements along the San Andreas fault say that it is characterised primarily by vertical displacement. As for the horizontal displacement, that took place only in recent geological time.

The answer to the problem of the scale and age of the horizontal displacement along the San Andreas fault will obviously only be obtained after additional detailed, special investigations.

Another example of a major transcurrent fault is the Great Glen fault in the Caledonian fold zone of Scotland, which has a north-eastern strike (see Fig. 39). The reference points here are two granite massifs, one located south-east of the fault (Foyers) and the other north-west of it (Strontian); the distance between them along the fault is as much as 100 km. It is believed that they are the sections of one massif which was cut by a fault and its parts have undergone a left off-set with an amplitude of 100 km. This displacement occurred at the end of the Lower Palaeozoic (before deposition of the Devonian red suite).

Major doubts, however, have arisen of late as to the adequacy of these ideas (1) because more than two similar granite massifs have been found along the fault and it has become unclear which pairs of them exactly comprise the separate parts of a single whole; and (2) because certain petrographic differences have been discovered between the rocks of the two massifs indicated above.

In the USSR a number of researchers consider the Talas-Ferghana fault in Central Asia a large transcurrent fault (see Fig. 40). This fault is about 400 km long. In general it has a south-eastern strike. In its northern part it separates the Caledonian and Hercynian Tien Shan geosynclines, and to the south it cuts obliquely across the east-west folded structures and individual tectonic zones of the Central Tien Shan. As these structures and zones approach the fault from both the west and the east, they stop short and do not continue directly across the fault. In the opinion of the advocates of the transcurrent nature of this fault, however, a continuation can be found, but with a dextral lateral displacement of 200 km. Just such a displacement the position indicates as it were by the facies zones of the Middle and Upper Devonian, Car-

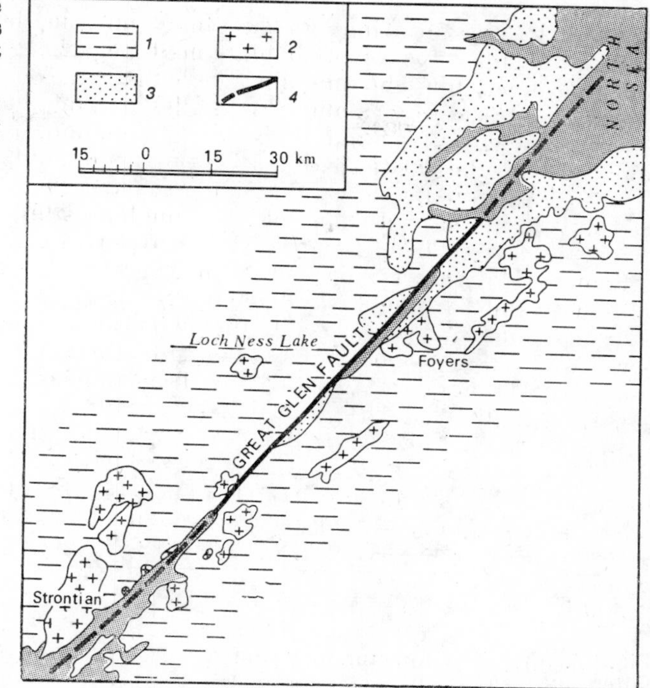

Fig. 39. Geological scheme of the Scottish Highlands and the Great Glen fault (after Kennedy, 1963):

1—Moine schists (Proterozoic); 2—Caledonian granites; 3—Old Red Sandstone (Lower Devonian); 4—fault

boniferous, and Lower Permian on both sides of the fault. The shear fault would have had to occur after the Early Permian. Later, in the Jurassic, only vertical movements were taking place along the fault, which led in particular to the formation of the above mentioned peri-fault Ferghana trough. In very recent geological time, however, already in the Quaternary, the dextral lateral displacement recommenced and led to a break in ravines intersected by the fault with an amplitude of 30 to 50 m.

Once again we come across vagueness and different points of view here. Thirty years or so ago no one doubted that all the observed peculiarities of the structure of the region around the Talas-Ferghana fault were explicable by vertical displacements. Since then one indisputable fact in favour of a transcurrent fault has been found, the displacement of the ravines, but this points to a Quaternary displacement with an amplitude of only a few score metres. A number of researchers still stick to the opinion that the facies of the Middle and Upper Palaeozoic sediments on both sides of the fault are so different that it is impossible to consider them mutually continuous. They could have been formed independently in different conditions and different basins, whose location on both sides of the fault is related to the primary distribution and not to a subsequent displacement. A difficulty also arises in "accommodating" material shifted over 200 km away. In the previous examples (San Andreas and Great Glen), the faults disappear under the sea and their further course can be guessed at however one wishes. The Talas-Ferghana fault is completely on land and both its ends, northern and southern, can be directly observed. It can hardly be imagined that, with a fault 400 km long and an amplitude of horizontal displacement of 200 km in the middle, the displacement of rocks entailed have disappeared through gradual dispers-

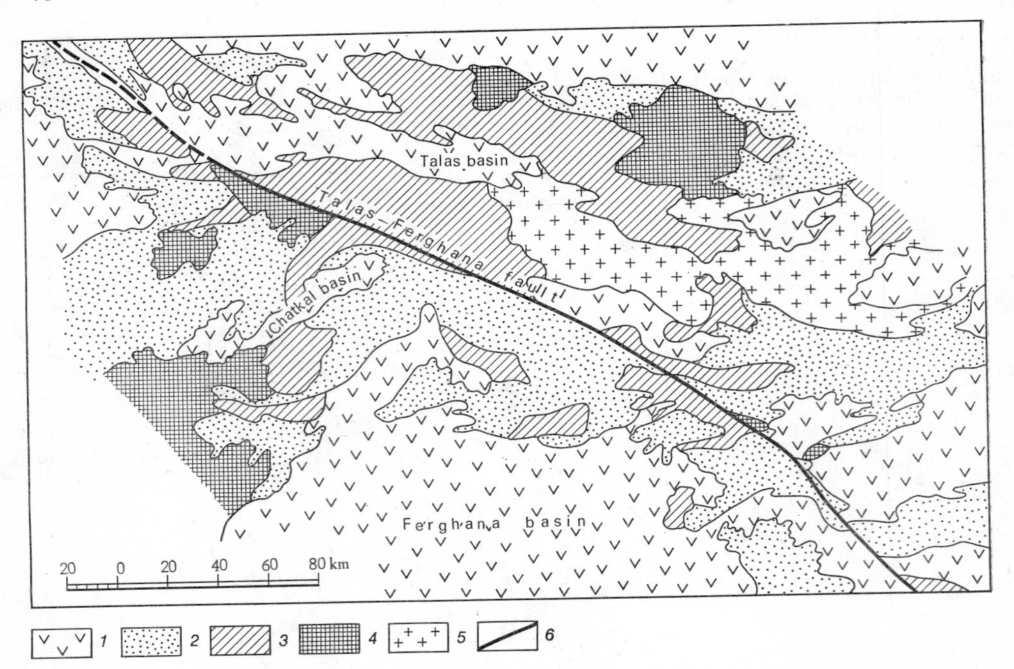

Fig. 40. Simplified structural map of the Tien Shan near the Talasso-Ferghana fault (after Burtman, 1969):

1—Meso-Cenozoic sediments; 2—Permian, Carboniferous, and Devonian sediments; 3—Silurian, Lower Palaeozoic, and Precambrian sediments; 4—Hercynian intrusions; 5—Caledonian and Precambrian intrusions; 6—fault

al along the fault in folds and fractures of normal dimension. One would expect much more intensive deformations, which means that in this case we come up against a number of obscure points in interpreting the nature of the fault.

In all of the cases mentioned, it must be noted, no doubts arise as to the present-day and recent faults during the Quaternary. The present-day ones cause earthquakes registered in our day by direct observation. The recent ones have left traces in the local landforms. The total amplitude of these displacements is measured in tens or hundreds of metres and does not exceed 1000 metres. As regards older and much broader faults, however, there are serious doubts about the unambiguity of the available data. The whole problem of deep transcurrent faults involving the whole crust, which is extremely important as regards general questions of the development of the Earth's crust as a whole, cannot therefore be considered enough and calls for further investigation.

All-Crustal Overthrusts

When it is possible to trace the position of the surfaces of deep slashes to a depth of several kilometres it can be observed everywhere that a fault vertical in depth becomes inclined under a relatively raised block as it approaches the surface. The vertical fault changes upward into an overthrust, the surface of which can become a gently sloping, horizontal, or

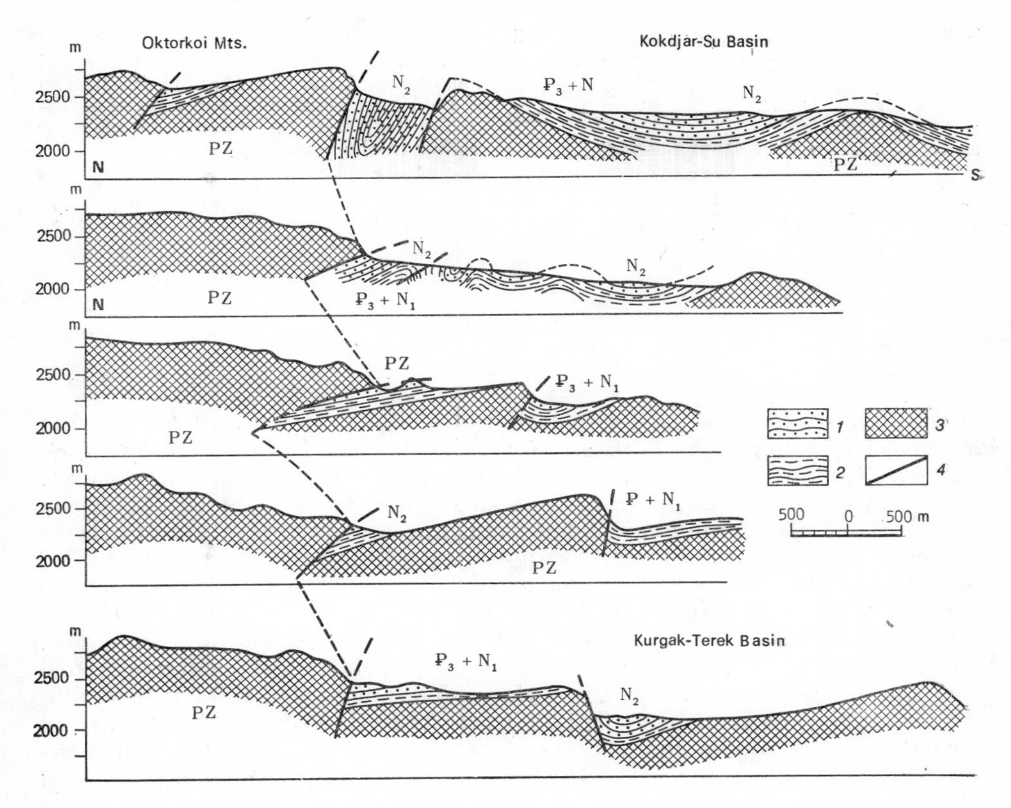

Fig. 41. Schematic geological profiles of the Kokdjar-Su basin (Northern Tien Shan), showing changes in the pitch of tectonic faults along the strike (after Goryachev, 1959). The various sectors along the strike correspond to the depth of the erosion section: *1*—Pliocene; *2*—Oligocene-Miocene; *3*—Palaeozoic; *4*—tectonic faults

even reversely dipping one (Fig. 41). This phenomenon is common in the Tien Shan, where the slashes separating zones of neotectonic uplift and subsidence pass into overthrusts at the top. The so-called Main Caucasus overthrust, confined to the southern slope of the Great Caucasus Mts., also becomes vertical at depth. Very many of the faults in mountainous regions held to be overthrusts with a shallow dip slide going down to a large depth are undoubtedly, in fact, the result of secondary deformation of the faults plane due to sliding sideways under the gravitational force exerted by the upper parts of relatively raised blocks of the crust.

The structures described obviously remain in the category of slash faults and cannot be classed as deep overthrusts.

The only type of structure that can be interpreted as an all-crustal overthrust and even of the mantle is what are called Benioff zones. These are confined to the boundaries between a continent or an island arc and the ocean, more exactly to places where the ocean is bordered by a deep trench (see Chap. 20). Such conditions exist along almost the whole perimeter of the Pacific Ocean; in the Atlantic and Indian Oceans they are confined to island arcs, e.g., the Scotia Arc (South Sandwich Islands) and the Sunda.

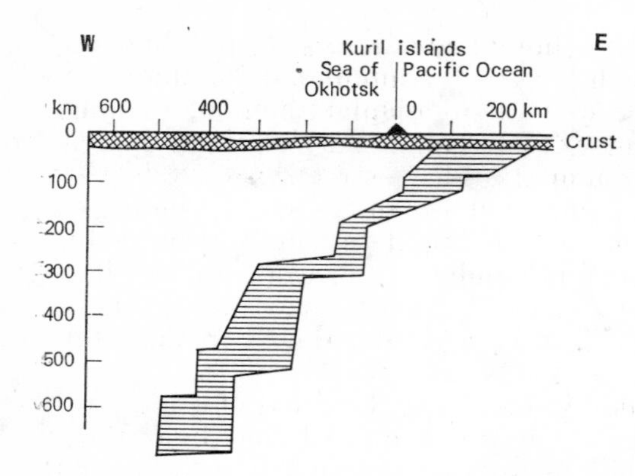

Fig. 42. Benioff zone beneath the Kuril island arc (after Tarakanov, 1972). The shaded area represents the zone of earthquake foci with a magnitude of 4 or higher in the period 1902-1966

Benioff zones form along the location of earthquake foci. In the Circum Pacific seismic belt, observations have shown that foci are concentrated in zones dipping beneath the continents or island arcs fringing the ocean. The average dip of these zones is apparently close to 45°, but the American seismologist Hugo Benioff concluded from processing the location of a large number of foci, that the dip in the upper part of the zone (to a depth of 300 kilometres) is gentler, at an angle around 30°, but to increase deeper to 60°. Since the deepest foci in this belt lie at a depth of 720 km, one must presume the Benioff zone to penetrate the mantle to such an immense depth. The thickness of the zone, measured by the deviation of earthquake foci above and below a certain average surface, does not exceed 100 km (Fig. 42).

Study of the mechanism of foci indicates that the earthquakes occurring in this zone are predominantly associated with displacements of an overthrust type, the hanging wall of the zone moving upward and the lying wall down; hence the idea of Benioff zones as overthrusts of the crust and upper mantle of continents on the crust and upper mantle of the oceans. More recently, however, it has come to be considered not as an overthrust of the continental margin over the oceanic one, but rather as an underthrust of the oceanic lithosphere beneath the continental margin. We shall discuss these points later (see Chap. 25). Geophysical observations only allow us to establish deep overthrusts for our time. On the basis of their permanent link with the deep trenches fringing continental margins of the Pacific type, their existence can probably also be extended to all geological times in which deep trenches exist. But, as we shall see later, the latter are undoubtedly very young, and are a neotectonic formation not earlier than the end of the Palaeogene. There are no evident geological data on the existence of overthrusts of the whole for earlier geological periods, and the views common in the current geological literature on Mesozoic, Palaeozoic, and older Benioff zones need to be considered as propositions lacking any factual basis

Deep (All-Crustal) Normal Faults

Deep normal all-crustal faults accompany tension in a stretching of the Earth's crust. Neighbouring sectors of the latter are pulled apart and inclined normal faults arise between them, where displacements compensate the tension. On the surface a system of faults rather than a single one, forms. The total displacement over the system leads to the development of complex rifts divided into numerous grabens and horsts of the second and the next orders. These complex structures, which stretch for hundreds and thousands of kilometres, and have an amplitude of vertical displacement of as much as several kilometres, are usually called rift valleys. Examples of continental rift valleys are the African-Arabian system of grabens stretching for 6500 km from the Dead Sea in the north to the Zambezi River in the south, the Baikal system of grabens 2500 km long (from Lake Kosogol to the Tokkin basin), the system of the Rhine grabens (600 km), and the aggregate of the fault structures of the Great Basin (or of Basins and Ranges province), lying in the centre of the North American Cordillera in California, Nevada, Utah, and Arizona (1000 km long).

The width of rift valleys varies greatly from tens to hundreds of kilometres. The Upper Rhine graben, for instance, is 40 km wide and the Great Basin 600 km. The vertical amplitude of the crustal subsidence in the Baikal rift is as much as 5000 m (according to some data 7000 m), while in others it does not usually exceed 3000 m. The extent of the stretching of the crust that led to the development of the rift can be calculated from the dip of the faults (usually 60° to 70°) and from their vertical amplitude. For the African-Arabian rift, it amounts to 2000 to 3000 m while for the Baikal and Rhine rifts it does not exceed 2000 metres, in the case of the Great Basin it attains 70 km. In the last case, however, the distension is distributed over a very broad zone occupied by fault structures and is roughly 10% of this width.

The development of rift valleys is accompanied not only by tectonic movements along the faults, but is also characterised by magmatic activity. Their location and time of their origin are connected with features of the preceding history of the crust. Rift-formation comes as a regular stage in the development of the crust and tectonosphere as a whole. We should therefore speak of a special *rift regime* consisting of a certain combination of various endogenous processes. The regime should be ranked with such others as geosynclinal and platform regimes. All endogenous regimes will be considered later, and then the rift regime will be discussed in greater detail.

Deep (All-Crustal) Separations

When tension in the Earth's crust leads not to faults, but to the development of open fissures in it, the structure arising is a separation. When there are grounds for thinking that the open fissure penetrates throughout the crust we are justified in speaking of it as a deep all-crustal separation of the whole crust.

It is clear that a deep fissure in the crust cannot remain open. It becomes filled with magmatic material, forming a dyke in the fissure. It is also difficult to expect that a separation or parting will be preserved in its pure form, that there will be no vertical displacement along the fault. Indeed, in most cases, fissuring is combined with vertical displacement, i.e., is a deep open slash.

A typical deep fissure is the open fault in South Africa, where the Great Rhodesian Dyke intruded. This dyke is a vertical intrusive body of basic and ultrabasic rocks 5 to 10 km wide and 500 km long, which intersects various Precambrian metamorphic and igneous rocks and extends north-north-east from the upper reaches of the Sabi River to the Zambezi. In South Africa the dolerite dyke at Matatieli (Cape Province) is 24 km long and 1600 m wide. The Breven dyke in Sweden is 25 km long and up to 1 km wide. Many dykes have a length measured in tens of kilometres and a width of tens and hundreds of metres.

In a number of places the total separation of the crust is distributed over a large number of individual fissures with which groups or swarms of dykes are associated. Each dyke may be as thick as a score of metres, but when there are hundreds of them, the scale of the total distension must be measured in kilometres. A swarm of dykes of predominantly dolerite composition of the Tertiary age (from the end of the Eocene to the Miocene) is known, for example, in West Scotland and neighbouring islands. The strike of the dykes is predominantly north-south and north-west. In some places they are so dense as to take up a fifth of the whole section.

The question of the space for such dyke swarms and on the extent of the stretching of the crust necessary for them has not yet been resolved. It is quite possible that the magmatic material intruding into the opening fissures pushes the enclosing rock apart mechanically compressing it. In that case the total width of dykes can be greater than the initial stretching of the crust. It is also unclear whether the stretching is compensated somewhere by compression and in what form, and only several hypotheses exist *a propos* of it.

Present-Day Manifestations of Deep Faults

Many deep faults are active at the present time. The commonest form of their activity is seismicity, seismic foci being confined to fault zones. This has already been noted in connection with the deep transcurrent San Andreas fault and with Benioff zones. Seismicity is also characteristic of rift valleys. The deep faults in regions of present-day intensive vertical crustal movements like the Tien Shan and Central Asia generally are seismogenetic. Their degree of seismicity is directly related to the contrast of the vertical movements; this pattern is employed in seismic zoning in order to predict the intensity of the earthquakes that can develop along one fault or another. Far from all seismic tremors, however, are perceptible to man and destructive. Most of them are only registered by instruments. These weak shocks (micro-earthquakes) occur extremely often, thousands of times more frequently than violent earthquakes. Along certain deep faults seismographs may record several minor shocks a day, whereas strong earthquakes occur along the same faults far from every year. The seismic method of tracing deep faults (granted that they are now active) is based on registering weak shocks by a network of seismographs.

Every seismic shock means a displacement along a fault. The slip is very small with weak tremors, but can be of great amplitude in strong earthquakes. We spoke about a displacement of 7 m during the catastrophic San Francisco earthquake of 1906 in California. In 1911, an earthquake occurred in a valley between ranges of Zaili Alatau and Kungei-Alatau during which a scarp up to 9 m high was formed on the surface. The greatest vertical displacement (12 m) occurred in Assam (India) during an earthquake in 1897. The displacements always involve only a certain area of the fault and therefore the latter does not "revive" on the surface all along its length, but over part of it. During the 1906 California earthquake the San Andreas fault revived over a distance of 450 km, in other strong earthquakes the faults have usually been activised over tens of kilometres. The zone of revival of faults, however, may remain deep down completely and then displacement will not become visible on the surface.

The regular confinement of seismic shocks, weak and strong, to deep faults permits us to think that movements do not occur along such faults smoothly and gradually in the main, but by jerks. Every jerk is usually of low amplitude and embraces only a small area of the fault plane. The slip plane migrates from one shock to another and the observed final major geological displacement comes as a result of accumulation of these minor shifts. The individual shifts, of course, may take a different direction, but one trend prevails and in the end, determines the geological nature of the fault.

Along some deep faults, however, also slow gradual displacements are now happening in addition to jerks. The array of instruments along the San Andreas fault indicates that a constant sliding is going on in it in the same direction, as is manifested during strong shocks. The rate of slow displacements varies from 1 to 3 cm a year, but displacement does not occur everywhere and in different years may involve different sectors of the fault.

Gradual subsidence of the Upper Rhine graben too, along its bordering faults, has been established by geodetic methods, it is occurring at a rate of 0.5-0.7 mm a year.

In concluding this section on deep faults we must note that although this type of tectonic movement plays a very important role in the crust's life and has great practical importance, much is still unclear as regards their origin, their geological history, and their relation with other endogenous processes.

SECTION B

Tectonic Movements Within the Crust

CHAPTER 10

Block Folding

The morphological and kinematic classification of foldings is considered in textbooks of structural geology. Here we shall follow the kinematic classification, which is based on differences in the nature of the movements of crustal material that lead directly to the formation of folding of one type or another. This classification envisages distinguishing between the folding of a block, injection folding, general crumpling, and deep folding. We shall consider block folding first.

Block folding includes those fold complexes that are connected in their development with the raising of some sectors of the Earth's crust in relation to adjacent ones. Morphologically this is idiomorphic or intermittent folding, quite variable in its individual manifestations. Its general features are the following: the presence only of anticlines in typical cases, while synclines are replaced by intermediate sectors of undisturbed occurrence of strata and independent development of each fold. The last feature is expressed in that the folds located alongside each other may have quite different heights, strike, and shape, both in profile and in area. Among folds of this type we find gently sloping domes with limbs dipping at small angles, massive horst-like box folds with a flat crest and steep limbs complicated by stepped flexures. Swells hundreds of kilometres long and tens of kilometres wide are also met, and very asymmetric swells with one limb sloping gently and the other relatively steep. Folds can be formed either by smooth bending of strata or be fringed by faults or broken up into separate blocks. The amplitude of the uplifts is measured in tens to over a thousand metres. Relatively big folds may be complicated by minor uplifts. Block folds are also known as "stamped", platform, or reflected folds.

Signs of protracted development proceeding simultaneously with sedimentation, i.e., signs of consedimentation, are characteristic of block folding. These sings are expressed as a change in the thickness and facies of the sediments within each fold: in that part of the section that corresponds to the time the fold developed there is a thinning of the thickness of the sediments toward the crest and a change in their composition in the same direction toward littoral and coarse facies.

The changes differ in degree from horizon to horizon. In some the thickness changes imperceptibly, in others there is a complete wedging out of one band of strata or another, and then a hiatus with signs of erosion of the subjacent rocks. In the same way the change in facies may also be greater or less. Hence it follows that block folds are manifested during their growth now as relative and now as absolute uplifts. It can be supposed that the crust subsides in a whole area and that folds grow on the background of this sagging. When the raising of a fold outstrips the general subsidence, the fold is manifested as an absolute uplift, and when it lags behind the general sagging of the area, the fold is expressed in the form of a relative uplift. A reduction in the thickness of sediments toward the crest of a fold leads to older strata being steeper in one and the same fold than younger strata.

Thicknesses and facies are consequently used both to reconstruct the history of oscillatory movements of the crust and to study the history of block folds. We can obviously obtain data on the development of such folds by comparing the thicknesses and facies of sediments in different parts of them, on their crests and on their limbs, and also in the area between folds. If there are no changes from the crest to the limbs in any suite, then the fold did not rise during the deposition of the suite, and if the thickness on the arch is thinner than on the limbs, the difference in thicknesses (where necessary with a correction for change in depth of sedimentation) is the measure of the uplift of the crest relative to the limbs. The ratio of the thicknesses, which is at the same time the ratio of the rates of subsidence of the crust on the arch of the fold and beyond it, can be substituted for the difference in thicknesses. When there is not only relative, but also absolute uplift on the arch of a fold, then the volume method can sometimes be employed, which helps determine the ratio of the volume rates of the raising of the fold's crest and the sagging of the limbs. This can only be done, however, when the detritus deposited on the limbs came wholly from the crest of the same fold and not from any other source.

Table 2. The Zhiguli swell during the late Devonian

Horizons	Thickness on arch, m (a)	Thickness on southern limb, m (b)	Ratio (a)/(b), percentages	Thickness on northern limb, m (c)	Ratio (a)/(c) in percentages
Shchigrovo-Semiluki	18	36	50	94	19
Voronezh	22	28	79	57	39
Evlanovo-Livny and Elets	132	137	96	208	63
Lebedyansk	216	236	91	215	100

An example of use of the ratio of the thicknesses of separate horizons on the arch and the limbs of a block uplift is given in Table 2. The data refer to Upper Devonian horizons on the Zhiguli swell.

The figures in the third and fifth columns indicate the rate of uplift of arch relative to limbs during the period from the Shchigrovo-Semiluki horizon to the Lebedyansk one decreased. The different ratios between the arch and both limbs is due to the difference in the distance of the boreholes on the northern and southern limbs from the arch.

An example of a change of sedimentation facies from the limbs to the arch is met in the Don-Medveditsa swell. In that area Middle Jurassic depositions lie on a heavily eroded surface of Carboniferous limestones. The Callovian sediments differ on the arch by being sandier than on the limbs and containing gypsum inclusions. The Neocomian (or Wealdian) stage is represented on the arch by coarser sands than on the limbs and lies on an eroded Upper Jurassic surface. In the Aptian stage there is a transition from clays to sands toward the arch. The Cenomanian sands are represented by a glauconite variety on the limbs and a quartz one on the arch. There are similar changes in the top stages of the Cretaceous and in the Palaeogene, all of which indicate that there was a rather shallower sea on the site of the Don-Medveditsa swell from the Middle Jurassic to the Palaeogene inclusive than in its immediate vicinity, and that an island emerged there at times. Because of the shallower depth the sediments on the arch suffered more vigorous erosion and more active grading by waves, with the result that the coarser material remained on the arch of the swell, while the finer particles were carried away to deeper places. The concentration of relatively coarse terrigenous material on the arch of the swell thus does not mean that it was produced by erosion of the swell itself; it could have been brought in from further away, the concentration being caused by secondary grading in the relatively shallow sector.

Study of the history of various block folds indicates that they develop very unevenly in time, their growth now accelerating and now slowing down. The Krasnokamsk swell (in the Volga-Urals region), for instance, went through periods of particularly rapid rise at the end of Late Carboniferous time and the beginning of Permian, at the end of the Artinskian Stage and beginning of the Kungurian, at the end of the Early Kungurian and beginning of the Middle Kungurian Stages, and so on. The uniform distribution of the thicknesses of the Givetian Stage on the Zhiguli swell indicates that it was not then rising; its most intensive uplift came later in Shchigrovo-Semiluki times, after which growth slackened and stopped altogether by the end of the Devonian (see Table 2). For the Tournaisian Stage there are no signs of growth of the swell. Uplift began again in Vereian time when the thickness of the accumulated sediments was 50 m thinner on the arch than on the limbs. During Late Carboniferous time, however, the swell did not rise for a long time, but in the Tertiary Period it was raised considerably and assumed its present form. Observations indicate that the growth of block folds accelerates and slows down synchronously for many folds over a wide area. The start of the growth of the various folds, however, falls into different geological epochs.

The Tuimaza swell in Bashkiria, for instance, began rising at the beginning of Devonian time. The Serafimov-Baltaev swell and the eastern half of the Bolshekinel swell began to rise at the end of the Givetian Stage. The Sok-Sheshma and the Vyatka swells began rising at the beginning of Carboniferous time. The western half of the Bolshekinel swell was formed at the end of Carboniferous and beginning of Permian time.

The length of the growth of individual block folds varies in the same way. Some folds in the east of the East European platform rose only in Devonian time (the Gusenki fold on the Saratov swell). Others rose until the end of the Palaeozoic (e.g., the Pristan and Shumeka folds on the same swell).

Fig. 43. Migration of the arch of a block fold (after Beloussov)

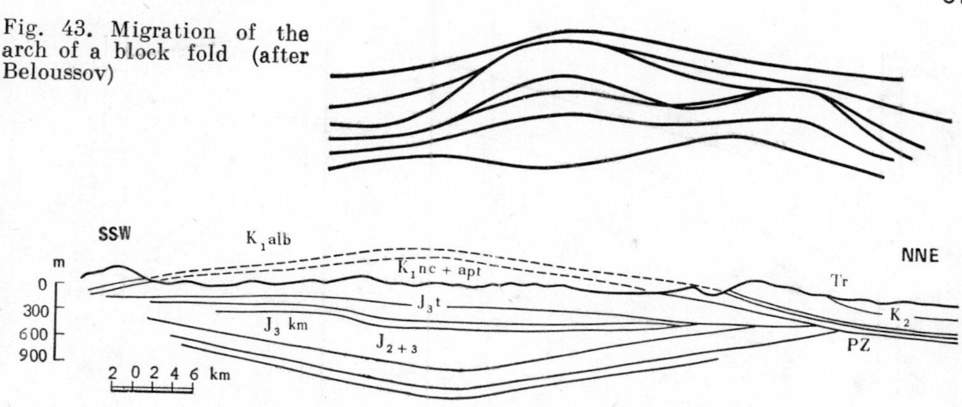

Fig. 44. Section through the Wealden swell (after Lamplugh, 1919)

A third type continued to grow in Mesozoic and Cenozoic time (e.g., the Vyatka swell).

Some block folds are still growing today.

Detailed study of the distribution of thicknesses and facies in the section of block folds often helps in many cases to reveal signs of migration of the arches over an area. One such case is illustrated in Figure 43. In each epoch the crest of the fold was where the thickness of the sediments of that epoch is thinnest. The practical significance of this phenomenon is that there may prove to be an accumulation of oil and gas to one side of the crest of the fold that is visible on the surface, i.e. where the crest was earlier at the time the oil or gas accumulated, rather than vertically below the apparent crest.

In some cases it has been found that there is a thickening of the sediments on the arch of a block fold below certain stratigraphic stages rather than a thinning. In that case the bedding of strata becomes gentler with depth rather than steeper, and the anticline may give way to a syncline.

This specific type of structure was noticed first on the Wealdian swell that separates the London and Hampshire basins in southern England (Fig. 44). Boreholes revealed that the anticlinal bedding of the Tertiary and Upper Cretaceous sediments observed near the surface changed at greater depth to a horizontal bedding in the Lower Cretaceous and a synclinal one in Jurassic sediments. The sedimentary formations thus formed a big lens there. The Jurassic and Lower Cretaceous sediments beneath the axis of the fold are as much as 1500 m thick, but wedge out on the limbs, where the Lower Cretaceous directly overlies the Palaeozoic. The Upper Cretaceous sediments alter little in thickness, while the Tertiary ones become thicker from arch to limbs. All the sediments are shallow-water ones.

Clearly the crust subsided at the site of the swell during Jurassic and Early Cretaceous time. Downwarping ceased in Late Cretaceous time, while a swell began to rise in the Tertiary period on the site of the former trough. A reversal or inversion of the vertical movement is consequently noticeable in the history of this swell.

Many such "inversion" block folds are found in the east of the East European platform. Folds are distinguished in whose sections there is a smoothing out of strata with depth beginnning in the Predevonian, Devonian, or Car-

7—0601

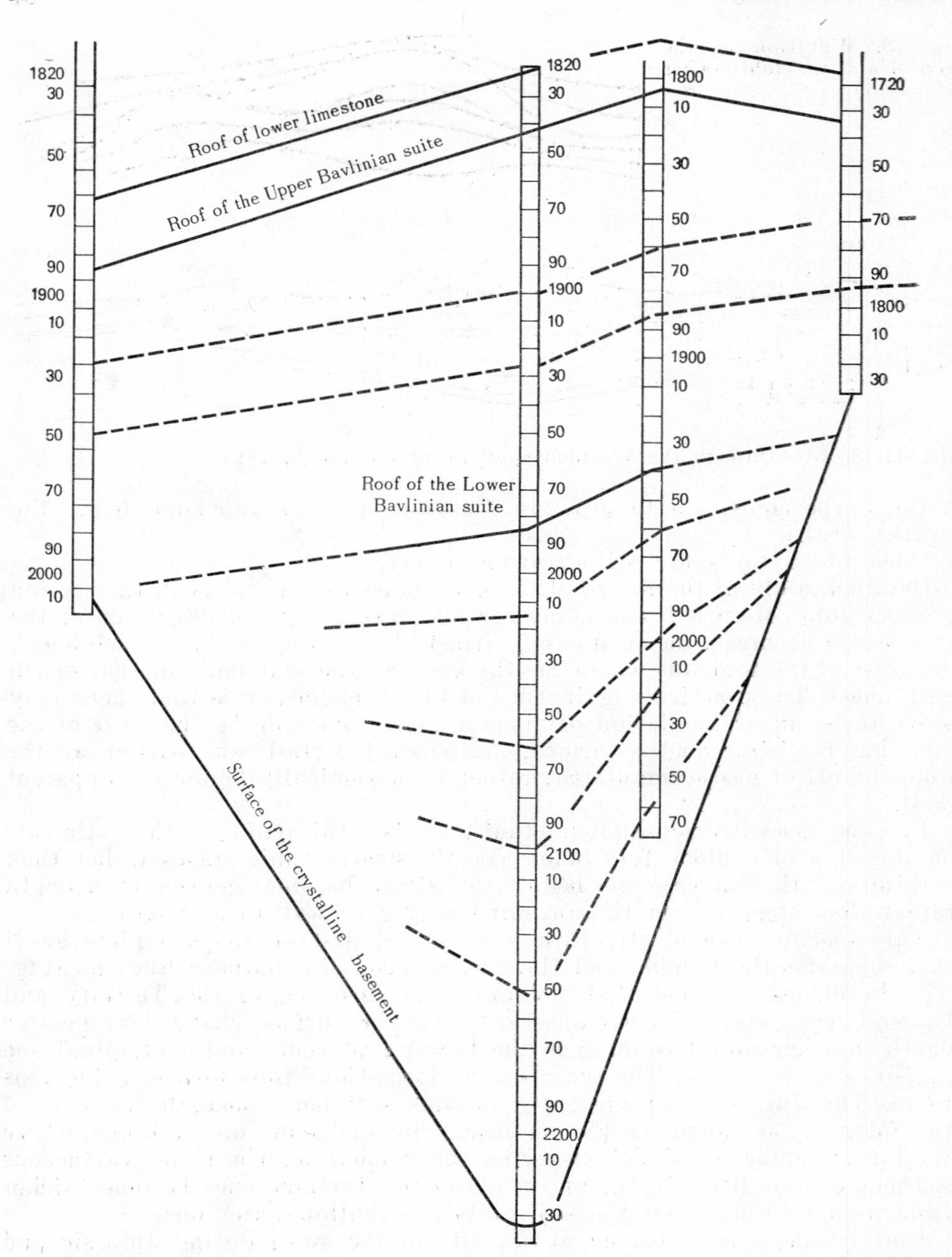

Fig. 45. Profile through the Bavly fold (after Rozanov, 1957). Inversion occurred after deposition of the Bavlinian suite

boniferous sediments, depending on the age of the inversion. A profile through the Bavly fold is shown in Figure 45, in which a smoothing out of strata with depth can be seen at first and then a transition to an opposite synclinal flexure. The inversion occurred after Bavlinian time and, consequently, a trough existed on the site of the present-day fold in Late Proterozoic time.

Block folds are characteristic of platforms. They are common, for instance, in many areas of the East European platform, and are no less so on the North American platform and others, both ancient and young. They are also found, however, in orogenic zones, mainly concentrated in foredeeps and intermontane troughs. They are known, for instance, in the Terek-Kuban foredeep, the Kura-Rioni intermontane trough in the Caucasus, and the Cisuralian foredeep.

The shape and history of block folds and the autonomy in development of each of them leave no room for doubt that they are formed as a result of uplift of separate of earth crust's sectors by forces acting vertically upward. The size and shape of the folds at the surface are determined by the size, shape, and amplitude of the uplift in deeper zones of the crust, but it must be borne in mind that the shape of an uplift can be altered significantly with depth by purely mechanical causes. Most block folds (and possibly all) are undoubtedly linked with vertical faults and slashes. These faults sometimes delimit raised crustal blocks on different sides, and the blocks act like a die on the overlying rocks. In other cases the uplift may be demarcated by a fault on one side but have a gentle slope on the other side.

Very many block folds are a surficial reflection of stepped swarms of faults on the slopes between the uplifts and troughs formed by oscillatory movements. The block folds on a platform, for example, are often distributed along the perimeter of syneclises and reflect the stepped vertical faults surrounding these major structures.

The causes of the uplifting of block folds are not known. It can be thought that they are processes of an increase in volume taking place at various depths in the crust, mainly in its deep zones, such as, for instance, the serpentinization of ultrabasic intrusive rocks or the diaphthoresis (amphibolisation, biotitization) of highly metamorphosed rocks through the effect of water circulating along deep faults in the crust—hence, possibly, also the link between block folds and faults.

Study of the history of block folds has practical interest as well as theoretical. Oil and gas fields are associated with many of them—swells and domes. But some folds generally located within petroliferous districts are devoid of accumulations of oil and gas. Their productivity proves to depend on the time they were formed. In the Saratov Region on the Volga, for instance, commercial concentrations of gas are discovered in Devonian sediments only in folds that existed at the end of Frasnian time. In the same region oil and gas are concentrated in Carboniferous sediments in structures that were formed before Mesozoic time. There are no oil or gas pools in folds raised later. In Western Siberia oil is found in structures that began to rise in Jurassic time and continued to grow during the Early Cretaceous; structures that arose only at the beginning of Jurassic time, however, or only in the Upper Cretaceous, do not contain productive pools.

In concluding this chapter we must recall that certain structures of non-tectonic origin are morphologically very similar to gently sloping block folds, structures, for instance, that reflect the unevenness of buried relief, or structures of non-uniform consolidation of sediments by the weight of the overlying rocks. Information on such structures will be found in textbooks on structural geology.

CHAPTER 11

Injection Folding

If a thick suite of rocks of high plasticity occurs at a certain depth below the surface in a sedimentary series, the rocks of this suite may flow in certain conditions and be squeezed out of one place and injected into others. In the area of squeezing the plastic series becomes thinner and increases in thickness in the area of injection. The less plastic sedimentary rocks lying above the plastic ones are deformed, yielding to the latter's movement. These strata subside above the zone of pinching and are raised above the areas of injection of plastic material. The regions of pinching usually occupy broad areas and the strata lying above them are not in the main disturbed, while the regions of injection are sites of concentrated swelling of the plastic material and local convex structures are formed above them in the overlying sedimentary series, either dome-shaped or brachyanticlinal depending on the form of the injection. The convex structures may vary in degree of expression from very gently sloping uplifts to pronounced folds with steep, and even overturned bedding of strata on the limbs, broken up by numerous faults. As for the rocks that occur below the plastic series, they may not be involved at all in the deformations, or movements in the "basement" may affect the location of the zones of pinch and swell in the plastic series to some extent. The mechanism of this effect will be dealt with later.

Taken together the deformations described form *injection folds.* Where it happens three complexes of rocks can be distinguished in the vertical section, as will be clear from the foregoing, namely, a lower complex, underlying the plastic series; a middle, plastic series; and an upper complex covering the plastic series.

As we have just said, the lower complex is simply the basement of the injection folds and its effect on them may be only indirect. Its internal structure is disharmonious with that of the overlying strata.

The middle complex, i.e., the plastic series, is the active one in injection folding. The processes of pinching, flowage, and injection take place in it. In the swell area an "injection core" is formed which may vary in shape from a gently convex lens to a narrow column 6000 to 8000 m high, with a diameter of hundreds of metres, and intricate branching or overhands into the overlying rocks.

The upper complex of stratified, less plastic rocks deforms "passively", submitting to the shape of the injection core. The relation of the middle and upper complexes above the swell is not limited to raising of the upper complex, but is often expressed in the plastic rocks of the middle complex piercing the strata of the upper complex, rupturing them and pushing them aside. The injection core then becomes a piercement or diapiric core; folds of this sort are called diapiric folds (or diapiric domes).

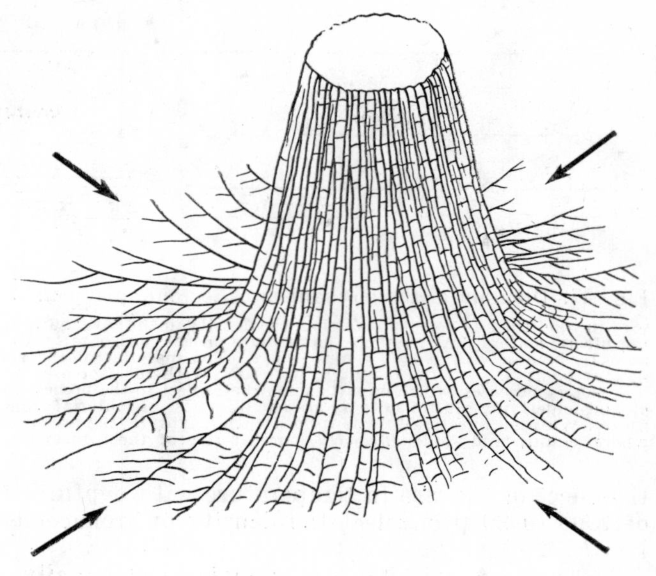

Fig. 46. Idealised salt dome (after Balk, 1949). Salt was pressed radially into the diapiric core, which is characterised by folds with vertical hinges

The plastic rocks that form the middle complex may be salt, gypsum, clay, marls, or thin alternations of them. Salt is most often the material of injection and piercement cores, but small diapirs with cores of coal and peat are encountered.

When the stratification of the plastic rocks enables one to see their internal structure, the cores are often found to have a very complex structure, which is particularly intricate when the material is layered salt. The salt is buckled into strongly compressed isoclinal, bizarrely bent folds. In columnar cores the hinges of these internal folds are vertical. Such folds are formed as a result of the salt's flowing into the site from all sides, and then being squeezed up the column. In the first stage, when the salt is flowing horizontally toward the column, it gathers into folds with horizontal hinges converging radially to the base of the column. With subsequent rise these folds are sucked into the column and their hinges become vertical (Fig. 46).

Squeezing and injection go hand in hand with disruption of the plastic series' mechanical stability which is caused in turn by the following:

a) density inversion;
b) an uneven load on the plastic series;
c) the squeezing out.

We shall consider the first mechanism from the example of an injection fold with a salt core.

The density of salt is around 2.2 g/cm^3. The mean density of the loose sand and clay sediments occurring on the surface is around 2.0 g/cm^3; their density increases with depth, however, under the load of overlying strata to 2.5-2.6 g/cm^3. At the same time the density of salt hardly alters at all under pressure. At a depth around 900 m its average density is equal to that of the overlying sediments. Deeper its density is less than that of arenaceous-argillaceous rocks, and deeper still the difference is greater. A situation of density inversion thus commonly sets in at depths around 900 m. There is a certain, additional, though small, reduction in salt's density through

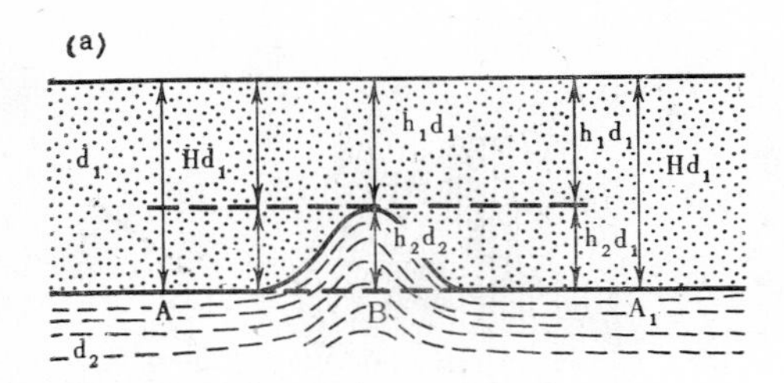

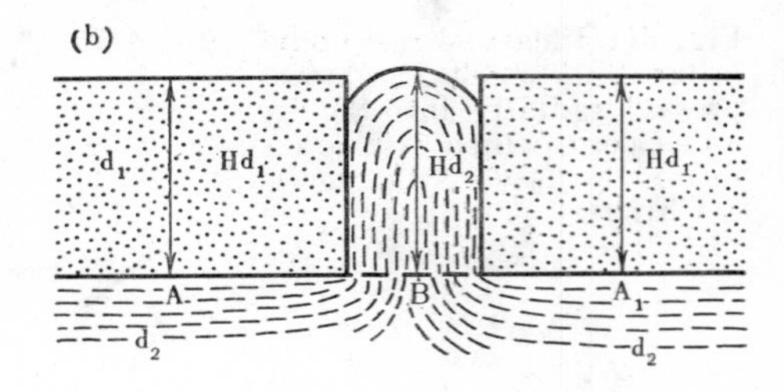

Fig. 47. Diagram of the mechanism of tectonic piercement (after Beloussov):

(*a*)—in conditions of density inversion: the density of the cover suite d_1 is higher than that of the underlying plastic suite d_2. At point *B* the pressure on the horizontal plane ABA_1 is lower than in sectors *A* and A_1, because $h_1d_1 + h_2d_2 < h_1d_1 + h_2d_1$. The material of the suite under pressure must therefore flow at point *B*, and the height h_2 will rise. With the increase in height of the fold, the difference $h_2d_1 - h_2d_2$ increases, which imparts a snowball effect to the process; (*b*)—the densities of the cover suite d_1 and underlying plastic suite d_2 are equal. A rise of material from d_2 is only possible along an open fissure. When an injection fold is formed with height *H*, the pressure along the plane ABA_1 is everywhere equal, and since $Hd_1=Hd_2$ further rise of the fold is impossible

the effect of the rise in temperature with depth. The temperatures at depths of 1000 to 2000 m affect the density of arenaceous-argillaceous rocks much less.

The state of density inversion is mechanically unstable. Any deviation from the horizontal plane on the interface between a denser material lying above and a less dense underlying one is sufficient to give rise to a tendency for this deviation to snowball; the greater the unevenness, the greater is the difference in pressure beneath and beyond it on any horizontal plane within the salt (Fig. 47*a*).

Whether this tendency causes movement of the denser material downwards, however, and movement of the less dense material upwards, depends on their capacity to move and deform, and on their viscosity and strength. At room temperature salt is not sufficiently plastic to flow under its own weight. But at higher temperatures its viscosity and limit of elasticity rapidly diminish. Salt becomes quite fluid at temperatures above 200 °C. At that temperature, too, its limit of elasticity is halved (from 9 g/cm^2 to 4.5 g/cm^2). In areas with a thick cover of sedimentary rocks this temperature is reached at a depth of 4 or 5 km, which is the depth most favourable for flowage of salt in a density inversion situation. As to the deforming properties of the overlying rocks, both their viscosity and their strength are important. They have been found usually to be low enough to let the upper layers of the complex bend and rupture under the pressure of the plastic series.

From the physical aspect the forming of injection folds in density inversion conditions is similar to convection, but it is not thermal convection in which one and the same material heats and rises to the surface, then cools and sinks. In our case it is gravitational, unidirectional convection, consisting in the sinking of the denser material and the rising of the less dense underlying material; the reverse movement does not occur. Such unidirectional convection is also known as advection.

Experiments have established the "flow patterns" that regularly arise with convection. The movement breaks up into streams that group into cells. Each cell consists of an ascending stream formed by rising material of lower density and a descending stream of material of higher density. In the ideal case, when there are no horizontal non-uniformities in the movement of the media, cells have a hexagonal cross-section. Depending on the ratio of the densities and viscosities of the two media, the ascending stream occupies the periphery of a cell and the descending stream the centre, but the arrangement may be the opposite. The horizontal extent of cells depends on the thickness of the layers involved, mainly that of the lower layer with the lower density. The thicker the lower layer the bigger are the cells. At a very rough approximation their diameter can be taken to be more than three times the thickness of the lower layer.

In a number of cases it is possible to observe a grouping on natural diapiric domes into irregular rings or polygons. That is the arrangement, for instance, of certain groups of salt domes in the Caspian syneclise (Fig. 48). In such a grouping we apparently can see a manifestation of convection cells. In these natural cells the lighter material rises to the surface on their periphery, while the sinking heavier material is concentrated at the centre. The geophysical data indicate that the domes are joined together at depth to form a solid enclosing framework of salt around the inner part of the cell. The separation of these frames into individual columns at the top is due to the fact that, since a cylinder is the geometrical figure with the minimum lateral surface, the penetration of salt into the overlying rocks in the form of cylinders is most economical of energy, because a cylinder encounters less frictional resistance from the host rocks. Where the deep section of a structure similar to that of the Caspian syneclise is visible on the surface the salt frameworks can be seen directly. An example is the Ishimbai area of the Urals (Fig. 49). There the convection cells are completely surrounded by salt frames of the Kungurian Stage, while their interiors are made up of stratigraphically overlying sediments of the Ufimian Stage, which form synclines.

In accordance with what has been established for convection the size of the cells formed by diapiric domes is directly related to the thickness of the salt series. In the central areas of the Peri-Caspian syneclise, for instance, where the salt series is thicker the cells are bigger than on the periphery of the basin, where the salt-bearing series is thinner.

The limit of unidirectional convection (advection) should be complete overturning of the substances involved, the denser one finally lying at the bottom and the less dense at the top. Such a state is then stable, since the centre of gravity of the system is then at its lowest position. Complete overturning is always observed in fact when diapiric domes are modelled by means of the tectonic models usual for the purpose, employing equivalent low-viscosity materials like bitumen (imitating salt) and dense sugar syrup (imitating the cover rocks). The separate stages of such an experiment are illustrated in Figure 50.

In natural injection folds, however, complete overturning does not happen. The growth of diapiric domes usually ceases at some depth or other beneath the surface. The reasons for incomplete advection should be clear from the foregoing description.

1. As salt rises, the density relation between it and the cover rocks alters in such a way that the buoyancy forces causing its upward movement diminish. When salt penetrates overlying rocks, the latter become enclosing rocks rather than covering. The closer they are to the surface, the lower is their density, while the density of salt alters little. As a result the difference between the densities diminishes as salt moves upward and the buoyancy force falls. In the last stages of salt's ascent it tends to spread out

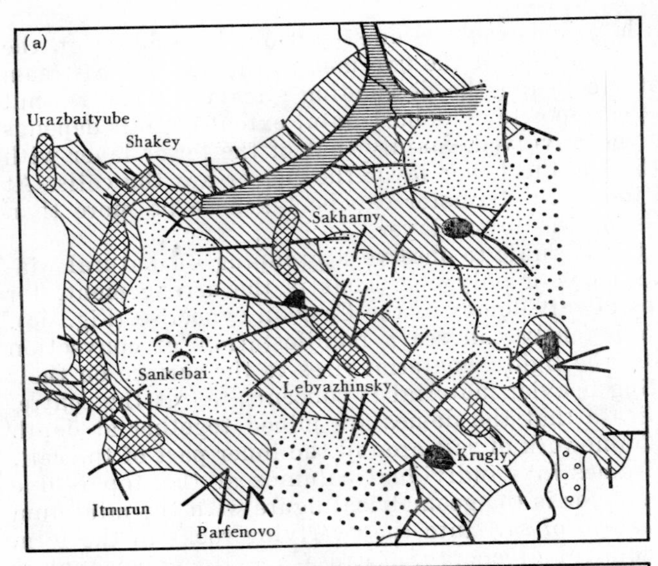

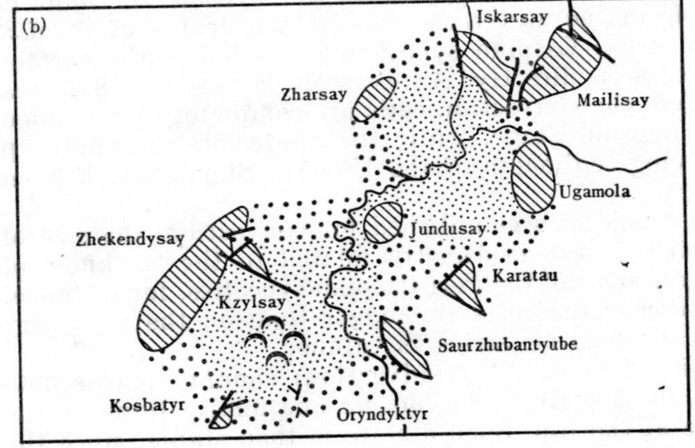

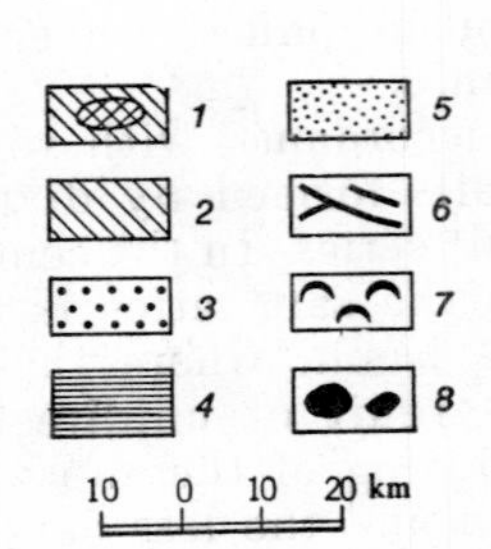

Fig. 48. Grouping of diapiric domes in irregular rings and polygons in the Caspian syneclise (Pre-Upper Miocene shear surface) (after Sycheva-Mikhailova, 1973):

a—nearly unbroken ring massifs (central zone); *b*—broken ring massifs (Khobdin zone); 1—piercement salt domes; 2—cryptopiercement salt domes; 3—salt isthmus; 4—salt-dome grabens; 5—primary compensation troughs; 6—faults; 7—second generation domes; 8—supradome subsidence troughs

more between strata than to pierce them and rise. Overhangs then develop at the top of the diapiric core, which assumes a mushroom shape.

2. As salt rises it reaches zones of lower temperature and its viscosity and limit of elasticity increase so that flowage becomes ever more difficult.

Another cause stopping uplift of a diapiric core can be depletion of the source through pinching out of the layer feeding salt.

The activity of the injection process also depends on the initial thickness of the salt. If it is thin, squeezing cannot develop. The point is that, since the size of convection cells depends directly on the thickness of the salt, only injection cores of very small diameter can develop when it is thin, and small cells are unable to deform and pierce the overlying series. That can only be done by broad cores with high total buoyancy forces sufficient to upwarp the covering rocks, overcome their strength, and rupture them.

The mechanism of the formation of injection and piercement cores consisting not only of salt but of other rocks is the same in conditions of density inversion. Clays saturated with water possessing a high pore pressure, for

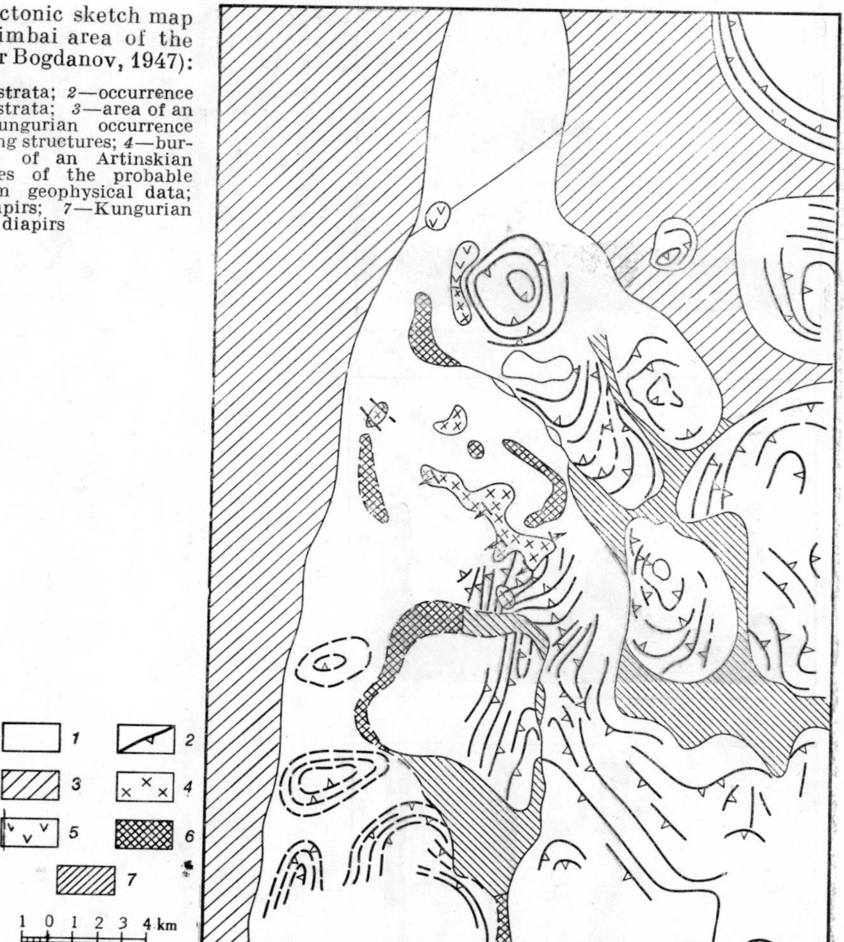

Fig. 49. Tectonic sketch map of the Ishimbai area of the Urals (after Bogdanov, 1947):

1—Ufimian strata; *2*—occurrence of Ufimian strata; *3*—area of an elevated Kungurian occurrence within sloping structures; *4*—buried massifs of an Artinskian bed; *5*—sites of the probable massifs from geophysical data; *6*—cryptodiapirs; *7*—Kungurian outcrops in diapirs

instance, are lighter than consolidated sediments, and the conditions for the formation of injection and piercement cores are similar to those for salt cores.

The whole process of the formation of injection folds, and above all the shape of the cores, are most vitally affected by the structure of the upper complex of rocks. The existence of faults in them naturally affects the sites and shape of the injection of rocks of the plastic series. Faults facilitate rise of the plastic series so much that they may often be squeezed out onto the surface through the crest of the injection fold and form small "pinched" overthrust nappes on the latter (Fig. 51).

The lower complex (the "basement" of the injection fold), as we said above, may not be involved in the dislocations at all, but its relief often has a stepped character associated with vertical block movements. The injection cores in the covering plastic series then lie mainly above the basement ledges. With block movements of the basement, weak zones, at least, if not fissures, develop above the faults separating the individual blocks, and

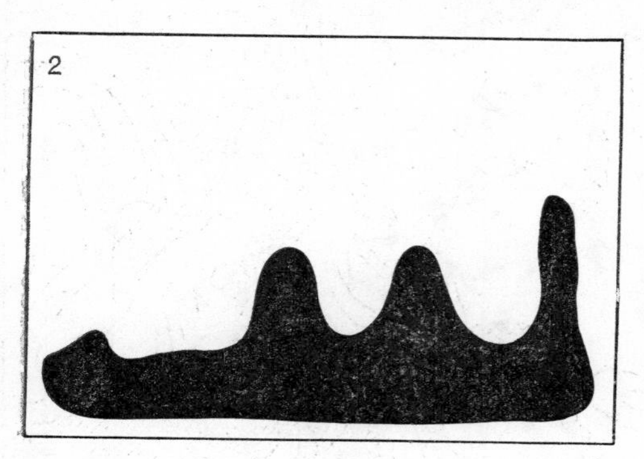

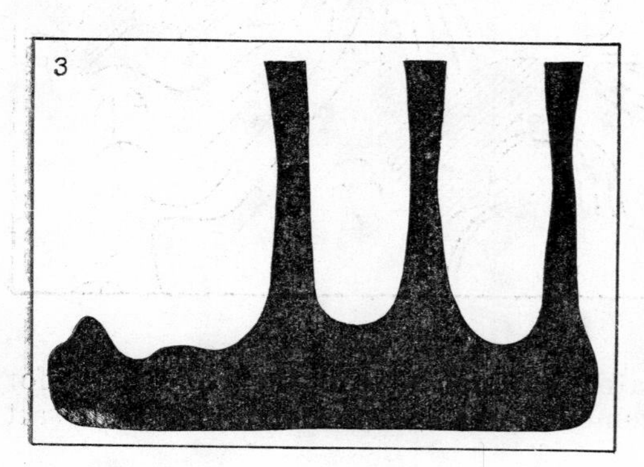

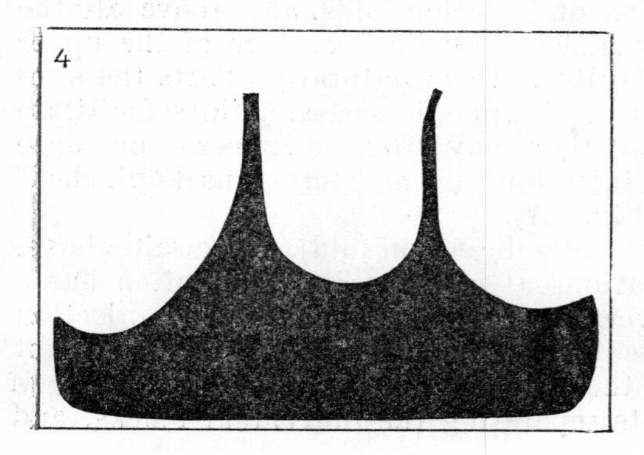

Fig. 50. Stages in the migration of two viscous liquids (after Sycheva-Mikhailova, 1973):

Composition: bottom—55 per cent bitumen (1 cm); top—syrup (12 cm)

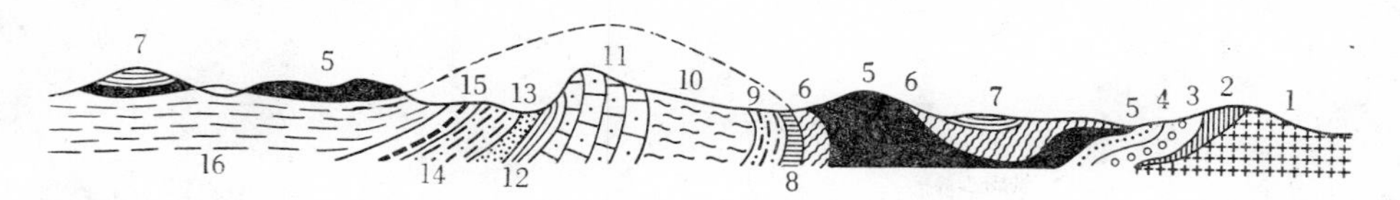

Fig. 51. Mass overthrust of diapiric origin (the Gros Cerveau anticline in Provence) (after Corroy and Denizot, 1943):

1—metamorphic rocks; *2*—Carboniferous; *3*—Permian; *4*—mottled sandstones; *5*—shelly limestones; *6*—Keuper; *7*—Infraliassic; *8*—Lias; *9*—Bathonian; *10*—Upper Jurassic; *11*—Hauterivian; *12*—Aptian; *13*—Cenomanian; *14*—sandy Turonian; *15*—Turonian with rudistids; *16*—Senonian

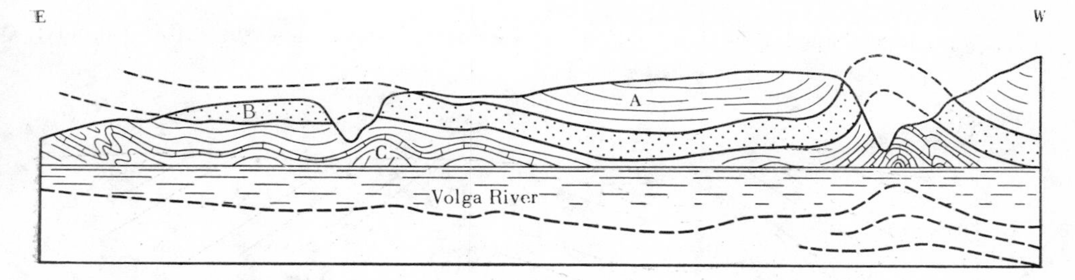

Fig. 52. Folds in Permian sediments formed by the squeezing of plastic rocks into valleys (Cheboksary district, Chuvash ASSR) (after Bronguleev, 1951). Letters denote different horizons of the Tartarian Stage

penetrate the plastic series and possibly even the upper complex of rocks. Injection of plastic material occurs along these weakened zones.

Now let us turn to the second mechanism of the formation of injection folds, namely through unevenness in the pressure of the load on the plastic series of rocks, without involvement of density inversion.

An unevenness of load can be produced by irregularities in relief. When a thick plastic series occurs horizontally not far below the surface, and the surface landform is broken up, the plastic rocks may be squeezed out from below watersheds, where the load on them is greatest, towards the valleys, where the load is least. For that to happen the difference of pressure must obviously exceed the limiting creep stress of the plastic rocks. The injection of plastic rocks will lead to an increase in their thickness beneath the valleys. They may at the same time be buckled into folds. The folds in strata of the Tartarian Stage, in some of the valleys along the Volga, for instance, developed in this way (Fig. 52).

There can be an unevenness of load in connection with a non-uniformity of density in the rocks overlying a plastic series. Clays can be squezzed out from under lenses of heavy sandstones and then folded concentrically around the lenses.

Non-uniformity of load on plastic rocks is often associated with faults in overlying series. Beneath each fault and below the whole more or less broad zone of attrition accompanying the fault the pressure on the underlying series is less than under the undisturbed rocks, so that material from the plastic series may be injected into them. Examples are the injection folds in the South-eastern Caucasus, where the cores consist mainly of clays of the Barremian and lower Aptian Stages, while the cap rocks are sandstones, limestones, and other compact rocks of the upper Aptian and Albian Stages, the Upper Cretaceous, and Palaeogene. In plan these

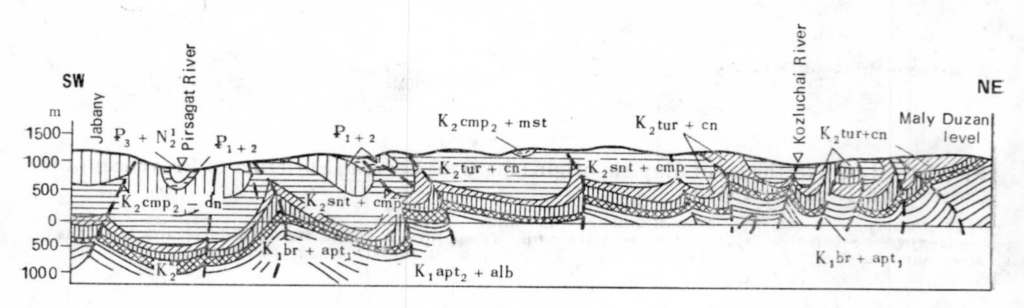

Fig. 53. Ridge-shaped injection folding in the South-eastern Caucasus (after Vikhert, 1966)

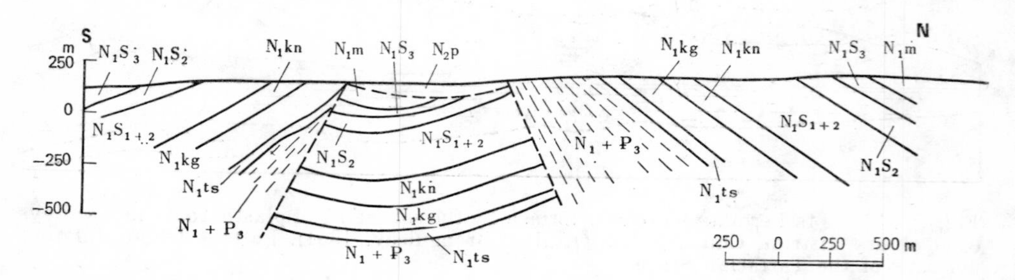

Fig. 54. Profile through the Andreev anticline (Kerch Peninsula) (after Beloussov). The diapiric core is complicated by a "depressed trough":

N_1+P_3—Maikop formation; N_1cs—Chokrak stage; N_1kg—Karagan stage; N_1kn—Konka stage; N_1s_{1+2}—lower and middle Sarmatian stages; N_1s_2—middle Sarmatian; N_1m—Meotian stage; N_2p—Pontian stage

folds stretch along faults and morphologically belong to the type of crest-form folding (Fig. 53).

What is common to injection folds whose mechanism does not involve density inversion is the absence of conditions for a snowballing of the injection process. On the contrary it slows down as the cores rise and must cease when the pressure in any horizontal plane within the plastic series beneath and beyond the core is equalized.

The combined effect of relief, faults, and density inversion was apparently manifested during the formation of clay injection folds on the Kerch Peninsula. The plastic series forming the cores there consists of clays of the Maikopian suite (Middle and Upper Oligocene and Lower Miocene). The cores are confined to faults in the overlying series of Middle and Upper Miocene sandstones, clays, and limestones. Squeezing out of the clays was promoted by a reverse relief (which is higher above synclines on the Kerch Peninsula than above anticlines), and may also be attributed to the lower density of the Maikopian clays as a result of their attrition and saturation with water and gas during their rise to the surface. The clay cores of these Kerch injection folds are complicated in several places by "depression troughs", circular synclinal basins formed by layers of Miocene sandstone and limestone (Fig. 54). These synclines occurred because the clay cores were being squeezed more strongly along their peripheries under the weight of the rocks directly overlying them than in the centre of the core further from the cover rocks. The upward movement of the middle of the core, consequently, lagged behind and the Miocene deposits, already eroded on the periphery, survived

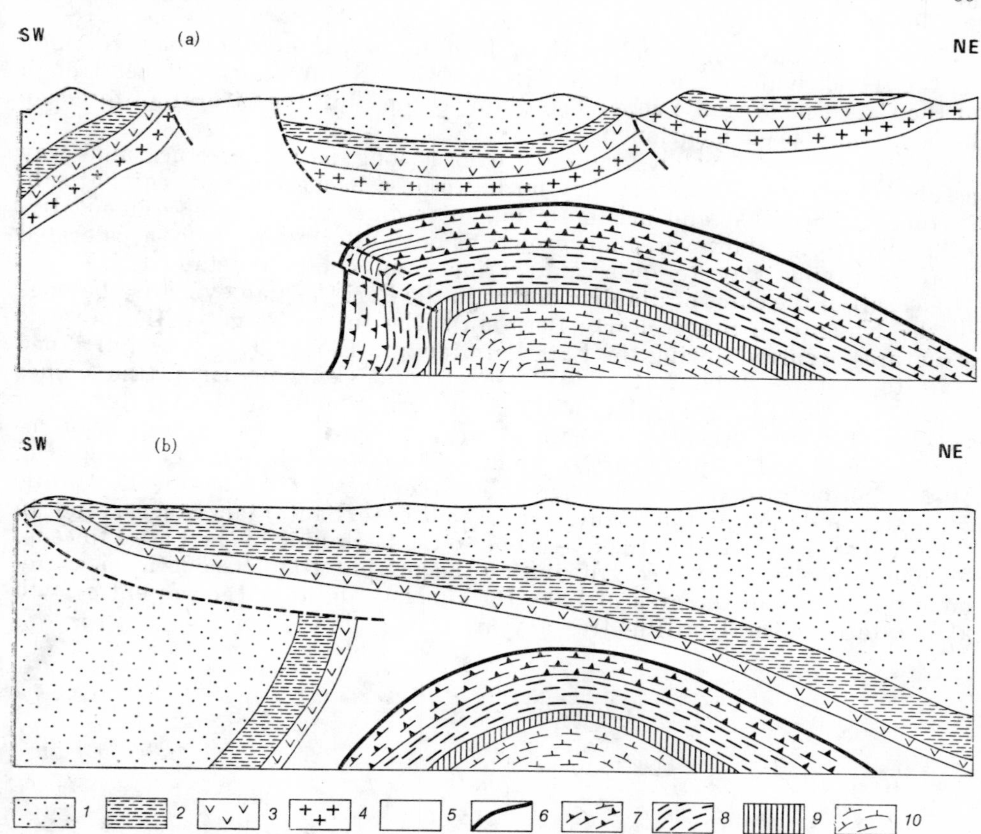

Fig. 55. Profiles through folds in South-eastern Iran (based on data of British Petroleum).

a—Lary fold; *b*—Aga-Jari fold; *1*—Upper Farsian; *2*—Middle Farsian; *3*, *4*, *5*—Lower Farsian; *6*—cap rock; *7*—Asmarian limestones; *8*—Eocene-Oligocene; *9*—Upper Cretaceous; *10*—Middle Cretaceous and older rocks

there. The blocks of Miocene rocks, separated from the surrounding rocks in this way, gradually sank into the relatively lighter Maikopian clays and squeezed them aside so that the blocks acquired the shape of round saucers (since the friction of the underlying rocks is minimal with that shape).

This mechanism of the formation of pinches and swells can be seen from the example of the salt folds of South-western Iran (Fig. 55). There the lower complex underlying the plastic series played a more active role. Block folds developed on its surface and as a result the plastic salt-bearing series was squeezed between the rocks of this complex and the hard overlying arenaceous-argillaceous series. The salt was pressed aside from sites of vertical flattening and broke through to the surface, where the upper complex was weakened by faults.

Injection folding is common in regions of thick accumulation of sedimentary series that have remained unmetamorphosed and contain plastic rocks in their section, especially salt-bearing strata. Such regions are the foredeeps and intermontane troughs of orogenic zones, and the deep syneclises of platforms. On the East European platform such folds in the form of diapiric domes are widely developed in the Dnieper-Donets syneclise, as well as in the Peri-Caspian syneclise. Crest-form folds with clay cores are common in the Terek-Kuban foredeep and in the Kura-Rioni intermontane trough of the Caucasus orogenic zone. The diapiric domes in the Cisuralian foredeep have already been mentioned.

Injection folds develop gradually over a long period, in the same way as block folds. On the whole they build up on a background of regional subsidence of the crust and accumulation of sediments, e.g., consedimentally. Like block folds, however, their growth is not uniform, epochs of accelerated increase giving way to quiescent periods and complete standstill.

The salt of the cores of diapiric domes in the Peri-Caspian syneclise belongs to the Kungurian Stage (Early Permian). The initial stage of the diapiric structures occurred during the Late Permian and Early Triassic and attained a quite mature state. Later, in Mesozoic and Cenozoic time, the domes continued to rise, generally without interruption, but with alternating periods of a strengthening and weakening of growth. The first stage of intensified uplift of diapiric cores is confined to times before the Late Triassic. More vigorous growth periods were noted later in the time before the Middle Jurassic, and before the Hauterivian and Late Santonian Stages, and in the period from the end of the Oligocene to the beginning of the Late Pliocene.

Some injection folds are growing at the present time. Geodetic measurements of domes in the area of Lake Baskunchak indicate the present growth at a rate of several millimetres a year.

CHAPTER 12

The Folding of General Crumpling

General Characteristics of Crumpling Folding

General crumpling or true folding (morphologically—complete or holomorphic) is on the whole the result of the bending of strata by the effect of longitudinal compression, i.e., tending parallel to the strata. Unlike block and injection folding, for which individual development of each fold is characteristic, which determines their morphological diversity, general crumpling folding is characterised by the subordination of big groups of folds to a single deformation common to them all. This implies reduction in the size of the whole massif in one direction and an increase in its dimensions in another. The main trend of the contraction is horizontal and of the increase vertical. In a homogeneous medium such deformation would have a regular character; each medium of any size would be deformed in the same way, contracting horizontally and stretching vertically. When, however, the medium is a layered, heterogeneous one and there is a possibility of the layers slipping relative to each other, the compressive deformation parallel to the layers is only partially expressed in their equal thickening. The rest of the deformation takes place through bending of the layers into folds. The ratio of the two depends on the thickness of the layers and their mechanical properties. (See V.V. Beloussov, *Structural Geology*, 2nd Ed. Mir Publishers, Moscow, 1978; in French.)

Although the original trend of compression and contraction is horizontal and of stretching vertical, the initial vertical position of the axial planes may be distorted by additional influences. If, for instance, a zone crumpled into folds rises above neighbouring ones, gravity may cause flowage of the top of the raised zone. The folds thereby become inclined fanwise and overturned.

Methods of Reconstructing the History of Crumpling Folding

The collective nature of general crumpling folds is reflected in the criteria by which their age can be determined and their history reconstructed. The criterion used to study block and injection folds was the change in thickness and facies of sediments related to each fold separately; for general crumpling folds it is unsuitable. These folds are not consedimentary, but postsedimentary and are formed after the strata involved have already been deposited. If the strata within folds vary in thickness, the changes were caused not by the original conditions of accumulation but through secondary flow of their material layer by layer during deformation.

The age of general crumpling folds is ascertained from the age of the rocks involved and the age of the rocks bedded on them with an angular unconformity. Folding was obviously confined to the time interval separating these two age limits, for it developed after the buckled strata were deposited and before fresh strata were laid down on the erosion surface that truncated the folds. It must be specially emphasised that the uplifting of the folded strata, their erosion, and fresh subsidence, during which the upper series of sediments accumulated, are necessary for an angular unconformity to develop. All the geological evidence indicates that the formation of general crumpling folds is closely and regularly linked with uplifting of the crust, which leads to erosion and the development of a denudation surface on the truncated folds. The subsequent subsidence necessary for the accumulation of a new series, however, is an independent event that may or may not take place. When it is missing we lack the possibility of defining the upper time limit of the folding. In many areas, however, the chain of events indicated is manifested, which enables us to decipher the history of fold structures.

Since subsidence after uplift and erosion is not linked with what went before, the geological time interval concealed in any angular unconformity may be of different duration. The age of folding can be determined most accurately from angular unconformities that involved the shortest interval of time. Such short-period unconformities also present interest in solving the rate of crumpling folding, whether it happens slowly geologically speaking, rather like block and injection folding, or takes place rapidly in "shocks" or "pulsations".

Cases when the time interval that includes an angular unconformity is geologically very short are particularly valuable for solving this problem. In the South-eastern Caucasus, for example, there is an angular unconformity of 35° between the beds of the top of the Tithonian Stage and the bottom of the Valanginian Stage. In the Little Caucasus there is an angular unconformity between the Campanian and Maestrichtian Stages. In Abkhazia the bottom of the Callovian Stage discordantly overlaps an overturned fold of Bathonian sediments.

Such cases justify us in considering crumpling folding to develop quickly, geologically speaking. On the geological time scale such an event of deformation can actually be considered a "pulsation". It is known as the "folding phase". It is quite difficult to determine the length of the folding phase in years. It has been suggested that each phase lasts not more than a few hundred thousand years.

The folded structure usually observed is the result of several phases following one after another. Each successive phase complicates the folded structure formed by the previous ones. The short-period folding phases are separated by much longer quiescent periods during which sediments accumulate but folding does not occur. The length of the quiescent periods may vary considerably, depending on which stage of the formation of the fold zone it takes place. In Alpine fold zones, for example, the Neogene is the time most tightly "filled" with folding phases. The intervals between them have to be measured in several million years. During the Mesozoic, however, folding phases occurred much less often and the length of the quiescent intervals was already measured in tens of millions of years.

The Link Between Crumpling Folding and Oscillatory Movements

General crumpling folding is confined exclusively to geosynclines. This points to a link between it and the regime of block-undulatory oscillatory movements typical of them, i.e., a regime of movements of high amplitude and steep gradients. Within geosynclines, moreover, crumpling folding is particularly intensive in zones of the sharpest contrast in the development of vertical crustal movements. The contrasts may be marked by ample gradients of the thickness of the sediments involved in the folding or by steep gradients in the neotectonic relief (when the folding is young), or by both. In the Great Caucasus, for example, crumpling folding is developed along the southern slope of the range and in part involves its axial area. At the same time there is almost none on the northern slope, which is either a gently sloping monocline or a region of the development of block folding. There was intensive folding of the southern slope in Mesozoic sediments during the Palaeogene and Neogene. The distribution of Mesozoic strata indicates that there was a zone of rapid transition from the thick Mesozoic thicknesses in the present-day ridge to the thin thicknesses in the present-day Kura depression just where the southern slope of the range now is. This is a zone of steep gradients of Mesozoic thicknesses. At the same time, the thinning of Mesozoic thicknesses from the axis of the present-day range towards the northern foothills (towards the Scythian Epihercynian platform) is gradual, without a high gradient. On the other hand, the southern slope of the present-day range is steep, while the northern is gentle, i.e. the distribution of neotectonic movements, as well, created great contrasts of landform on the southern slope and minor ones on the northern (see Fig. 56*A* and *B*).

The strike of the crumpling folds has been noticed, from many examples, to run parallel as a rule to the trend of the isopachs or to the contours of the main forms of the neotectonic relief. Because of that the folds extend along zones with a different regime of oscillatory movements.

That this is not simple coincidence can be seen from the development of bunches of folds running across the strike of those zones where they close up along the strike and where, consequently, the isopachs or contours of the neotectonic relief run across the general strike of the geosyncline. Such transverse crumpling folds, complicated by overthrusts, which developed on the western subsidence of the Pyrenees where the series involved in the folding quickly pinch out in a westerly direction along the strike of the whole zone of folding, are illustrated in Figure 57.

The Origin of Crumpling Folding

Since general crumpling folding takes place *directly* through the effect of horizontal compression, the question of its genesis boils down to that of the origin of horizontal compression in the crust.

The contraction hypothesis that predominated in geology in the second half of the nineteenth and beginning of the twentieth century treated geosynclines as plastic zones that were being crushed between the converging rigid blocks of platforms located on both sides of them. This idea of external pressure on a geosyncline was also adopted by subsequent geotectonic hypotheses, including those popular today. The hypothesis of continental drift, for instance, both in its old form and in its present-day modifications, pre-

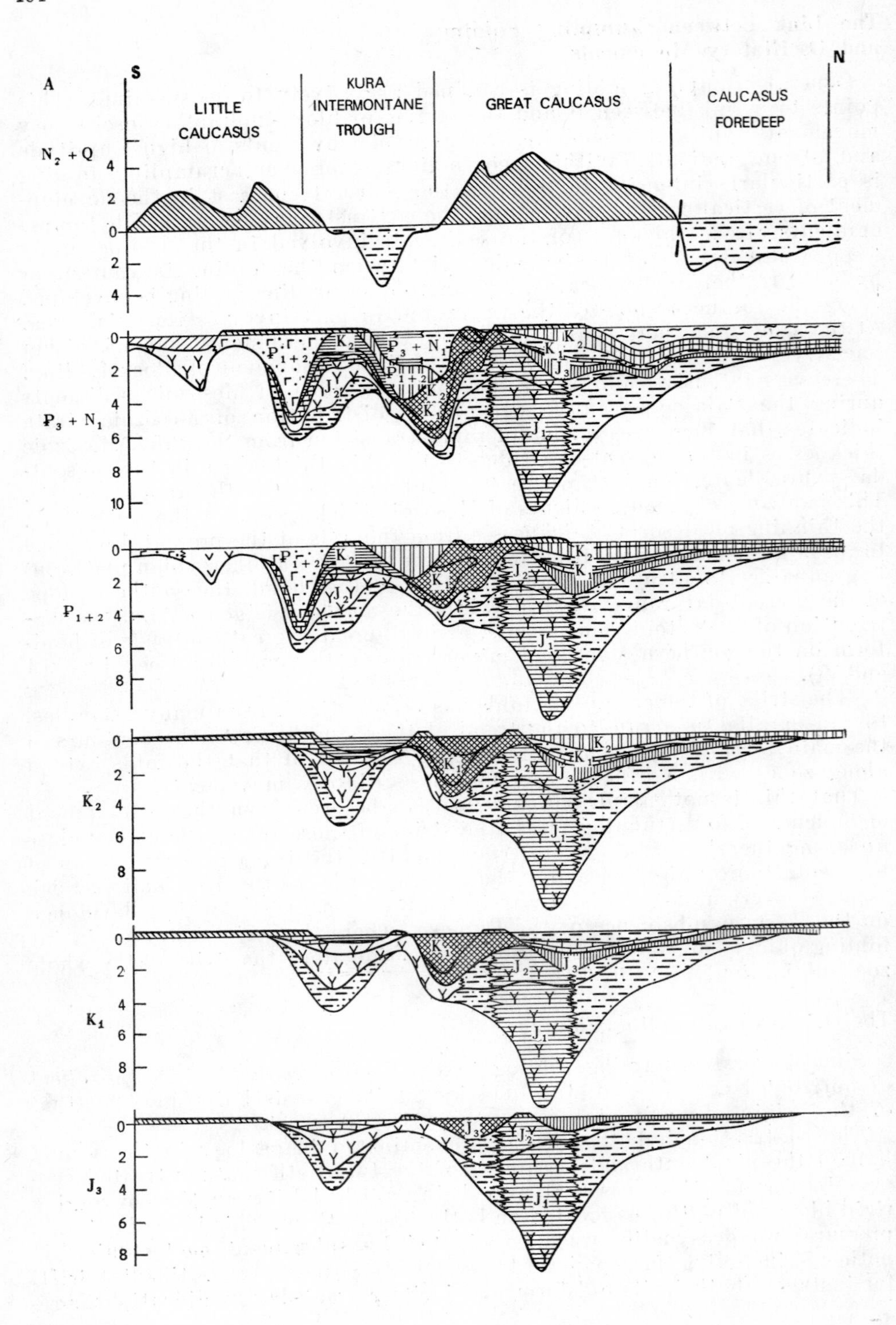
A
S
N
LITTLE CAUCASUS
KURA INTERMONTANE TROUGH
GREAT CAUCASUS
CAUCASUS FOREDEEP
$N_2 + Q$
$P_3 + N_1$
P_{1+2}
K_2
K_1
J_3
J_1
J_2

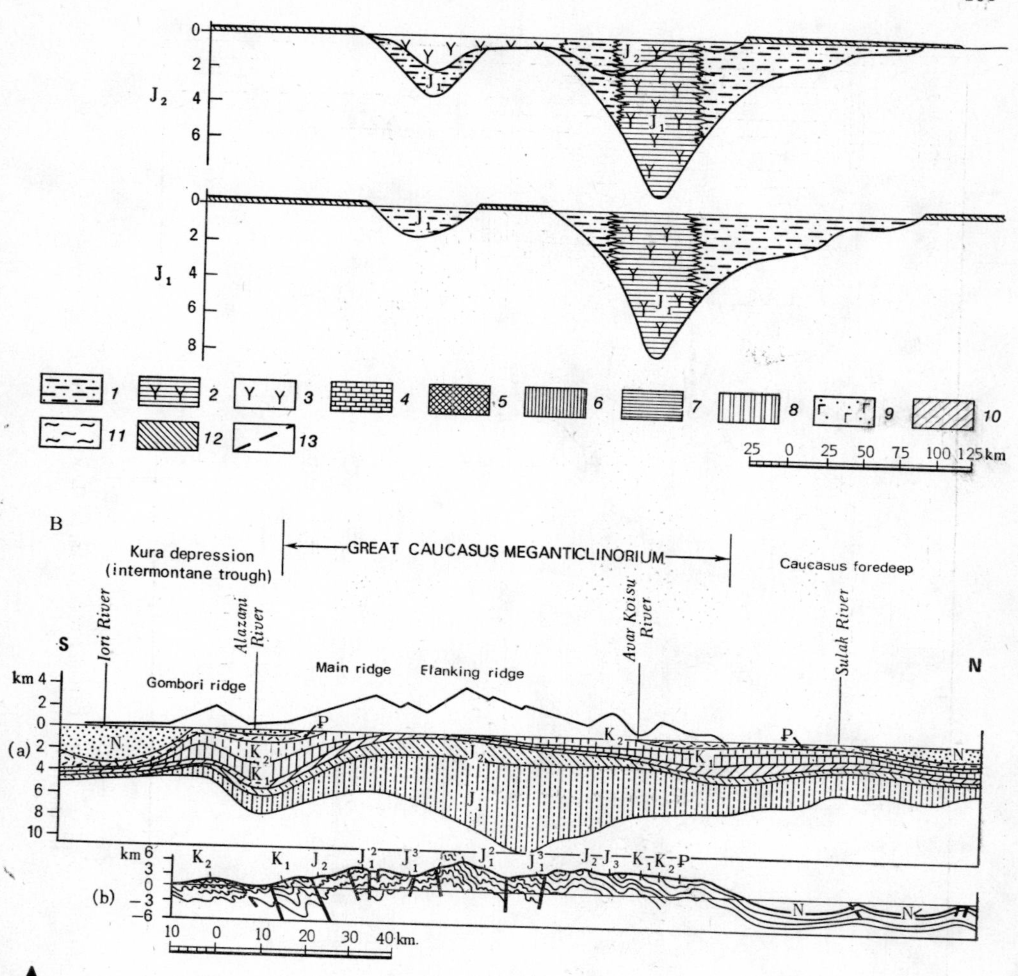

Fig. 56. *A*. Development of oscillatory movements in the Great Caucasus during the Alpine cycle (palaeotectonic profiles) (after Sholpo, 1978):

1—arenaceous-argillaceous sediments; *2*—clay series containing diabase dykes; *3*—volcanogenic series; *4*—limestones; *5*—carbonate terrigenous flysch; *6*—dolomitised limestones; *7*—carbonate series interbedded with volcanogenic material; *8*—marls and argillaceous limestones; *9*—tufas and tufagenous rocks; *10*—arenaceous limestones; *11*—clays; *12*—regions of uplift and erosion; *13*—faults

B. Relationship of folding (profile *b*) to the total thickness of the Meso-Cenozoic and to the neotectonic relief of the Great Caucasus

sumes that folding in geosynclines is caused by squeezing of the latter between convergent blocks.

This simple idea of an external origin of horizontal compression forces in relation to geosynclines, however, is refuted by a number of features specific to the structure and history of fold zones, of which the following are of basic significance:

a) the shape of the fold zones;
b) the history of their development;
c) the distribution of folds of different kinematic type over a fold zone.

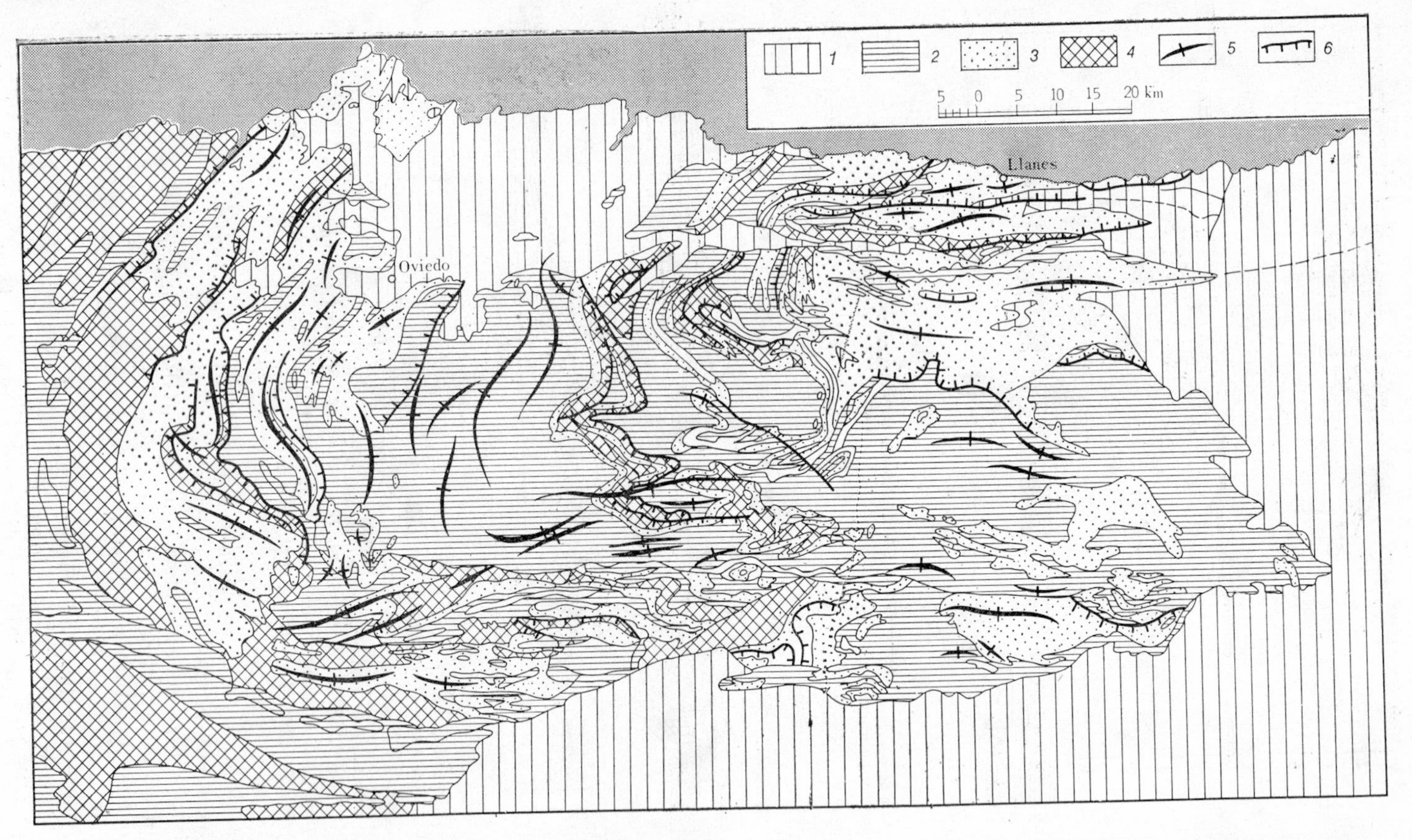

Fig. 57. Transverse folds in the western dip of the Pyrenees (after Casteras):

1—Tertiary and Mesozoic rocks; *2*—Upper Carboniferous rocks; *3*—Middle Carboniferous rocks; *4*—Lower Carboniferous and Lower Devonian rocks; *5*—axes of anticlines; *6*—overthrusts (strokes on the side of the thrusting)

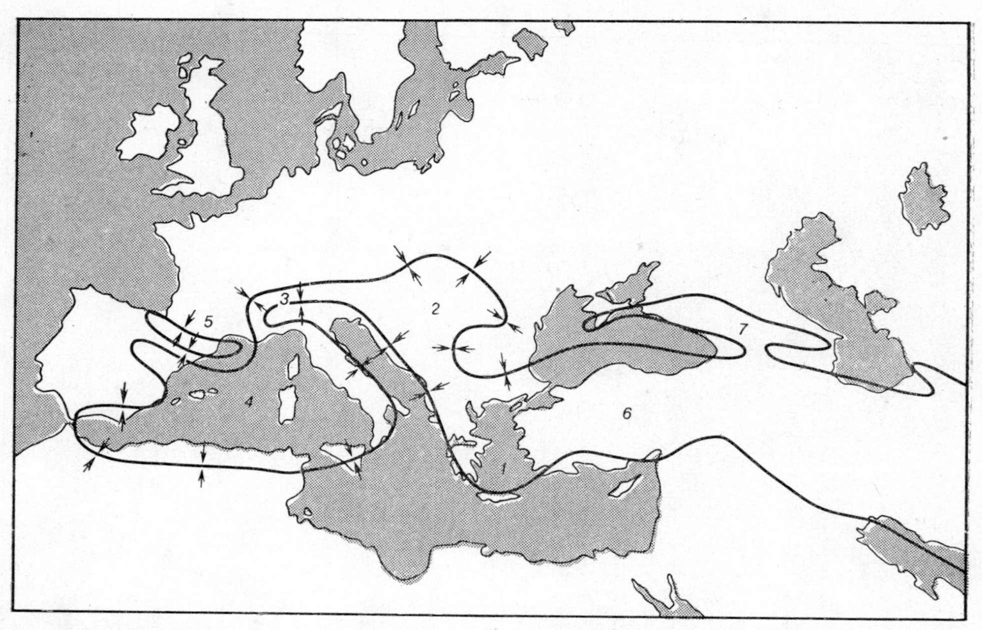

Fig. 58. The shape of the Alpine folded zone in Europe (after Beloussov). Ovals: *1*—Aegean; *2*—Hungarian; *3*—Alpine; *4*—West Mediterranean; *5*—Pyrenees; *6*—Asia Minor; *7*—Caucasian

Let us consider these features.

The Alpine fold zone in Europe has complicated outlines. It forms a number of arcs, viz., the Carpathian, Alpine, Gibraltar, Apennine, and Balkan. These arcs are grouped into several ovals (see Fig. 58).

These complex outlines contradict the idea of an external application of compressive forces to the geosyncline. For the fold arc of the Carpathians to have been formed through pressure from the East European platform it would have been necessary for the latter, in fact, to have pressed against the fold zone in different directions: from north to south in the north of the arc, from east to west in the middle, and from south to north in the south. It is, however, impossible, since according to this hypothesis the platform should be monolithic and rigid, and since there are no deformations in its body pointing to the possibility of independent movements of its separate pieces. For the same reason it is impossible to explain any other folded arc from this standpoint.

The development of folding in a geosyncline always begins in its interior, from which it spreads to the periphery. The zone involved by folding in a given time interval is thereby raised and a trough forms in front of it. As the zone broadens the trough is displaced to the periphery of the geosyncline and finally, in the orogenic stage, "rolls" over the edge of the platform in the form of a foredeep. This history does not tie up with pressure being applied to a geosyncline from the outside in which the first dislocations would have to arise on the geosyncline's periphery and spread into it from there.

Special observations indicate that general crumpling folding, despite the widely held view, by no means involves the whole of a geosyncline, but

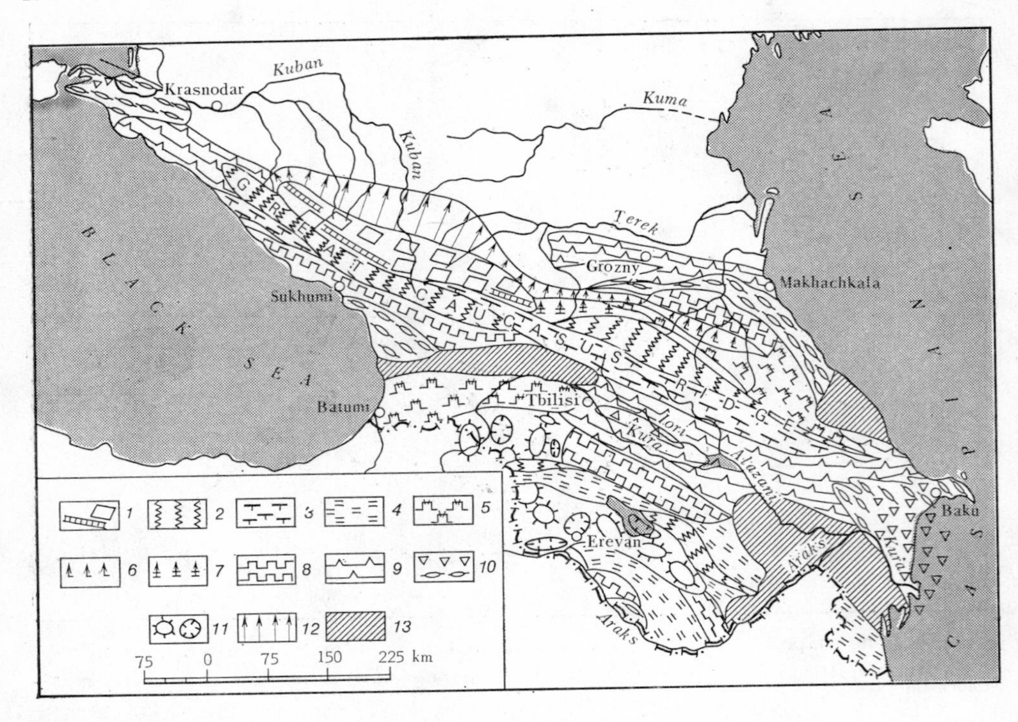

Fig. 59. Morphological types of Caucasian folding (after Sorsky, 1964):

1—zone of Prealpine block-fold tectonics; *2-4*—zones of crumpling folding [*2*—complete disharmonic folding with cleavage; *3*—tight disharmonic folds overturned to the south (southern slope of the Great Caucasus ridge); *4*—moderately compressed V-shaped folds]; *5*—zone of box folds complicated by faults and minor folding; *6*—zones of stepped flexures, complicated by minor folds and faults; *7*—steep monocline; *8*—zones of simple box folds; *9*—zones of narrow ridge-type anticlines separated by broad flat synclines; *10*—zones of brachyfolds, diapiric domes, and mud volcanoes; *11*—zone of major brachyform and dome-shaped arched uplifts expressed in the present-day relief; *12*—monocline complicated by small flexures and minor faults; *13*—zone of horizontal position of Neogene and Quaternary sediments

only certain bands of it, all of which are always fringed by zones of the development of block and injection folding (see Fig. 59). These two types of folding predominate on the peripheries of geosynclines, but have nothing to do with horizontal compression. This circumstance also contradicts the idea of external application of horizontal compression forces to a geosyncline, for such forces could not by-pass its peripheral areas in order to reach the bands of development of crumpling folding.

The signs enumerated indicate that the horizontal compression that causes the formation of type of crumpling folding could not be applied to a geosyncline from the outside. It must be "internal", i.e., must arise as a result of processes of some kind occurring within the geosyncline itself. These processes would have to consist in a displacement of material within the geosyncline, in which longitudinal compression of strata would occur in certain zones. Since the breadth of the fold zone does not alter, compression in one place must be compensated by stretching somewhere else.

It can be supposed, for instance, that longitudinal compression of strata arises through their displacement by gravity along the slope of an uplift formed during undulatory oscillatory movements. In that case buckling into folds would be expected on a slope where the further downward movement of strata encounters resistance from rocks lying quietly below the slope, while the zone of stretching of the same strata would be higher up the slope.

One can imagine, however, that the buckling of some strata or other into folds occurs as the result of their being penetrated by rocks occurring at great depth that push them aside and create horizontal compression in them. Such penetration could have a diapiric character and be caused by the same force of gravity in a situation of density inversion.

The special research into the problem of the origin of folding carried out in the USSR over the past 15 or 20 years indicates that the main driving force in the forming of general crumpling folding is in fact gravity. The concrete mechanisms of the formation of folds are as follows:

a) flowage or sliding of rocks down the slopes of uplifts;
b) spreading of the upper part of a relatively raised block of rocks, and its pressure on the strata of a neighbouring, relatively subsided block;
c) penetration by a "deep diapir" and its pressure on the surrounding rocks.

The first mechanism has long been known, and is most clearly manifested in the formation of tectonic nappes of Helvetian type.

As an example we may cite the Ubaye-Embrunais nappes in the French Alps (see Fig. 60). With flow of a water-soaked plastic series of Palaeogene flysch down a slope, rocks of another age were entrained, in particular Cretaceous and Jurassic limestones. On the whole this flow nappe is a huge recumbent anticlinal fold extending down the slope, complicated by smaller recumbent folds and faults.

Fold deformations requiring constriction are largely compensated in flow nappes by stretching and thinning of the strata in its upper part. But they can also be related to a nappe's complete detachment from its roots and the formation of a gap between them and the body of the nappe. In such cases the nappes are packets completely torn from their roots and sliding freely down the slope.

An extremely low gradient, measured in a few degrees, is sufficient to move nappes of Helvetian type. Movement proceeds slowly, through creep, which is much facilitated by saturation of the rocks with water and high intrapore pressure. The range of horizontal displacement can be as much as tens of kilometres; it does not happen, however, in one motion but develops gradually, as the region of uplift enlarges and its slope is moved further and further forwards.

Spreading of the tops of uplifted blocks of crust is quite common. When we were considering deep faults above, we said that the vertical faults became inclined toward the top and ran below the relatively uplifted block. That is also the result of gravitational spreading or slumping of the crest of an uplifted block. The spreading is compensated either by fissuring or by plastic horizontal stretching of the rocks comprising the crest of the block. When a raised block is fringed on two sides by relatively subsided zones spreading may occur on both sides; but in geosynclines, in which the processes described mainly develop, the slopes of the anticlinoria usually have a stepped character, which is determined by a series of parallel deep faults. In these cases the spreading on one limb is unilateral; each higher-lying block presses on the neighbouring lower one.

The fold deformations caused by collapse of a relatively uplifted block weaken and pinch out completely with distance from the source of pressure, i.e., from the edge of the raised block. But if the slope of an anticlinorium is

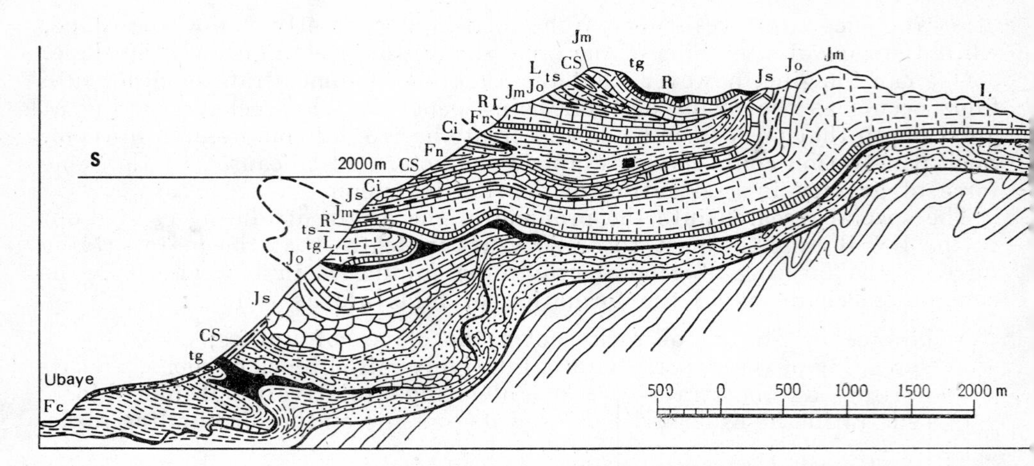

Fig. 60. The Ubaye-Embrunais gravitational tectonic

1—Quaternary sediments; *2-4*—flysch (2—upper Priabonian; *3*—Lutetian-Priabonian; *4*—Palaeocene); *8*—Middle Jurassic (limestones); *9*—Lower Jurassic (limestones); *10*—Rhaetian Stage; *11*—Upper Triassic Stage

divided into a great many steps, the whole slope may be covered with folds, the degree of constriction of which, however, differs at its various levels, being stronger at the top of each step and attenuating toward the bottom.

The mechanism of the effect of a raised block on the surrounding rocks is not limited to their moving apart. It also includes the squeezing of the surrounding rocks out from beneath the overthrust margins of the block where the overthrusts have sufficient horizontal amplitude for that. Squeezing out occurs particularly intensively when the overthrust rocks are heavier than the underlying ones.

Signs of the spreading of the tops of raised blocks are quite plain in the Tien Shan (see Fig. 41).

The change in shape along the strike of the Sugano fault in the Southern Alps is illustrated in Figure 61. The fault, which is vertical at depth, becomes horizontal and even plunges near the surface. The raised anticlinorium to the north of the fault is composed predominantly of granites and metamorphic rocks, and in part of Mesozoic formations. The collapse of this anticlinorium toward the south caused folding of the Cretaceous and Tertiary deposits of a neighbouring synclinorium.

The profile through the Polar Urals traced by H.P. Kheraskov and A.S. Perfiliev makes it quite probable that the folding of the Lemva synclinorium was caused by collapse of the Kharbei anticlinorium toward it (see Fig. 62).

In the preceding cases horizontal pressure was exerted on the surrounding rocks by blocks composed of much older rocks than those of the buckled strata around them. The deformations and metamorphism in the rocks of the raised blocks were completed before they were uplifted and began to exert a mechanical force on the enclosing series. We shall provisionally call these blocks "passive".

There are also other uplifted massifs within the fold zones that exert a deforming effect on the country rocks. Some of them are composed of rocks of the same age as the enclosing series and others of rocks which, though older,

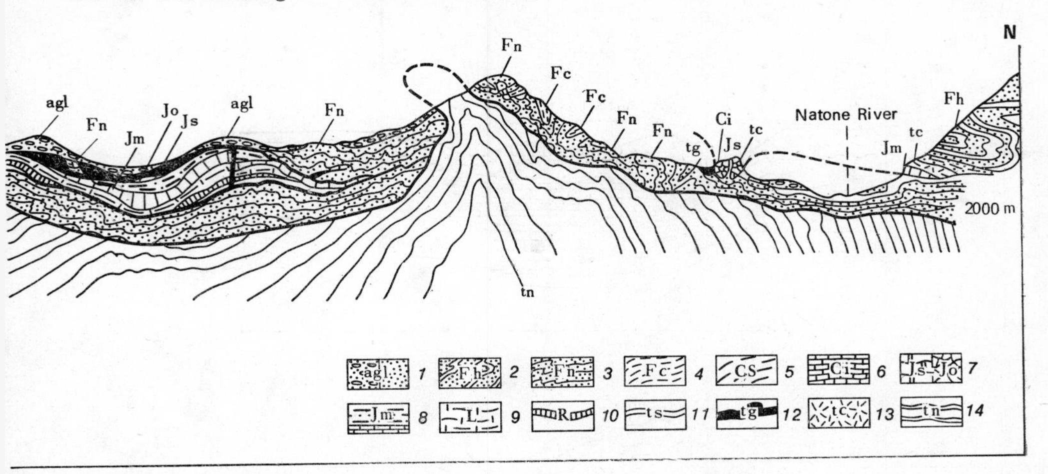

nappe (French Alps) (after Schneegans, 1938):

5—Upper Cretaceous (limestones); 6—Neocomian Stage (limestones); 7—Upper Jurassic (limestones); (argillites); 12—Middle Triassic (gypsum); 13—Lower Triassic (limestones); 14—Bathonian-Oxfordian (clays)

have suffered metamorphism and granitization simultaneously with the uplift. Still others consist of granites, granite-gneisses, or gneisses formed simultaneously with uplifting of the block. These are "active" massifs or the so-called "deep diapiric folds". They differ in their mechanical properties from the "passive" blocks of the previous type in having been much more plastic during their rise, as a consequence of which, in particular, they not only forced the rocks apart, through which they rose, but spread out mushroom-wise, even more than the "passive" massifs over the enclosing series which were squeezed out from beneath this cover.

W. Bucher explained the folding of the Appalachians by the uplifting of a central metamorphic massif now largely hidden beneath the young sediments of the Piedmont lowland, its creeping over the adjacent sedimentary series and squeezing out of the latter from beneath it (Fig. 63).

In the axial belt of the eastern half of the Great Caucasus Lower Jurassic shales crumpled into tight isoclinal folds in which there is marked cleavage outcrop separately along faults from the country rocks. Farther away from them the folding in Lower and Middle Jurassic arenaceous-argillaceous suites rapidly attenuates on both sides. There are grounds for seeing a deep diapir of complexly dislocated Lower Jurassic shales in this zone, penetration of which into the overlying series caused the latter to crumple (Fig. 64).

The tectonic nappes of deep or Pennine type known in the Alps belong to the category of deep diapirs. They consist of metamorphic rocks that have undergone a certain degree of granitization. The fact that Pennine nappes are the result of diapiric rising is confirmed by their shape, which takes the form of thick tongues and mushrooms, linked with the deeper interior by narrow, vertical "roots", a shape that can hardly have arisen in any other way. The nappes, which lie on top of one another, crushed each other and crept over normally stratified Mesozoic and Palaeogene series on the northern slopes of the Alps, squeezing them out from under them, crumpling them into folds, or forcing them to slide to the periphery of the fold zone in the form of Helvetian nappes (Fig. 65).

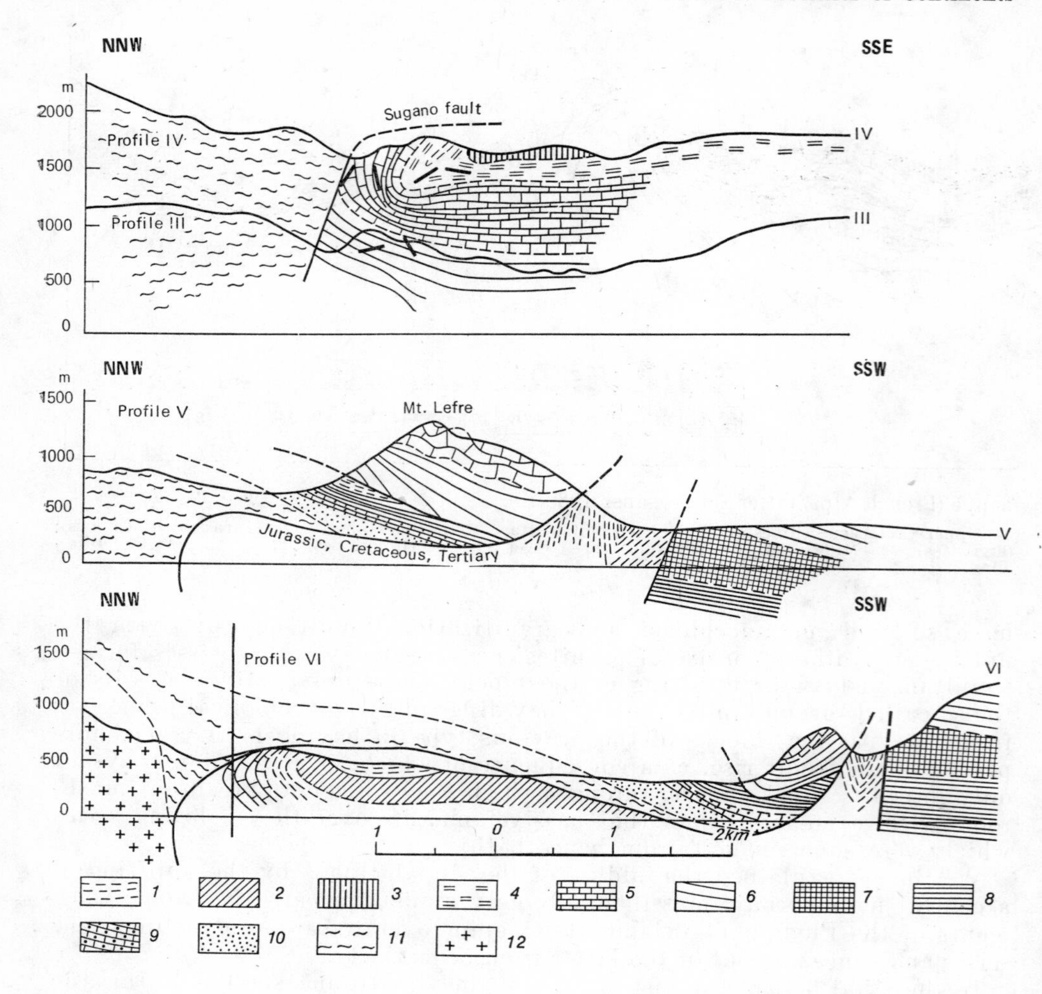

Fig. 61. Profiles through the Sugano fault (Southern Alps) (after Agteberg, 1961):
1—Miocene; *2*—Oligocene and Eocene; *3*—Upper Cretaceous; *4*—Lower Cretaceous; *5*—Jurassic; *6*—Upper Triassic; *7*—Middle Triassic; *8*—Lower Triassic; *9*—Upper Permian; *10*—Lower Permian; *11*—quartz phyllite; *12*—granite

Where, thanks to the stratification of the rocks, we can decipher the internal structure of Pennine nappes, it proves to be quite similar to that of the salt domes described above. Here also, tight isoclinal folds are developed inside the deep diapirs (Fig. 66).

The idea of deep diapiric folds provides an explanation of the complicated fold structure of the axial parts of fold zones often observed. The two previous mechanisms required there to be stretching in the axial area, but it is just there that very tight vertical isoclinal folds are often observed. Given the presence of deep diapirism these become understandable.

The problem of the origin of deep diapiric folds will be considered in the next chapter.

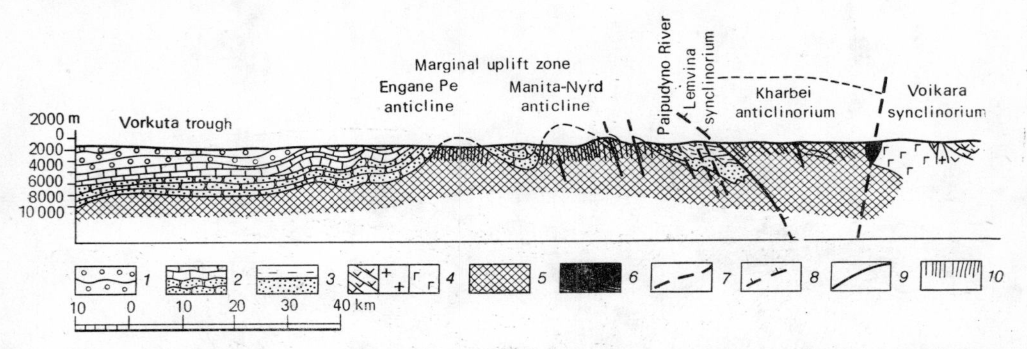

Fig. 62. Schematic geological profile through the Polar Urals (after Kheraskov and Perfiliev, 1963):

1—sediments of a periclinal trough (Upper Carboniferous and Triassic); *2*—sediments of the zone of marginal uplifts (*a*—lower structural stage; *b*—upper structural stage); *3*—sediments of the West Urals synclinoria (*a*—lower structural stage; *b*—upper structural stage); *4*—formations of the zone of greenstone synclinoria (*a*— lower structural stage; *b*—plagiogranites; c—amphibolites and gabbro-amphibolites); *5*—protuberances of ancient structures into the cores of anticlinoria (complex of preuralites); *6*—hyperbasites; *7*—faults; *8*—overthrust of the western edge of the Kharbei anticlinorium and its analogues; *9*—Main Urals deep fault; *10*—preuralite internal structure

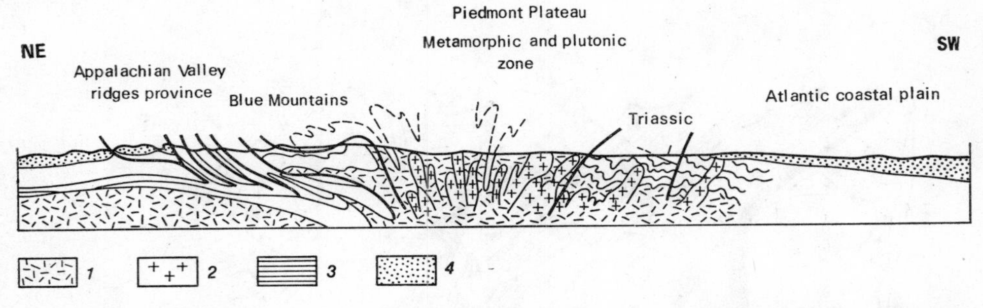

Fig. 63. Schematic profile through the Appalachians from Kentucky to South Carolina (after Bucher, 1957):

1—metamorphic rocks; *2*—Hercynian granites; *3*—Lower and Middle Palaeozoic sedimentary deposits; *4*—Triassic and younger deposits

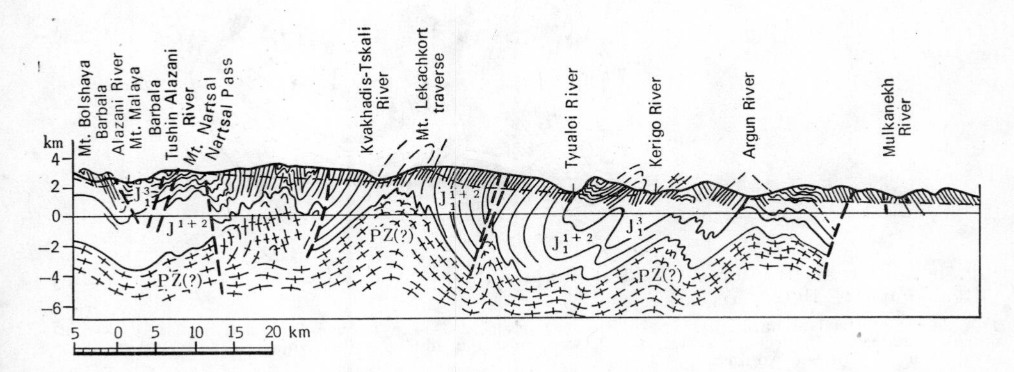

Fig. 64. Cross-section through the eastern part of the Great Caucasus (after Sorsky, 1964)

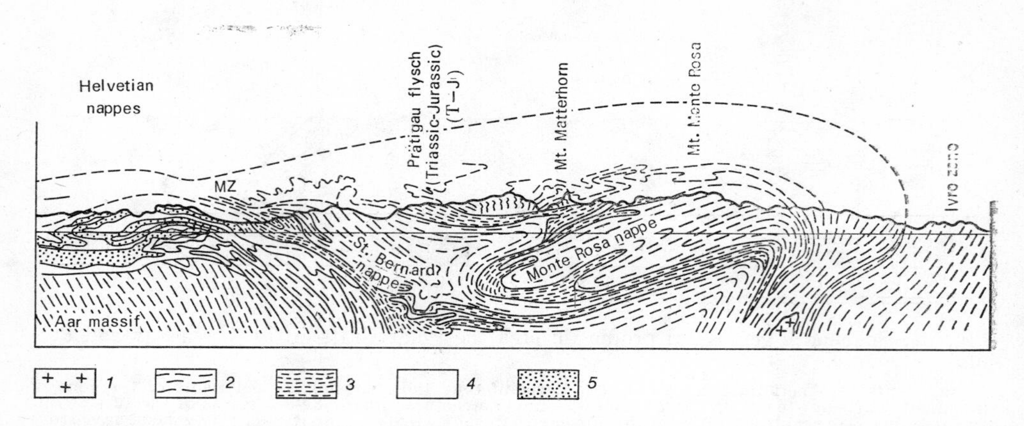

Fig. 65. Pennine nappes in the Swiss Alps and the squeezing of Helvetian nappes from beneath them (after Heim, 1922):

1—Tertiary granites; *2*—Palaeozoic crystalline rocks, activated by Alpine metamorphism; *3*—metamorphosed Lower Mesozoic rocks; *4*—non-metamorphosed Mesozoic; *5*—Eocene flysch

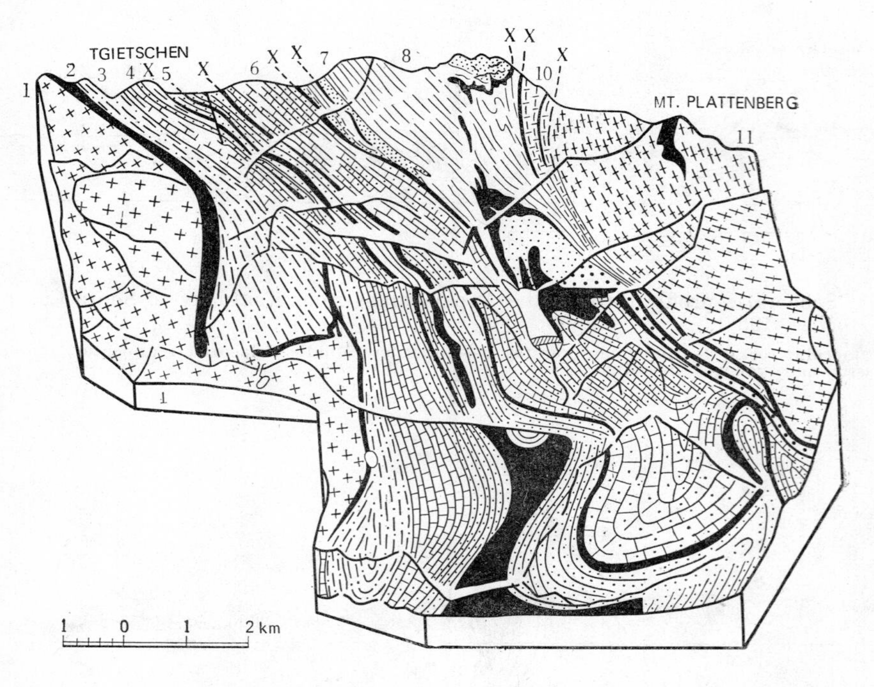

Fig. 66. Internal structure of a "deep diapir" in the Swiss Alps (Upper Blanco Valley) (after Leupold, 1962):

1—gneisses of the Gotthard massif; *2*—Triassic system; *3-5*—Lias of the Gotthard massif; *6-8*—metamorphic shales; *9*—Verrucano suite; *10*—Lower Mesozoic metamorphic shales; *11*—gneisses of the crystalline core of the Adula nappe

CHAPTER 13

Deep Folding and General Considerations About Folding Movements

Deep Folding

In earlier chapters we mentioned a complex folding typical of the metamorphic and granitized rocks that forms the cores of deep diapirs. Now we must add that tectonic deformations in all heavily metamorphosed, granitized, and migmatized rocks have a complicated, specific character. The features of the folding observed in the cores of diapirs are thus only a special case of a more generally specific kind of deformation inherent in rocks that have undergone metamorphic and ultrametamorphic transformations.

The best examples of the special nature of these complex deformations can be seen in crystalline shields where Archean and Lower Proterozoic, highly metamorphosed, granitized, and migmatized rocks are exposed over wide areas. We call the complex deformations observable in rocks that have been strongly metamorphosed and ultrametamorphosed *deep folding*.

A feature of the structure of series involved in deep folding is the broad development of dome-like structural forms or uplifts with intricate irregular outlines. These forms may consist of granites, granite-gneisses, gneisses, or migmatites, and are commonly called granite, granite-gneiss, gneissose, and migmatic domes. They are formed during the rise from great depths of the material composing them. All are varieties of deep diapirs.

The laminated or banded metamorphic rocks within deep diapirs are buckled into folds. A combination of folds of different orders, superimposed on one another, is typical of deep folding. From folds measured in hundreds of metres we descend via folds of decreasing size to fine plications of only one or two centimetres. All these orders of folds form highly disharmonic combinations. Foliation and boudinage are common. Isoclinal folds with very drawn-out noses are characteristic, but are quite imperceptible so that bundles of them assume the form of very thick monoclines. A superimposing of folds of different systems onto one another is also characteristic: the axial fold planes of one system may be buckled into the folds of a second system, and the axial planes of these into the folds of a third system. The schistosity and cleavage generally characteristic of deep folding may also, accordingly, be represented by several systems, each linked with a corresponding system of folds. Finally, it should be noted that folds with steep, vertical hinges are common (see Fig. 67).

All the features of the morphology of deep folding enumerated point on the whole to its having been formed in conditions of high plasticity and low viscosity of the rocks. The fold forms observed could not in the aggregate have been formed through longitudinal compression of the strata by forces applied somewhere at their ends. The whole picture witnesses to their being the result of the application of body forces causing general flowage of the rocks. In that case, however, the folds are simply "flowage patterns" reflect-

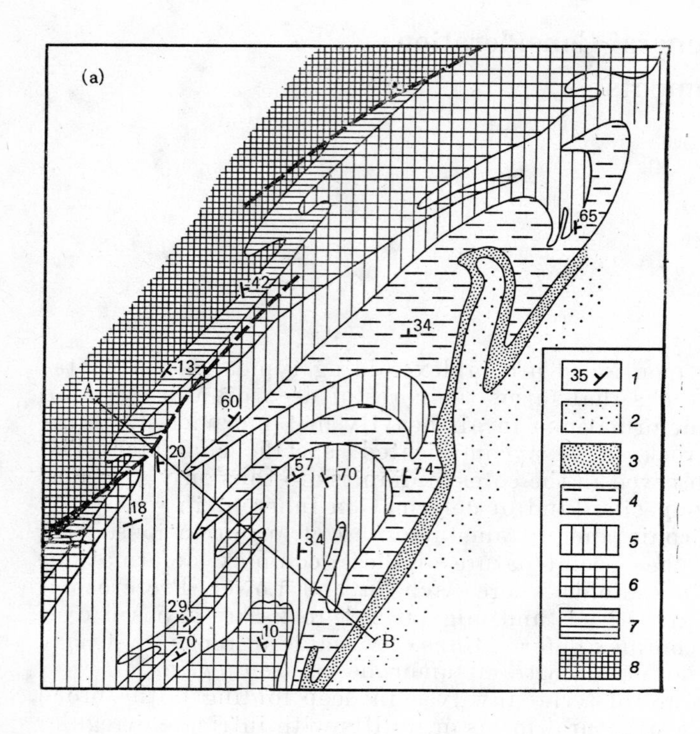

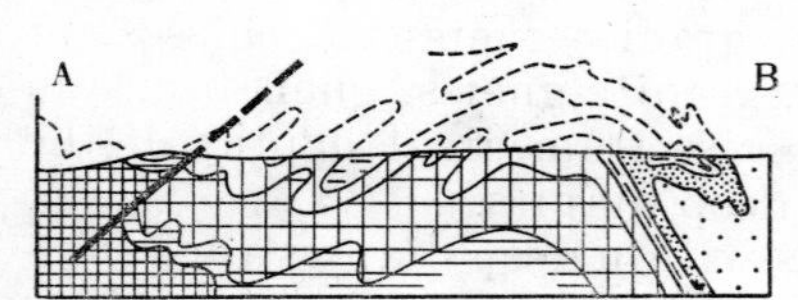

Fig. 67. Complex superimposed folding developing in metamorphosed series:

a—superimposition of folding in the Lower Palaeozoic of Quebec (Canada) (after Osberg, 1955): *1*—bedding of strata; *2*—dolomites; *3*—upper phyllites; *4*—limestones; *5*—lower phyllites; *6* — quartz-muscovite-chloritic schists; *7*—quartzites; *8*—epidote-chloritic schists; *b*—block-diagram of a structure formed by mutual arrangement of folds of three generations. Archaean in Karelia (after Kukley). Axial surfaces of folding: *1*—first generation; *2*—second generation; *3*—third generation

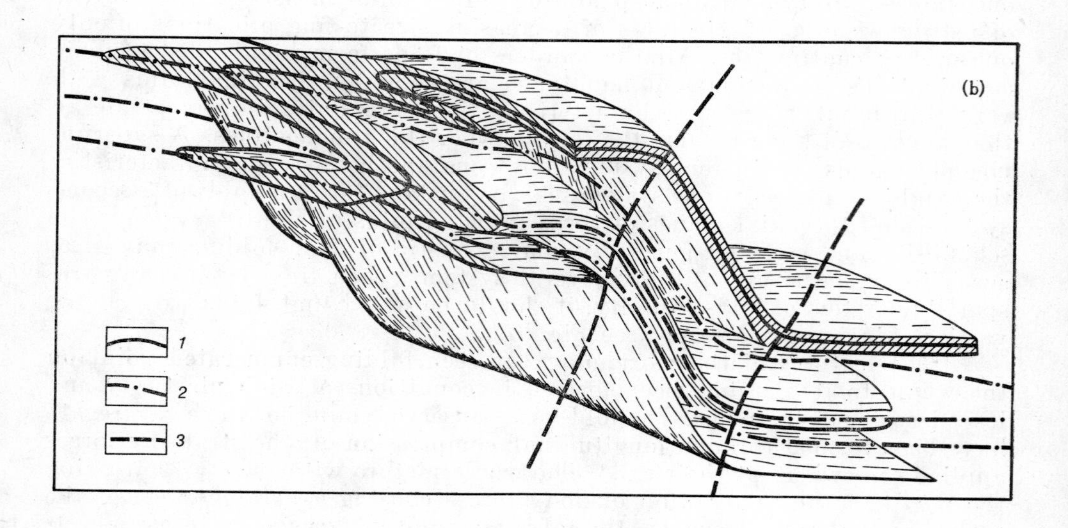

ing the unevenness of this flowage, broken up into a great many large and small streams; and it is these streams that determine the flexures of the strata. A substantial role is also played by processes of squeezing and injection; a bundle of strata is squeezed in a certain sector and the material pinched out is injected into a neighbouring sector where, being in a confined space, it gathers into folds.

In order to understand the origin of deep folding, let us recall the general conditions in which the metamorphic and ultrametamorphic processes develop.

All these processes require a high temperature. They also depend on pressure, the presence of water, and an influx of alkalis and silica from deeper regions. Regional metamorphism, as we know, is manifested in different "facies" that reflect metamorphic transformations of different degree, the main ones being the greenschist, amphibolite, granulite, and eclogite facies and the ultrametamorphic facies or granitization.

Mainly metamorphic rocks of the greenschist and amphibolite facies (plus subfacies) are involved in the structure of massifs in which deep folding has developed, and also the products of ultrametamorphism, i.e., granites, granite-gneisses, and migmatites. The granulite and eclogite facies are encountered in rather less volume.

Laboratory experiments to obtain metamorphic rocks artificially indicate that the temperature at which greenschists develop varies between 200 ° and 500 °C under a lithostatic load corresponding to depths of 3000 to 5000 m. The presence of a large amount of water is also needed.

The amphibolite facies of metamorphism includes various gneisses of granitic composition and amphibolites. It is formed at a temperature of 500 ° to 600 °C and a pressure corresponding to the lithostatic pressure at depths of 5 to 15 km. The rocks contain water but rather less than the preceding facies.

Ultrametamorphism develops first in the form of migmatization and then of complete granitization within a zone of amphibolite facies. Given enough water at a pressure corresponding to a depth of 5 to 15 km, granitization takes place at a temperature of 650 ° to 750 °C, and the rocks partially melt.

By comparing temperature with the probable depth of the formation of these metamorphic and ultrametamorphic facies we see that the geothermal gradient for the average conditions needed for these processes must be 70 ° to 100 °C per kilometre, which is two or three times higher than the mean present-day gradient of 36 °C per kilometre.

The temperature at which metamorphic and ultrametamorphic reactions take place affects the viscosity of rocks, reducing it. Where granitization develops the lowering of viscosity is particularly great, since rocks undergo at least partial melting. The viscosity of the melt is usually of the order of 10^6 poises, whereas the viscosity of a crystalline rock is 10^{18} to 10^{20} poises.

The viscosity of metamorphosed rocks is greatly diminished and their plasticity increased by the presence of water in them. Water enters during the development of regional metamorphism in part from the mantle (see Chap. 23), but most of it comes from the rocks themselves. The majority of metamorphic rocks were originally sediments with a preponderance of muds, which contain much pore water, frequently more than 30% of their volume. Most of the water is pressed out during compaction of the muds under the pressure of overlying sediments, until they turn into clay, but even in dense argillaceous rocks there is up to 5% water. When the heating of the rocks begins that subsequently leads to metamorphic transformations, the original pore water is joined by water released from the surface of minerals and the crystalline lattice. Adsorbed water from the surface of minerals, which volume comprises from 2.5 to 20% of the total bulk of a rock, depending on its composition, is released. Then constitutional water is released during the transition of minerals rich in water into ones poor in water or anhydrous. When muscovite, for example, turns into common potassium feldspar, the water released is 14.5 per cent of the solid phase.

The water content of rocks at the stage of metamorphic changes thus rises steeply.

The higher plasticity of rocks during metamorphism is also associated with recrystallization. When metamorphism occurs concurrently with the action of deforming forces (and in the vast majority of cases that is what happens) recrystallization is oriented in accordance with Riecke's principle: flat or elongated crystals are formed whose short axes are parallel to the direction of maximum compression and long ones perpendicular to it and parallel to the maximum elongation of the medium, i.e., deformation occurs largely through the mobility of ions and their regrouping in accordance with the stress field.

In all these cases the lowering of viscosity and raising of plasticity are temporary phenomena. As soon as the epoch of regional metamorphism ends, the temperature declines, all the rocks that underwent partial melting recrystallize, and the water that filled the pores is squeezed out from the rocks to the surface; all this leads to an increase in viscosity and a reduction of plasticity to the levels normal for solid rocks.

Parallel with the changes in viscosity and plasticity during metamorphic and ultrametamorphic reactions there are also changes in the rocks' density. Since partial melting occurs with ultrametamorphism, the rocks' transition from the solid state to the liquid one is accompanied with lowered density and a corresponding increase in volume. The density of solid acid igneous and metamorphic rocks is usually around 2.8 g/cm³. The melts of these same rocks have a density around 2.2 g/cm³, which is thus less than that of most sedimentary, and all dry metamorphic and igneous solid rocks.

Density also changes as a result of the release of water from the surface and crystalline lattices of minerals into the pores. When muscovite and quartz alter as a result of metamorphic reactions into potassium feldspar and sillimanite, with release of water, at a lithostatic pressure of 2.5 kbar and a temperature of 600 °C, the total volume of the solid and liquid phases increases by 14%:

$$KAl_2(OH)_2[AlSi_3O_{10}] + SiO_2 = K[AlSi_3O_8] + Al_2SiO_5 + H_2O$$

Molecular volumes:	Muscovite	Quartz	Potassium feldspar	Sillimanite	Water
	140.5	22.7	108.7	49.9	27.0
	163.2		185.6		

Density diminishes accordingly.

This effect is also transient. The water released from the rocks will, of course, rise continuously to the surface in the form of thermal streams and, when its release from minerals ceases, the rocks will become nearly anhydrous, increasing in density and diminishing in volume to the values normal for dry, solid rocks.

In order for water to be released within a big massif of rocks and escape from it, however, a time is needed that depends on the amount of water and the hydraulic permeability of the rock. It has been estimated that, with the permeability characteristic of argillaceous shales, and the most probable volume of water, it would need at least 10^7 years for complete release of water to the surface. The process, of course, might be accelerated by fissuring of the rocks, but even if the figure mentioned were reduced by one order of magnitude, the phenomenon of a temporary lowering of density and increase in bulk of a massif of rocks that takes place during metamorphism would last around a million years, which is a perceptible interval of time even on a geological scale.

The phenomenon, however, even though temporary, plays an important tectonic role. It leads to a temporary state of density inversion in the Earth's crust; rocks occurring at a depth of 5 to 15 km below the surface are subjected to the effect of metamorphism and ultrametamorphism and may prove to be temporarily lighter than the sedimentary and older metamorphic rocks covering them.

Hence it follows that where metamorphism and ultrametamorphism occur conditions are created in the course of time that promote the formation of deep diapirs composed of granitized and metamorphic rocks. The density inversion, it must be especially emphasised, arises right in the earliest stages of the heating of the rocks, when recrystallization has not yet begun and only adsorbed water is being released. This circumstance explains the cases when there prove to be shales in the core of a deep diapir, which have not passed beyond the facies of slates or phyllites in degree of metamorphism. The complex folding characteristic of metamorphic rocks is a reflection of the complicated streams of material that are created within a deep diapir during its formation.

The general character of the deformations in metamorphic series is such, as was mentioned above, as can only be understood by attributing a fundamental role in their mechanism to volumetric forces applied directly to each volume of rocks, even the smallest. It is volumetric gravitational forces that cause the rise of deep diapirs and are responsible for the complex deformations of their cores.

When penetrating the overlying rocks deep diapirs also deform them by their bursting forces. Granite, granite-gneiss, and metamorphic deep diapirs are therefore usually surrounded by a folded fringe of country rock.

General Considerations About Folding Movements

The types of folding under consideration are confined to different tectonic conditions and levels of the Earth's crust. Block folding in its pure form is manifested in the most quiescent state of oscillatory movements, where their amplitude and gradients are not too great. It is confined to platforms and to the quietest zones in geosynclines and orogenic regions.

The similarity between the formation of block folds and general block-undulatory oscillatory movements of the crust suggests a certain degree of affinity between them, the more so that a reversal of the direction of vertical movements (reverse block folds) is observed in the history of the development of block folds.

Injection folding is characteristic of regions where the oscillatory movements are of considerable scale and where thick sedimentary series therefore

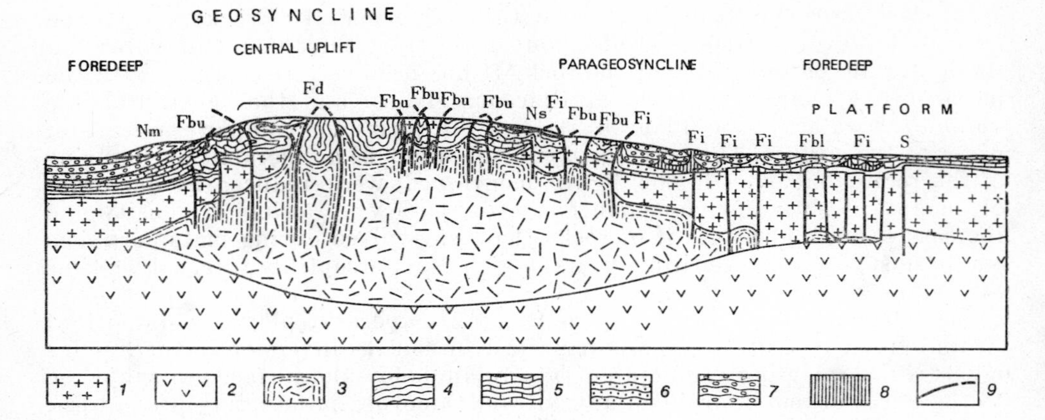

Fig. 68. General scheme of types of folding (after Beloussov):

1—ancient folded basement; 2—"basaltic" layer of the crust; 3—young granites and migmatites; 4—lower terrigenous formation; 5—limestone formation; 6—upper terrigenous formation; 7—molasse and lagoon formation; 8—salt; 9—tectonic fault. Nm—tectonic mudstream nappe; Ns—tectonic slide nappe; Fbu—buckle folding; Fi—injection folding; Fbl—block folding; S—swell of block origin; Fd—deep folding, granitic-gneissose domes, tectonic nappes of pennine type

accumulate. But these should remain unmetamorphosed and should include rocks of high plasticity and low density. Such conditions exist in deep syneclises on platforms and in the foredeeps and intermontane troughs of orogenic zones with their sections containing salt-bearing series and thick series of plastic clays saturated with water. To some extent injection folding can be considered as a complication of block folding, when we take into account that block movements in the crystalline basement predetermine the localization of injection folds.

General crumpling folding with its three formation mechanisms develops in zones of high-gradient oscillatory movements, i.e., in a more complicated situation, from this aspect, than the two previous types. Its confinement to geosynclines is therefore quite natural. It is essential to note that it develops in the upper layers of the crust where flow or sliding of bundles of strata occurs on the slopes of uplifts, and where buckling of strata also occurs through the effect of spreading of the tops of relatively raised blocks of crust and the pushing aside of strata by the penetration into them of deep diapirs.

Deep folding develops n deeper layers of the crust, although it is not so much depth as metamorphism that is mainly significant in it. The deep diapirs that originate from this folding penetrate overlying normal sedimentary series and deform them.

The relationship between the different types of folding is shown in Figure 68 as they are depictable in the light of the opinions expressed here. In this scheme the vertical movement of blocks, like the formation of deep diapirs, is related to metamorphism and granitization occurring at deeper levels; we assume that the movements of blocks are caused by the uplifting of deep diapirs that do not, however, penetrate the overlying series of normal sedimentary rocks but push blocks of the basement out from under them. The latter, in spreading sideways at their tops through the force of gravity cause the formation of general crumpling folds in surrounding strata.

From this survey of the formation conditions of the various types of folding, it follows that injection folding, general crumpling, and deep folding are related to a disturbance of gravitational equilibrium of one sort or another. Equilibrium can be disturbed through the formation of a slope between an uplift and a trough, when the rocks lying on the slope prove to be in an unstable state and tend to flow or slide down it. Similarly equilibrium is disturbed on the margins of crustal blocks raised above the surrounding area along faults. It is also disturbed when a density inversion develops and less dense material in the crust lies below denser material.

The causes of density inversion may be: (a) sedimentation (the case with a salt series below other sedimentary rocks); (b) metamorphism of various stages from the initial heating of rocks when adsorbed water begins to be released into pores from the surface of minerals, to the amphibolite facies and ultrametamorphism. To these processes leading to density inversion we must add the development of secondary ones in connection with the displacement of tectonic nappes when the denser rocks of the sheet are thrust over less dense autochthonous rocks. Strictly speaking the instability of the rocks on the slope of a tectonic uplift or the margin of an upraised block can also be considered a kind of density inversion, because the hard rocks also occur, in these cases, lying above a less dense aqueous or atmospheric environment (depending on whether it takes place on the sea floor or above sea level); and since, when rocks slide down a slope, they displace a certain volume of this less dense medium, the process is in principle quite analogous to the sinking of heavier rocks and ascent of lighter ones during the formation of injection folds or deep diapirs.

In all cases, consequently, the formation of injection, crumpling, and deep folding is caused by a disturbance of gravitational equilibrium, through the development of a density inversion in some way or another. The fold deformation itself is the structural result of movements tending to restore equilibrium. But equilibrium is only restored by the rocks sliding down the slope or by the tops of raised blocks spreading, or by the rise of diapirs, both surficial and deep-seated.

We said above that the movement developing during density inversion is similar to convection, and pointed out that it is possible to use the term "advection" for such unilateral convection, in which there is no recirculation of one and the same material, but the light material rises and remains at the top while the heavy one sinks and remains at the bottom. We have come to the conclusion that all tectonic deformations in general caused by the crust's tending to gravitational equilibrium is the result of advection or advective movements.

The main kinematic elements of advective deformations are squeezing and injecting. The rocks are squeezed out from some places and injected into others, piling up in masses of greater or smaller size.

Squeezing and injecting take place in different directions, although vertical displacements should on the whole predominate, since the general tendency of the system consists of upward displacement of the light material and downward displacements of the heavy. In this both the light material is being injected upwards and squeezing heavy material out and the heavy material is being injected downwards and squeezing out light material. These movements take place on a different scale; the squeezing out and injection of various orders are nested, as it were, in one another. From the movement of a whole tectonic nappe we descend, in passing from one order to another, right down to the boudinage of individual strata. Gravitational forces, however, it must be noted, have a greater direct effect on deformation, the greater are

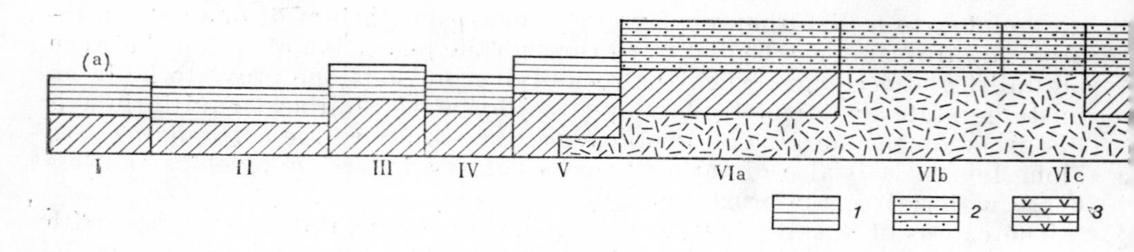

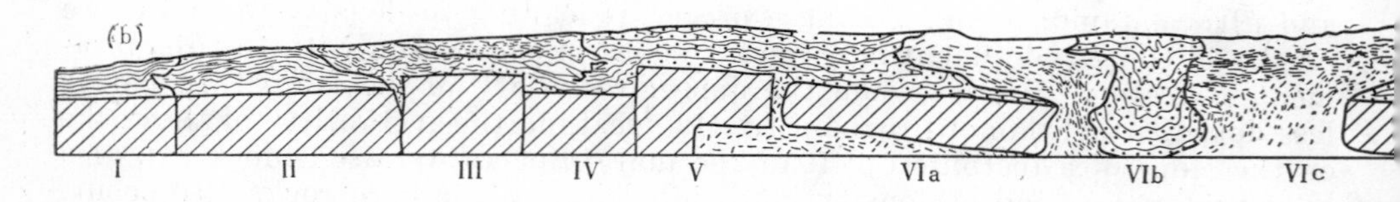

Fig. 69. Model of a fold zone

a—the model's initial form; *b*—at the end of the experiment (drawn from photographs); *1*—70% colo-
pentine; *4*—petroleum jelly and sand; *5*—73% colophony, 27% turpentine (with fine hairs); *6*—faults;

the bodies of the rocks involved in them. In passing to deformations of small volumes the effect of gravity becomes more and more indirect, reflecting only the degree of the link between the minor deformation and the bigger one underlying it.

In conclusion let us turn to the model of a folded zone created in the laboratory by N.B. Lebedeva (1966) (see Fig. 69).

The modelling material was resin mixed with machine oil in an amount that made it possible to control the density and viscosity of the material. The model was built of adjoining blocks of different height initially separated by partitions. The blocks had a layered structure and a density inversion existed in some of them. A particularly marked contrast of densities was created in the middle of the model, where the lower layers had a density of 1.0 g/cm^3 and the upper ones of 1.8 g/cm^3. There, too, the thickness of the light material was maximum and not layered, imitating rocks in a state of metamorphism and granitization. The different heights of the blocks represented the result of prior vertical block-undulatory movements of the crust.

When the partitions between the blocks were withdrawn, the model went into motion, for it became mechanically unstable because of the density inversions and different heights of the individual blocks. In the middle the light material, in floating up, formed two "deep diapirs" (see Fig. 69*b*, middle), between which the layered material initially lying on top of it was squeezed and buckled into folds by the expanding diapirs. The tops of the diapirs spread out and formed a Pennine type of nappe above which the layers were stretched. These deep diapirs pushed the layered series of adjacent blocks ahead of them. The material of the raised blocks, moreover, flowed aside toward the lower ones in the form of surficial sheets of Helvetian type. As a result a complex folded structure with sloping overthrusts was formed on the left of the model. On the right there was a certain spreading of the tops of the raised blocks and folds developed in adjacent blocks under their

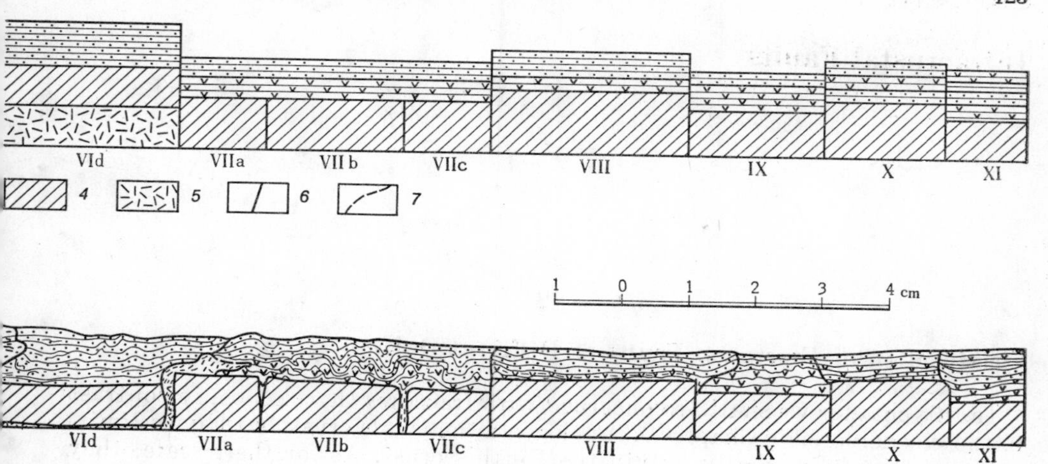

(after Lebedeva, 1966):

phony, 30% machine oil; 2—63% colophony, 30% machine oil, 7% clay; 3—73% colophony, 27% tur-
7—boundaries between the members of the different blocks; I-XI—blocks of various height

pressure. An injection structure was also observed there along the boundaries between blocks.

A model of a fold zone with a complex variety of deformations was thus obtained, with a complete absence of external horizontal compression. All the deformations were caused, on the one hand, by upward and downward circulation of the material in accordance with the local density inversions, and, on the other hand, the original uneven surface became almost horizontal after the gravity flow, as a result of "shrinkage" of the model, the thickness of the material of the raised blocks diminished while the layers lengthened, which caused buckling.

In accordance with the theory of physical similarity the model described satisfies certain dimensions of natural fold structures. Its overall length corresponded to that of a natural fold zone 150 km wide. The height of the raised blocks before the beginning of movement corresponded to natural mountains approximately 5000 m high. The horizontal amplitude of the overthrusts, on the same scale, amounted to 10 or 15 km.

This model has indicated that gravitational processes in stratified plastic series could, through advective displacements, lead to extremely complex deformations, including nappes and folding of different types.

CHAPTER 14

Intracrustal Faults

Classification of Fractures. Faults of Different Orders

In this chapter we shall be concerned with fractures that do not, in contrast to those considered in Chapter 9, cut through the whole crust, but have a limited extent and lie wholly within the Earth's crust. As for their scale, they are extremely varied (from some that stretch for tens of kilometres to cracks of centimetric dimensions).

Tectonic intracrustal faults (which we shall refer to simply as faults) are described in detail in textbooks of structural geology, in which, too, the tectono-physical and geological conditions of their formation are considered. Here, therefore, we shall confine ourselves to a survey of certain problems only.

Fractures are divided into two groups: *cracks or fissures* and *ruptures with a displacement or faults*. This division is conventional. Where the boundary is drawn between them depends on the scale with which we approach examination of a given structure. The scale determines what amplitude of displacement can be disregarded; then fault displacements of smaller amplitude than that are classed as fissures without amplitude. On the ordinary scales of geological mapping fractures with displacements of a few centimetres, and even of tens of centimetres, are classed as fissures, whereas even such displacements have to be taken into account in special, detailed investigations.

The group of fractures with displacement (faults) is divided in turn into partings or joints, shear faults, reverse or upthrown faults, cuts or slashes, thrust faults and faults. In this morphological classification singling out upthrusts and overthrusts is conventional. All faults with an upthrown hanging limb can be called upthrown faults if the dip is 60° or steeper. In that case all steeper faults with an upthrown hanging wall will be reverse faults.

An essential property of the generally accepted classification is that all the displacements are considered in relation to the horizontal and vertical planes and are taken to originate in either the horizontal or the vertical plane.

Experience of geological investigations indicates that these features of the classification are quite acceptable when major faults are being studied. Large-scale fault displacements in fact occur mainly either in a nearly vertical or an almost horizontal direction. It can be supposed that the force of gravity significantly affects the direction of the displacements of big massifs of rocks. It acts either directly, causing displacement of the massifs predominantly along the vertical (e.g., in the formation of gravitational faults), or supplementing other forces and altering the direction of a displacement carried by a force of another trend and bringing it nearer to the horizontal,

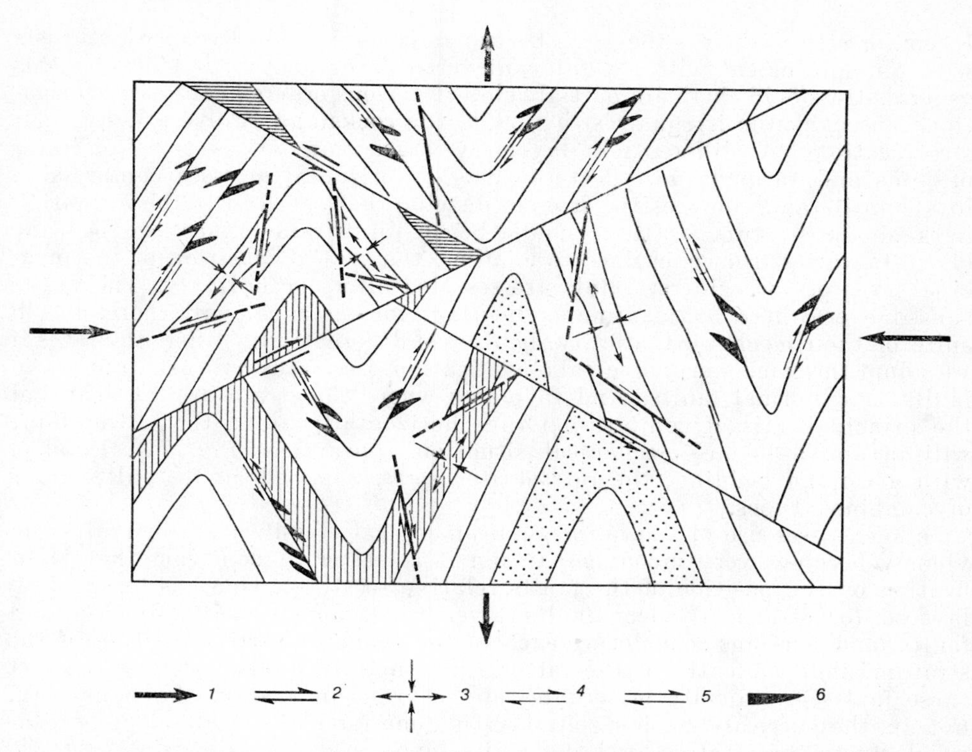

Fig. 70. Stress fields of various orders in a folded series of strata (vertical section) (after Beloussov):

1—compression and tension of the first order caused by crumpling and overthrusts; 2—displacements along overthrusts linked with a stress field of the first order; 3—a stress field of the second order caused by laminar flow; 4—stress field caused by strata sliding over one another; 5—displacement along faults connected with a stress field caused by laminar flow; 6—fissures of a fracture connected with the sliding of strata along one another

since the latter needs the minimum expenditure of energy (e.g., in case of the collapse of raised blocks of crust to one side).

As the size of the displaced massifs diminishes, however, the role of gravity also decreases and with a fall in the weight of the displaced blocks ultimately becomes an insignificant factor. In passing from large massifs to smaller volumes, therefore, mechanical non-uniformities in the structure of a series of rocks due, for instance, to stratification become increasingly important. Partial stress fields connected with such non-uniformities play an increasing role—hence, the loss of the regularity in the direction of displacements along faults observed in large massifs. Movements relative to small volumes of rock take place equally in any direction. When we retain the morphological classification, that classes fault displacements in relation to vertical and horizontal axes, we must then speak of multiple or complex faults.

To propose to this end some system of terms, however, proves to be highly difficult in cases when, by virtue of irregularities in the rock's structure, the spatial position of the principal axes of stress that directly causes some group of faults alters from one spot to another. The most general stress field in the formation shown in Figure 70, for example, is characterised by a horizontal principal axis of compression and a vertical principal axis of tension

In conformity with this the largest shear faults are presented by overthrusts with a displacement with a significant vertical component, but besides this general stress field there are partial fields of various orders in the same formation caused by the irregular structure of the rocks and, above all, by their stratification. A differential layer by layer flow arises in the limbs of folds and through the effect of general horizontal stresses compression local tensile and compressive stresses develop that can lead to the formation in a single stratum or to a small bench of strata. Since the bedding of strata varies from one place to another, the spatial position of the principal axes of these intrastratal stresses also alters accordingly. The faults caused everywhere by maximum shearing, for instance, will therefore, in spite of their mechanical affinity, be classed differently in different places if we adopt the same morphological classification, i.e., as overthrusts, upthrown faults, and normal faults. And in places where the strata are vertical and the principal axis of compression runs horizontally along them, the fault will in an analogous mechanical situation appear (see Fig. 71). Finally, with a sloping bedding, the intrastrated stresses may lead to fault strains of combined types.

In such cases use of the normal morphological classification of faults obviously becomes very inconvenient and it is better to construct the classification on the position of the faults relative to strata. One has to speak of layered (or stratified) shear faults, overthrusts, upthrown faults, normal faults, and partings considering each in the plane or vertical section of the stratum, independently of the latter's position. It is also useful to depict these faults graphically on profiles and charts, in exactly the same way, not in the vertical and horizontal as is done for major faults, but perpendicularly to the stratum or lying in its plane.

All this also fully applies to faults without displacement. In strata buckled into folds cracks are observed perpendicular to the strata and breaking up into systems trending along the dip and strike, and obliquely to both. When the stratum bends around a pericline or centrocline, these systems dip correspondingly (Fig. 72). They should clearly be distinguished precisely by their position relative to the stratum, rather than by their strike and dip measured by a surveyor's compass.

What has been said can be generalized as follows: in series irregular in structure and mechanical properties there are always stress fields of different orders that, as it were, nest into one another.

Each new structural feature, every new fault generated by strain of a non-homogeneous series in turn causes the development of new local stress fields. Figure 73 shows stress fields of various orders that accompany slipping along the master fault. They arise with displacements along faults "feathering" the master one.

Any unevenness in the plastic flow of rocks also gives rise to its own stress field. When an initially short normal fault has been formed in conditions of tension (see Fig. 74), the displacement along it will be uneven—faster and greater in amplitude opposite the middle of the fault and dying out at its ends, which leads to shearing faults at the ends of the fault; if it continues to lengthen, the process will proceed in arcuate fashion, as shown in the figure, rather than in a rectilinear manner.

All these examples indicate that one must distinguish between faults linked in origin with fields of different orders and consider them separately.

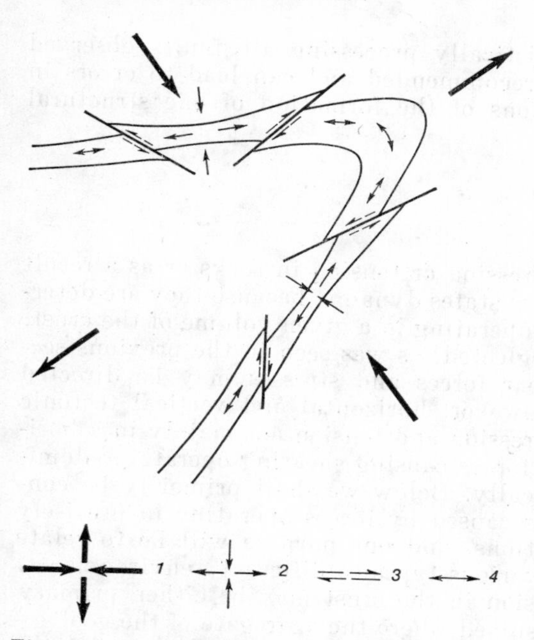

Fig. 71. Shear faults in a vertical layer caused by laminar flow (horizontal scheme) (after Beloussov):

1—principal axes of the major stress field; 2—stress field connected with laminar flow; 3—displacement along shear faults; 4—bedding of the layer

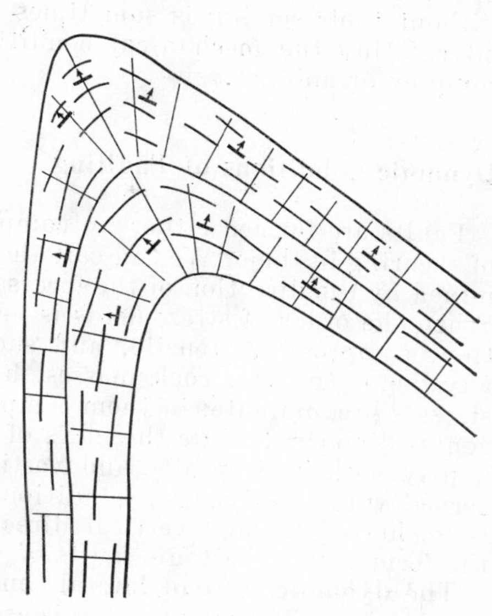

Fig. 72. Bending of fractures on the pericline of a fold; one system tends along the dip of the strata, the other along their strike (after Beloussov)

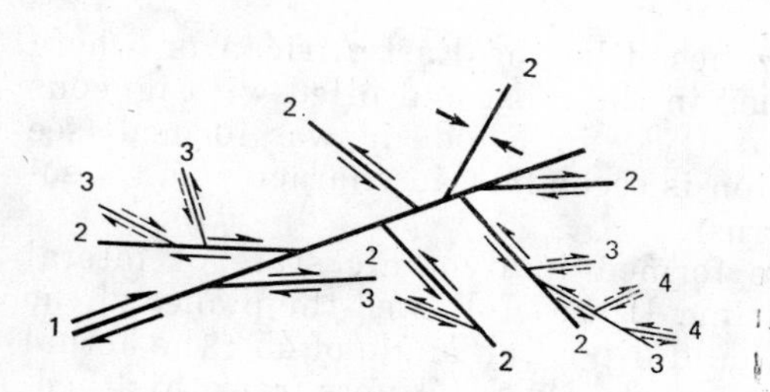

Fig. 73. Master fault (*1*) and its feather of different orders (*2-4*) (after Beloussov). The higher the order of the fault, the thinner is the line

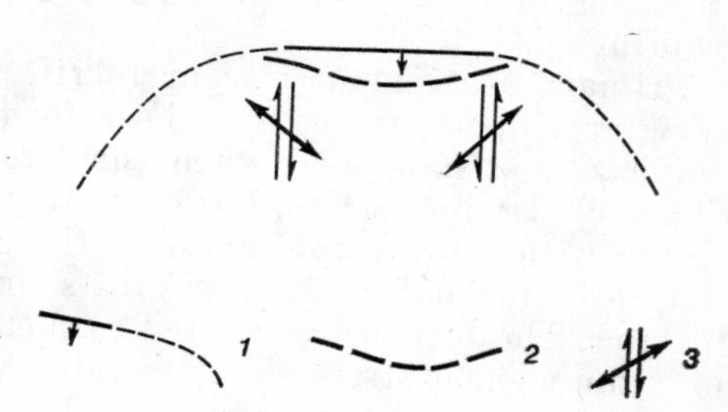

Fig. 74. Unevenness of movement along a fault leads to arcuate bending in plan during its further growth (after Beloussov):

1—fault (arrows point toward the footwall, dashed lines indicate the track of the fault's growth); *2*—curve conventionally depicting unevenness of displacement of the footwall; *3*—shear stresses caused by uneven displacement of the hanging wall and associated axis of maximum tension

The practice of recording and statistically processing all faults observed without their sorting is sometimes recommended and can lead to errors in interpreting the mechanical conditions of the formation of one structural complex or another.

Dynamic Conditions of Faulting

Faults are formed either by compression or tension in rocks or as a result of shearing in them. We can call these states dynamic because they are determined by the direction of the forces operating in a given volume of the crust. When the order of stress fields is neglected, as was seen in the previous section, compressive, tensile, and shear forces and stresses may be directed anywhere. In large rock masses, however, horizontal and vertical tectonic stresses predominate; in them compression and tension act mainly in a horizontal direction, while the pairs of forces causing shearing operate predominantly both horizontally and vertically. Below we shall primarily be concerned with the dynamic conditions caused by forces operating in precisely these horizontal and vertical directions, and our purpose will be to relate the formation of tectonic faults of various types to different environments.

The dynamic state of lateral tension in the crust may be either primary or secondary. Primary tension is assumed where the aggregate of the geological evidence suggests that the crust is being subjected to tension by the direct action of deep-lying forces, whatever their nature. Secondary tension is produced by upwarping either of the crust as a whole or some of its layers under the action of forces operating vertically. Such a situation arises on any convex structure generated by vertical pressure from below, e.g., the upper complex of diapiric domes, anteclises, anticlinoria, and orogenic uplifts. In these cases the tension is caused by the increase of surface during bending.

Primary tension may occur in rift zones, like the East African or where very large open fissures have been formed in the crust and filled with igneous materials, as happened when the Great Dyke in Rhodesia was formed (see Chap. 9). In most cases, however, tension is secondary in character and associated with a local upwarping of the crust.

Mainly normal and open faults are formed in dynamic state of lateral tension. The former are related to shearing. If they followed the planes of the maximum shear stresses, their planes would dip at an angle of 45°; the actual angle of shearing, however, as we know, nearly always differs from this theoretical one. Because of the friction forces that develop on the fault plane which grow with increase in the compressive component on it. The angle of shearing is the position of the fault plane at which the best dynamic "compromise" is established between the shearing stresses, which must be as large as possible, and the friction, which must be as small as possible. Observations indicate that the planes of big normal faults in the crust mostly dip by around 60°.

The main structural feature in areas of lateral stretching is a graben, but horsts also develop between grabens. The division into grabens and horsts is linked with a redistribution of stresses within the crust or at the bottom of the crust. Inclined faults cut the crust up into blocks that narrow toward either the top or the bottom. The pressure on the base of blocks that taper downward is greater than that on the base of blocks that widen downward.

The former therefore sink squeezing some volume or other of the plastic subcrustal or intracrustal material aside. There grabens are formed; the other type of block rises in compensation, forming a horst.

Normal faults are viscous ruptures formed by a concentration of previous plastic flow. In this case the preceding plastic deformation is a flexure — hence the close association of normal faults with flexures. Because deformations in the earth's crust develop unevenly, when in one place a fault has been already formed, in some neighbouring place the deformation is still at the stage of a flexure. When a fault is traced along the strike or dip of the fault plane, it can be seen in favourable cases to pass into a flexure through the branching of a single major fault into an increasing number of smaller and smaller faults.

The distribution of normal faults on an uplift is primarily determined by the tensile stress pattern, which depends on the shape of the uplift, but any irregularity in the structure of the rocks has a very great influence, and sometimes forces fault planes to deviate very strongly from their theoretical positions.

Open faults or separations are open tear fissures extending vertically deep into the crust. Some are filled with either igneous or neptunean dykes. Their position, as with normal faults, is dictated by distribution of tensile stresses and can be altered by the effect of local inhomogeneities. This type of fault is the structural form that has been little studied. It is still not known, in particular, how far the width of an igneous dyke that fills an open fault corresponds to the amplitude of initial opening, and how far the fault was widened further by the intruding magma. Extremely dense clusters or swarms of dykes are known in various areas, in which the total thickness of all the dykes suggests that the area was widened by 20 % of its initial width. Such a scale of stretching has been found, for example, on the western coast of Scotland in a zone of Tertiary igneous dykes. Study of such tension fields is no less interesting and important than study of the crust's compression zones.

A state of lateral tension can also give rise to shear or wrench faults (in the geological sense of the term, i.e., to horizontal displacement along a vertical fault). In the absence of other effects the angle between the direction of shear and the axis of maximum tension should be close to 60°, but in this case, too, inhomogeneities of structure can cause substantial deviations from this position.

Displacements formed in a dynamic state of lateral tension are shown in Figure 75.

In a dynamic conditions of horizontal compression the main type of fracture displacement is the thrust fault. Since general buckling takes place in the same situation, a close relationship arises between this type of folding, characteristic of geosynclines, and thrust faults, which are also a structural form common in folded zones.

Thrust faults are the result of shearing. When the same mechanical patterns operate during their formation as during normal faulting, the average inclination of the thrust planes should be 30°, but it is possible that this does in fact predominate. There are so many irregularities in a folded series of rocks that considerable local deviations from this angle should not surprise anyone. Thrust faults are known whose plane follows a horizontally bedded group of strata for a long way, because these layers can act as a good "lubricant" (Fig. 76). In other cases the dip of the thrust plane changes from

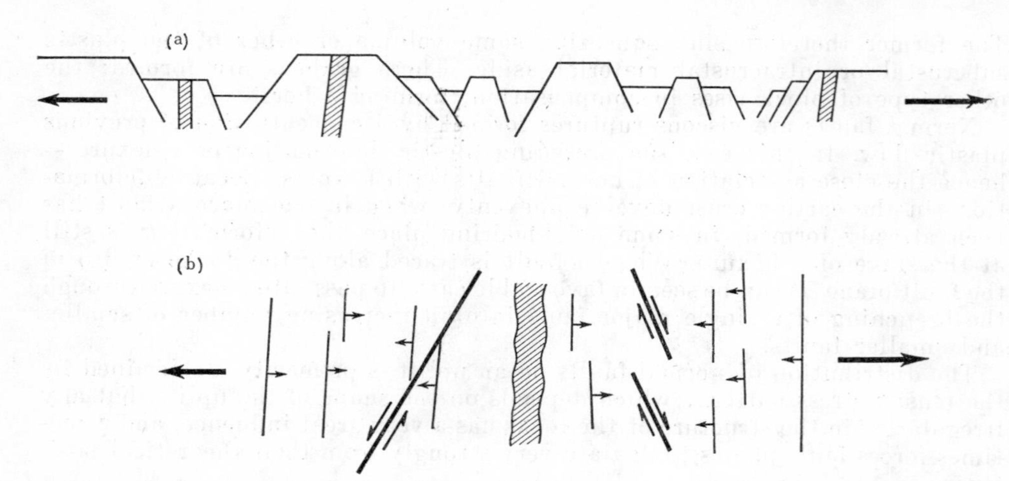

Fig. 75. Fracture characteristic of a dynamic state of horizontal tension (large arrows) (after Beloussov):

normal faults, separations (shaded), and wrench faults. The arrows show the direction of the dip of the fault planes, of displacement of the hanging walls of faults, and of displacement along wrench faults: *a*—profile; *b*—plan

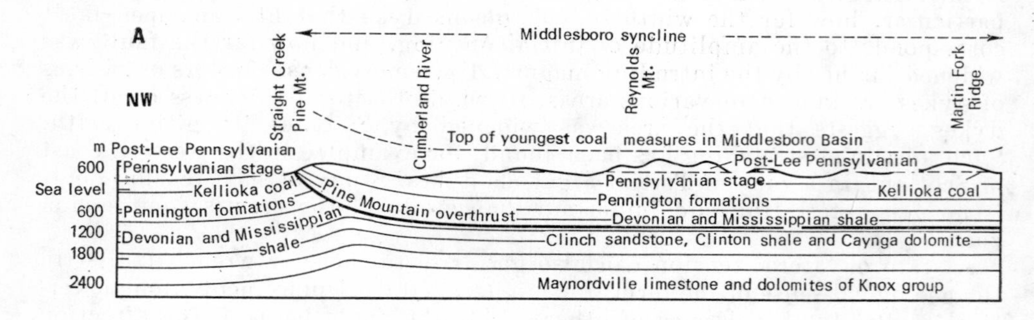

Fig. 76. Horizontal thrust fault in the Appalachians along the plane of a plastic bed of Devonian-Early Carboniferous clays (after Eardley). The displacement is about 9 km

group to group, since mechanically different rocks have different angles of shearing.

Like normal and wrench faults, thrust faults are viscous fractures, but whereas normal faults are preceded by flexures, thrust faults are preceded by folds, hence the systematic combination of thrust faults with folds, which is manifested particularly well in what is called imbricate structure, consisting of greatly tilted or overturned folds each (or almost each) of which is undercut by a thrust fault on its steep or overturned limb. By tracing the thrust plane, we can see a transition of the fracture into plastic deformation, as with a normal fault, and a branching of the major fault into smaller and smaller ones. It can thus be established that the thrust fault is immediately preceded by a thinning of the strata on the fold's overturned limb. In this case, however, the relationships between the plastic deformation and the succeeding fault are not so simple as in the faulting that follows the formation of flexures. While folding affects the position of the future thrust fault, this fault affects the course of the plastic deformation. Here it is a matter of a future but still non-existent fault and its influence is consequently directed backwards, so to speak; the plastic deformation, in developing, "knows", as it were, where the future fault is to be located. Of course, that is not the work of some fantastic "time machine" reversing the sequence of events, but simply an indication that the site of the future thrust fault is marked out by some phenomena or other long before the fault actually develops.

When a massif of rocks is compressed only at the beginning of deformation, the latter develops uniformly throughout it, but already in the early stages the deformation begins to be concentrated in certain zones, the orientation of which is determined by the angle of shearing (in this case it is 30° to the horizontal axis of compression). Such zones of increased flow are repeated along the axis of compression with a definite rhythm. (The reasons for such rhythms are discussed on p. 133.) During further concentration of the deformation they get narrower, and the rate of relative flow in them increases. It is these zones that determine the position of the overturned limbs of folds, and their dip that determines that of the folds; hence the folds in such imbricate structures are, so to speak, plastic manifestations of shearing.

With lateral compression wrench faults are also formed as a result of slipping along vertical shearing fractures, which must arise as two symmetrical systems at an angle around 30° to the axis of compression. They must consequently intersect the strike of folds and thrust faults obliquely, one of the systems being connected with the right displacement and the other with the left. Shear faults are also directed across the trend of folds and thrust faults, cutting transverse strips shifted relative to each other from the fold and thrust fault zone. This is a manifestation of irregularity in the horizontal movement of rock mass being folded. A cross-section and scheme of the combination of folds and faults characteristic of compression areas are shown in Figure 77.

A dynamic state of lateral shear arises over a considerable area near the big deep-seated shear faults considered earlier (see Chap. 9). It was also pointed out then that derivative compressive and tensile stresses were generated near such faults. These secondary stresses can lead to the formation of "feathered" folds and thrust faults (in areas of compression) and normal faults (in areas of tension).

A dynamic state of vertical shear is one of block tectonics, when the crust is divided into "keys" that are raised and depressed relative to one another. All the problems associated with it, as with the preceding one, were considered earlier in connection with the problem of deep slash faults. It only remains to add that this term (*vzrez* in Russian) can be also applied to intracrustal vertical faults.

It will be seen from the foregoing that each dynamic state is accompanied with a certain complex of faults, which are often systematically combined by definite folded structures. Knowledge of them helps to solve the "inverse problem", i.e., of reconstructing the previous dynamic state from existing structures. Acquaintance with them makes it possible to predict, after some of the structures and their mutual arrangement have been established, what other related structural features may also be found in the same area. Finally, such knowledge of dynamically related structural complexes helps to distinguish structures foreign to the given complex and so establish that a change took place there in the dynamic state. If, for example, thrust faults are intersected by normal faults, it is clear that conditions of compression earlier existing there where followed by tension (Fig. 78).

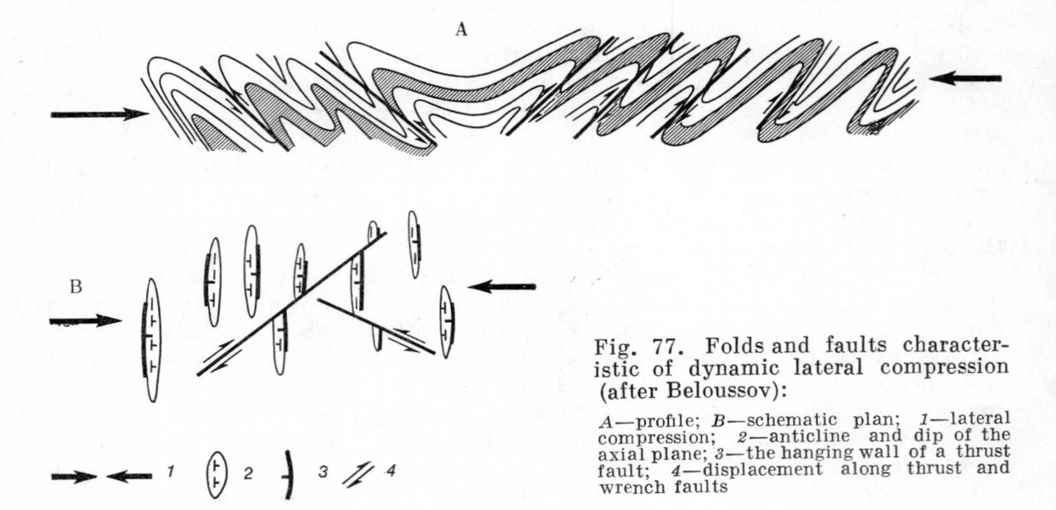

Fig. 77. Folds and faults characteristic of dynamic lateral compression (after Beloussov):

A—profile; *B*—schematic plan; *1*—lateral compression; *2*—anticline and dip of the axial plane; *3*—the hanging wall of a thrust fault; *4*—displacement along thrust and wrench faults

Everything we have said in this section about the dynamic states in big massifs of rocks holds for stress fields of the second and higher orders. In them, however, it would be more correct to speak of compression and tension in the plane of a layer or of some other element of structural inhomogeneity rather than of horizontal and vertical compression and tension, because, as has been pointed out earlier, the stresses in small volumes of rock can be oriented in any direction.

General Factors Affecting the Position of Faults

The shape and position of a fault observed today may differ essentially from its initial shape and position. The reasons for the differences, which give us no little difficulty in reconstructing tectonic stress fields from faults, belong wholly to the field covered by structural geology. We shall simply recall them briefly here.

Every fracture, whether brittle or viscous, is formed as a small crack or a group of small, isolated cracks, which subsequently join together in bigger and bigger faults, and displacement takes place along them. Viscous faults are the end result of an increasing concentration of plastic flow, which occurs not only before the faulting but also after it. The fault proves to be included in a massif that undergoes plastic deformation as a whole.

When the deformation is uniform and the fault is no longer lengthening, it turns during deformation. Shear faults, for example, that initially arise at an angle of 30° to the axis of compression, turn, with further deformation in the same stress field, so that the angle between them and the axis of compression increases; an acute angle between two systems of shear faults can become obtuse. If new shear faults develop during the same deformation, it must be emphasized, they will always be formed at the same angle of shearing, i.e., at an angle of 30° to the axis of compression. Thus, in the course of a single protracted deformation several generations of shear faults may be formed that will be located, according to their age, at different angles to the principal axes of stress (Fig. 79).

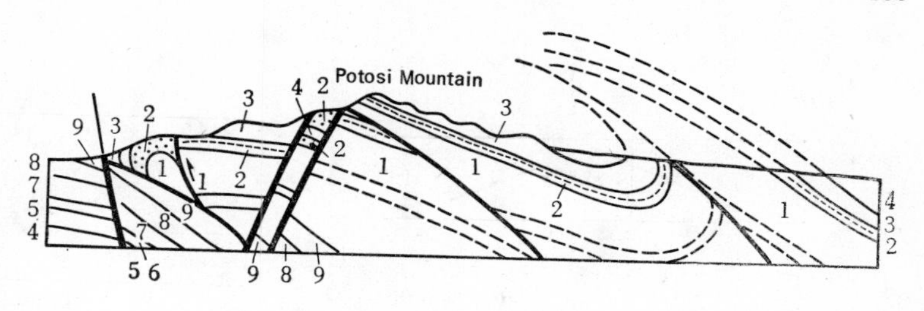

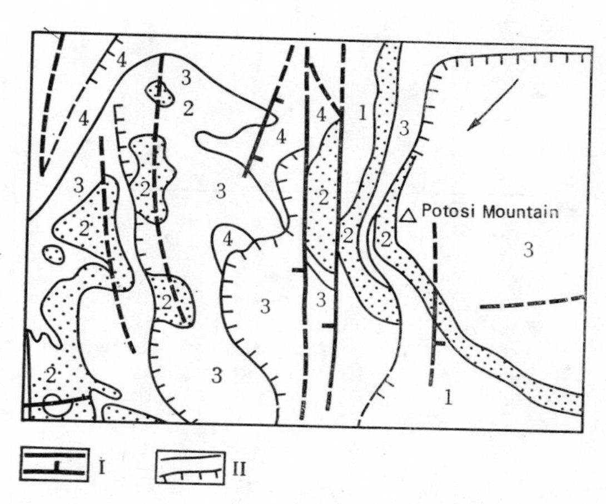

Fig. 78. Section and map of Mount Potosi in the U.S. Rocky Mountains (after Eardley): a combination of normal (*I*) and thrust (*II*) faults; *1*—Upper Cambrian-Devonian (?); *2*—Devonian; *3-6*—Carboniferous; *7-8*—Triassic; *9*—Jurassic (?)

An initially flat tectonic fault may subsequently be bent into folds.

When a fault continues to grow simultaneously with plastic deformation, its final shape will be arcuate: the older parts of the fault will turn and deviate from the initial position, while the newer ones will be formed at the same angle to the principal axes of stress (Fig. 80). All textbooks of structural geology give an example of the formation of S-fractures in a zone that has experienced plastic shearing.

An important feature of the structure of the earth's crust is that faults of one type usually occur, in groups rather than as single features within which they are repeated many times and often with a certain rhythm. The rhythmicity of the areal distribution of thrust faults in an imbricate structure was noted above. In the areas of tension normal faults also develop in groups, the total amplitude of the stretching not being concentrated in one fault, but being distributed over many.

The repeated character of areal occurrence of stress faults, i.e., primarily normal faults, is explained by distributed application of tensile forces to the crust (or to a group of layers within it). Such distributed tension applied to the whole underside of the group characterises the development of any convex structure associated with the action of vertical forces from below. For example, the upper complex of strata on a diapir is under pressure from below, applied to the whole bottom of the complex, while tensile forces are correspondingly applied to any point of it. Groups of faults are formed in rift zones because the stretching in them was also caused, as it may be supposed, by mushroom-like spreading of a plastic "pillow" beneath the crust which suffered stretching distributed over the whole area of the rift zone (for further details, see Chap. 23).

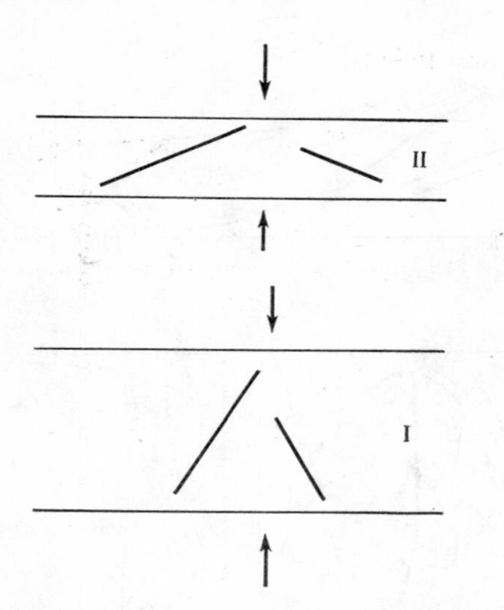

Fig. 79. Deflection of shear faults during plastic deformation (after Beloussov). Arrows indicate the direction of compression. Between stages *I* and *II* the massif's size was halved along the axis of compression

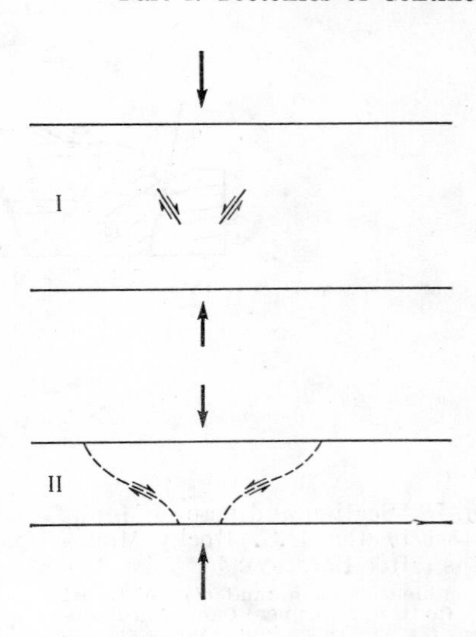

Fig. 80. Deflection of a shear fault during its growth in the course of plastic deformaton (after Beloussov):

I—initial fractures; *II*—growth of fractures (dashed line) to the stage when the massif's size was halved along the axis of compression (heavy arrows). Small arrows indicate displacement along shear faults

The fact that tension is caused by distributed tensile forces rather than their external application leads to the formation of faults anywhere independently of one another; if tensile forces were applied, however, only to the margins of the zone, one fault arising would be enough to relieve all the stresses. With distributed application of tensile forces, their scale and the strength of the Earth's crust or of the deformed massif naturally determine the distances between successive faults.

It could be supposed, by analogy, that the grouping of faults associated with compression is due to the same cause, viz., distributed application of compressive forces. That would be feasible, however, if the compression occurred in a situation of "subduction" of the crust beneath the compression zone somewhere in the interior. With such a mechanism, each segment of the crust would have to be compressed independently and faults could be repeated many times. We do not have sufficient grounds, however, to share this hypothesis, and there is no need of it, incidentally, because gravity can be an adequate cause of multiple, rhythmic repetition of thrust faults. When the first thrust fault is formed by external lateral compression applied from the margins, movement of its hanging wall upwards is limited to the amplitude at which the ever-increasing weight of the uplifteds rocks reaches some critical value. Upward movement along the fault then ceases, and the deformation spreads out farther from the margin of the massif, and a new thrust fault is generated there, and so on. Folds may also fill the area gradually, since gravity restricts growth of each of them by some critical size.

General Fracturing of Rocks (Joint Systems)

What is called general fracturing is ubiquitous in rocks. We shall touch on it precisely because of its commonness and because its origin is still an unsolved problem. General fractures develop not only in deformed strata but also in horizontally bedded ones in which they are particularly regular and systematic.

In quietly bedded sedimentary rocks general fractures are for the most part perpendicular to the strata and occur within a bed, i.e., individual fractures do not go beyond the bed. Fractures of the same strike arise again in the next bed but with a lateral shift. These fractures form systems, in each of which they are parallel to one another. The systems, having different strikes, intersect, orthogonal combinations of fractures of different systems being particularly common. The frequency of the fractures in each system in identical rocks, and in strata of equal thickness persists on average over large areas, in which connection the strata are cut up into joint systems of more or less regular shape and similar size. The usual forms are parallelepipeds. There is a regular relationship between the frequency of fractures, on the one hand, and the mechanical properties and thickness of a bed, on the other. In otherwise equal conditions, the thinner the bed, the closer the fractures are to one another. In the Carboniferous limestones of the East European platform, for example, the distance between the fractures of a system in strata 2.3 m thick is as much as 2 m and in strata 15 cm thick is 20 cm on average.

It would be most natural to see in general fracturing in quietly bedded sedimentary rocks an aggregate of tear fractures that could have been caused by a decrease in the volume of the rocks during lithification. In that case, however, hexagonal prismatic joints, sets like those formed, for example, when basaltic sheets cool or on the surface of drying clay would be much more likely than orthogonal ones. Even more surprising is the fact that general fractures in horizontal beds persist in their strike over vast areas. On almost the whole area of the East European platform, for example, "diagonal" strikes of the joint systems predominate—north-westerly and north-easterly.

Attempts to link the strike of fractures on a platform with major tectonic structures of some sort (anteclises and syneclises, or block folds) have not been successful. In some regions it has been noticed that the joint systems in a quiescent sedimentary cover follow the strike of ancient folds and faults in the underlying ancient basement. Such "duplication" of basement strikes is to be seen, for example, in the Hungarian depression. The mechanism of the transfer of strikes from the basement to the sedimentary cover, however, remains a puzzle. The origin of such fracturing in deformed rocks, in which they may survive from predeformation time, or arise during deformation, is an even more complicated matter.

We do not yet have a convincing answer to the problem of the origin of such common structures as joint systems. Meanwhile, general fracturing is not only of theoretical interest, but also has great practical importance for mining, geomorphology, hydrogeology, engineering geology, and other earth sciences.

10–0601

SECTION C

Patterns of the Evolution of Continents

CHAPTER 15

Endogenous Continental Regimes. General Features and Geosynclinal Regimes

A General Characteristic of Endogenous Continental Regimes

A survey of the main patterns of development of the continents indicates that the endogenous processes on them (tectonic, magmatic, and metamorphic) are combined in certain patterns, and that the intensification or weakening of these various processes, as a rule, runs parallel, which demonstrates an inner unity of the processes and warrants our saying that certain *endogenous regimes* operate in the Earth's crust. Each regime has a definite form, scale, and sequence of tectonic movements and magmatic and metamorphic processes, and operates in some one area during a period of geological time.

Features that determine the difference between regimes are:

1. **The character and extent of the permeability of the crust** to magma and its liquid and gaseous products. Permeability may be manifested in varying degree from very great to minimal, in which case we can speak of the crust's *impermeability*.

Permeability may be *diffused* or *concentrated*. In the first case, the crust is penetrated by a dense network of broad and very fine passageways, intersecting or parallel to the bedding. When filled with magmatic material, they transform the crust into a sort of mixture of intrusive and intruded rocks. In the second case, permeability is determined by rare separate faults.

2. **Character and extent of magmatism.** The crust's degree of permeability determines what type of magmatism will occur, whether intrusive or extrusive. The character of the permeability affects the shape of igneous bodies. It and the degree of permeability also affect the evolution of the composition of magma. With high diffused permeability, the broad surface of the magma's long interaction with the enclosing rocks leads to the evolution of a composition taking place largely through melting and assimilation of these rocks. When permeability is poor and concentrated, differentiation plays the main role in the evolution of the magmatic material.

3. **Regional metamorphism and granitization.** These processes, typical of certain regimes, are mainly important because they form the crystalline crust. They also affect the permeability of the crust by creating an impermeable

shell of metamorphic rocks and granites during its recrystallization where the Earth's crust was previously composed of pervious rocks. Strong, new tectonic movements are needed to break this shell.

4. **The degree of contrast between block-undulatory oscillatory movements of the crust.** The various regimes are characterised by the different degree to which the crust is divided into zones of block-undulatory uplift and subsidence and by different contrast between them, measurable, for example, by the vertical velocity gradient.

5. **The relationship between crustal uplifts and subsidence.** In some regimes uplifts predominate over subsidence, while subsidence, on the contrary, predominates in others. This factor determines the relief of the Earth's surface and the character of sedimentary formations.

6. **The character of dislocations.** The various regimes are characterised by various types and intensities of folding and faulting.

Classification of Endogenous Continental Regimes

The following classes of endogenous continental regimes are distinguished:
1) geosynclinal
2) platform
3) orogenic
4) rift
5) magmatic platforms activization
6) marginal.

Each of these classes is divided into separate regimes.

The class of geosynclinal regimes is divided into eugeosynclinal, miogeosynclinal, parageosynclinal regimes and the regimes of median masses.

Eugeosynclinal and miogeosynclinal regimes are sometimes combined in a general *orthogeosynclinal regime*, while the *parageosynclinal regime* is considered intermediate between geosynclinal and platform regimes.

The class of platform regimes is divided into the *regime of young platforms and that of ancient platforms* (*or cratons*), the latter being the quietest continental regime.

The class of orogenic regimes is divided into an *epigeosynclinal regime*, i.e., that which follows a geosynclinal regime, and an *epiplatformal regime*, which is the result of orogenic activation of a platform.

The class of rift regimes cannot at present be divided into regimes of different types.

The class of magmatic platforms activization regimes includes very diverse manifestations of effusive and intrusive igneous activity in areas that had previously been stable platforms. Within it the following can be distinguished: a *plateau-basalt regime* and a *regime of central intrusions* and *vents* (*pipes*) *or diatremes.*

Finally, the *class of continental margin regimes* can be subdivided into an *Atlantic* and a *Pacific type regimes.*

When familiarizing ourselves with the features of the various regimes, we continually find that they obey some rhythm in their origin and development that is determined by so-called *endogenous cycles*, which will be described in more detail later. Suffice it to say here that these cycles are the main stages in the evolution of the Earth's crust. The development of individual geosynclinal regimes in an area, for example, takes place in the content of one or more cycles. The replacement of one regime by another in ime is often confined to the boundary between two cycles. Some regimes have a certain place within an endogenous cycle. The orogenic and rifting regimes, for example, always occur at the end of a cycle. There were several endogenous cycles in Precambrian time, the last of which, the Baikalian in the Late Proterozoic, is quite well known. In Palaeozoic time the Caledonian and Hercynian (or Variscan) cycles are known and in Mesozoic-Cenozoic time only the Alpine cycle, or Cimmerian and Alpine (Pacific) cycles, are known.

Geosynclinal Regimes

The class of geosynclinal regimes is characterised by contrasted block-undulatory oscillatory movements of the crust, intensive folding and faulting, much igneous activity, both intrusive and extrusive, and manifestations of regional metamorphism and granitization. All these phenomena are most active in a eugeosynclinal regime, weaker in a miogeosynclinal one and weaker still in a parageosynclinal regime.

A **eugeosynclinal regime** is therefore the most "excited" endogenous regime. Its development falls into two stages: the ophiolitic and the inversion stage.

The following are the characteristics of the **ophiolitic stage.**

A. *Very great contrast between block-undulatory oscillatory movements.* The region where a eugeosynclinal regime is manifested is always divided into closely spaced zones of intense subsidence and uplift (in both rate and magnitude), known respectively as intrageosynclines and intrageanticlines. In area usually these zones take the form of greatly elongated ovals tens of kilometres wide and hundreds of kilometres long. The transition from an intrageosyncline to an intrageanticline is seldom gradual; in most cases zones of opposite vertical movement are contiguous along deep faults.

B. *Great preponderance of subsidence over uplift.* Intrageosynclines downwarp so rapidly that sedimentation lags behind. As a result, after a short period of compensation of the subsidence by accumulation, which occurs at the beginning of this stage when the sandy-clay sediments of the lower terrigenous formations (also known as shaly-greywacke or slate) are being depicted in the intrageosyncline, compensation is abruptly interrupted and the same intrageosyncline becomes a deep marine basin several kilometres deep. Evidence for this depth is the character of the sediments that succeed the slate-greywacke formation, in which siliceous rocks now predominate, i.e., radiolarites and cherts (the chert formation), and red non-calcareous abyssal clay; abyssal limestones occur in lesser amounts. The thickness of these rocks is less than the amplitude of the intrageosyncline's subsidence. These troughs are not compensated by sedimentation and are characteristic of the first stage of a eugeosyncline's development. (In the Western European geological literature they are referred to as *leptogeosynclines.*)

The preponderance of downward movements of the crust is also reflected in the evolution of intrageosynclines and intrageanticlines. During this stage, they expand at the expense of rising movements and the absolute uplift of an intrageanticline partly changes in places to relative uplift, and these sectors are altered from areas of erosion into zones of relatively slight downwarping which are compensated because of their slowness by sandy-clay or carbonate sedimentation. In that connection the deposits on many submerged intrageanticlines are thicker than in rapidly sinking intrageosynclines not compensated by accumulation.

C. *Great diffuse permeability* (*or perviousness*) *of the crust*. Pronounced magmatism is characteristic of this stage. It is manifested in various ways, as follows.

1. Submarine outpourings occur and extensive sills are formed near the surface. Basic (basaltic) magma is the main source of both. It solidifies in sills as diabase, but the magma that has flowed out onto the sea floor is altered (mainly enriched in natrum) to spilite. Among the extruded rocks, particularly those of the later stages, there are also acid varieties, keratophyres, which are regarded as differentiates of basalt.

2. Another form is a complex swarm of dykes, acting as channels for outpourings and sills. Dykes are composed mainly of diabases and form what is called the spilite-keratophyre, or spilite-diabase-keratophyre, formation.

3. A third form is a large discordant stock-like intrusion of gabbro and plagiogranites (the gabbro-plagiogranite formation), which are mainly concentrated in intrageosynclines.

4. There is "ophiolitic formation" which plays a very important part in "early geosynclinal magmatism" and after which the whole stage under discussion has been named. The rocks of the ophiolitic formation, which are formed mainly along the faults separating uplifts and troughs, are predominantly (ultrabasic) peridotite, which is usually largely altered into serpentinite. They also include gabbro, diabase, and norites in smaller quantities. All these rocks form very large bodies of the type of laccoliths of lenticular form which may be tens of kilometres across and up to a 1 km thick. Ophiolites are sometimes broken up into numerous intrusions along a dense network of fissures with the result that the enclosing rocks are mixed with the ophiolite and form a kind of breccia known as melange.

D. *Folding* is not characteristic of the first stage of the eugeosynclinal regime, but block movements occur in intrageanticlines and are expressed in the sedimentary cover as isolated block folds.

The second, inversion stage of a eugeosynclinal regime differs greatly from the first stage. Its distinctive features are as follows.

A. *Moderate contrasts in block-undulatory oscillatory movements of the crust*. The gradients and differences in rate of the vertical movements of the crust are much less than in the first stage. Dispersion of their rates also slows down, which is an indicator of that contrast. The role of deep faults is still high, however.

B. *A balance between the volume of uplift and subsidence*. Compensation of subsidence by sedimentation is characteristic of the second stage. The sediments consist mainly of terrigenous material—sands and clays—laid down in a shallow marine basin (the upper terrigenous formation). A terrigenous flysch formation is typical, frequently giving way to a paralic coal-bearing formation.

The generally terrigenous, medium to fine-grained, uniform sediments in troughs indicate that equilibrium was established on the whole between the subsidence of the troughs and uplift of the erosion zones feeding them with elastic material, but the rhythmic alternation of coarser and finer sediments characteristic of flysch and coal-bearing formations reflects rhythmic, though small, oscillations of the correlation between sinking and uplift around a mean level.

The rates of subsidence and uplift are balanced not only through slowing of the rate of downwarping of the intrageosyncline, but to a considerable extent through the formation of new uplifts within it. Such newly formed

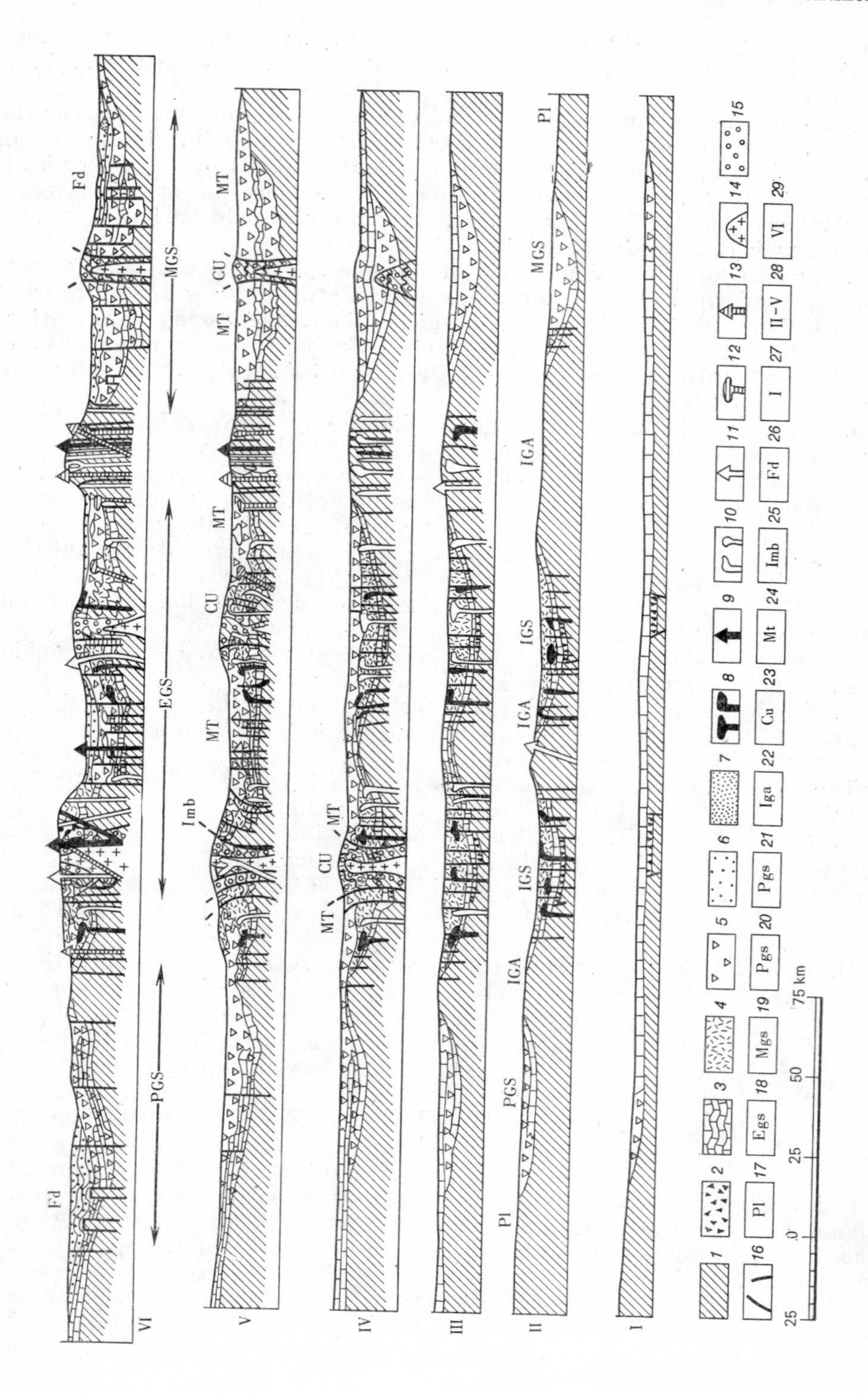

upwarps are called central uplifts, and their formation is an expression of what is termed partial inversion of the geotectonic regime. Because of this phenomenon, which plays a great role in the development of geosynclines, the whole stage is called an inversion one.

As we mentioned above, intrageosynclines expand during the first stage of the eugeosynclinal regime at the expense of intrageanticlines; this continues during the second stage as well. At the same time, however, central uplifts develop and grow rapidly within the expanding intrageosyncline. The whole process has a wave-like character; the margins of the intrageosynclines "roll over" the adjoining intrageanticlines, while uplifts form and expand within the former (see Fig. 81).

The process leads to a decrease of the area of the intrageanticlines and a slowing of their rise, and their conversion from regions of *absolute* uplift to regions of *relative* uplift, so that they supply less and less clastic material. The central uplifts become the main sources of detritus feeding the flysch formation. Accumulation proceeds in the marginal troughs on both sides of each central uplift. The troughs, being residual ones after the formation of the central uplifts, coincide in part with the margins of the intrageosynclines and in part encompass the areas previously occupied by the intrageanticlines. The distribution of areas of erosion and accumulation consequently mainly alters in connection with partial inversion within geosynclines, and largely becomes the opposite of what it was in the first stage. The sedimentary formations belonging to the different stages of a eugeosyncline evolution accumulate in different places, rather than on top of one another; as the eugeosyncline develops, the areas of accumulation migrate, and the sediments of various age form lenses overlapping like tiles.

C. *Regional metamorphism and the formation of granite massifs.* The development of central uplifts is accompanied by regional metamorphism, intrusion of granites, and granitization of the country rock. These phenomena are restricted to central uplifts and embrace new areas as they expand. When the full cycle develops, plagiogranite stacks arise first in the forming central uplift, then regional metamorphism takes place (greenschist or amphibolite facies), and the formation of granite batholiths follows in turn, either through granitization of sedimentary and ancient metamorphic rocks or through "rejuvenation" of ancient granites. The composition of the granites of this stage is "normal", i.e., calcareous-alkaline. Finally, in the last stage, discordant granite massifs of higher alkaline content invade. In particular areas, separate stages of the cycle may be absent.

The recrystallization caused by metamorphism and granitization, and the intrusion of large granite bodies lead to the "healing" of previously existing deep faults and the formation of an integral crystalline shell in the Earth's crust.

◄ Fig. 81. Scheme of geosynclinal development (after Beloussov):

1—the geosyncline's basement; *2*—continental coarse clastic sediments, partially coal-bearing; *3*—limestone formations; *4*—ophiolite and spilite-keratophyre formations; *5*—terrigenous marine formations, including flysch, and lower molasse, predominantly marine; *6*—upper molasse, mainly continental; *7*—salt in diapiric cores; *8*—basic and ultrabasic intrusions and basic effusive rocks; *9*—central volcano with basic lava; *10*—acid and intermediate intrusions; *11*—central volcano with acid and intermediate lavas; *12*—intrusions of alkaline magma; *13*—volcano with alkaline lava; *14*—granite; *15*—area of regional metamorphism; *16*—tectonic faults; *17*—platform; *18*—eugeosyncline; *19*—miogeosyncline; *20*—parageosyncline; *21*—intrageosyncline; *22*—intrageanticline; *23*—central uplift; *24*—marginal trough; *25*—intermontane basin; *26*—foredeep; *27*—pregeosynclinal undifferentiated trough (parageosynclinal regime); *28*—stages in the development of eugeosynclinal, miogeosynclinal, and parageosynclinal regimes; *29*—orogenic regime

D. *Deep folding, general crumpling folding, and associated faulting.* The time of partial inversion is also the main period of folding and associated faulting in a geosyncline. Folds and thrust faults are formed within a central uplift, in a narrow zone while the uplift is narrow, and in a broadening belt as it expands. In the axial zone of the uplift the folding is generally deep-seated, i.e., deep diapirism; further out from the axis, there is general crumpling folding.

The formation of deep folds and general crumpling is divided into pulses or phases, divided by longer quiet periods. The phases are closely associated with the unevenness of the central uplift's growth and expansion.

As a result of folding and accompanying tectonic faults and nappes, the geosyncline becomes a *folded zone.*

Examples of a eugeosynclinal regime are the Uralian geosyncline in Silurian, Devonian and Early Carboniferous times, and the Western Sayan and Central Kazakhstan geosynclines in Cambrian and Ordovician times.

The miogeosynclinal regime differs from the eugeosynclinal in the absence or weak expression of "primary magmatism", i.e., absence of the ophiolite stage, which is replaced by a stage of compensated downwarping during which subsidence predominates over uplift, but very much less than in eugeosynclines. All accumulation proceeds in conditions of compensation of the subsidence by sedimentation. The predominance of subsidence over upwarping expresses itself in the succession of formations: the lower terrigenous (or slate) formation with which the miogeosynclinal section begins gives way in time to a limestone formation composed primarily of shallow-water limestones, which is indicative of exhaustion of the sources of terrigenous material, which in turn is the result of intrageosynclines' expansion at the expense of intrageanticlines. In some cases, however, a limestone formation does not develop and the preponderance of subsidence leads only to the terrigenous sediments becoming finer with time.

The second (inversion) stage of the evolution of the miogeosynclinal regime does not differ in principle from the similar stage of the eugeosynclinal regime. It involves partial inversion, regional metamophism, granite magmatism, folding, and faulting. But these are all less intense than in eugeosynclines. The general crumpling, for example, may have a limited spread in miogeosynclines and be replaced by broad areas of injection and block folding.

A zone of a miogeosynclinal regime may border on one of a eugeosynclinal regime within a single geosyncline. In such cases, it usually lies on the periphery of the geosyncline. In the Palaeozoic, for example, a miogeosyncline occupied a strip on the western slope of the Urals between the East European platform and the eugeosyncline of their eastern slope.

The parageosynclinal regime, as indicated above, is intermediate between the geosynclinal and platformal regimes. The contrasts, rates, and amplitudes of the block-undulatory oscillatory movements are much smaller in it. Accumulation proceeds in conditions of compensation of subsidence by sedimentation. All the marine sediments are shallow-water ones. The section begins with a sandy-clay terrigenous formation above which comes a limestone formation. Limestones are generally common in the section of a parageosyncline as a result of poor supply of detritus from weak upwarps. There is no primary magmatism. There is also no partial inversion, regional metamorphism, or granitization in parageosynclines. Their absence makes parageosynclines to be very close to platforms, but all vertical movements within them are more contrasted and intense than on typical platforms. There is also no general crumpling folding, and in this sense again parageosynclines are similar to platforms, but injection and block folding is much more intensive in them than such folding on platforms.

The absence of partial inversion leads to there being no redistribution of areas of subsidence and uplift, which remain in their old sites, only their outlines altering, mainly through extension of regions of subsidence. All the formations filling a parageosyncline are consequently deposited one above the other.

Some signs of magmatism may be found in parageosynclines. Intrusive activity is expressed by dykes and sills of a basic composition emplaced early in the cycle and by laccoliths and igneous diapirs, in which an alkaline magma is characteristic. Small granite fissure intrusions are found in places.

Examples of parageosynclines are the northern slope of the eastern Great Caucasus (what is called Limestone Daghestan) and the Abkhazian zone on the southern slope of the western Great Caucasus. The Donets Basin also had this regime.

The regime that often historically precedes eugeosynclinal and miogeosynclinal regimes can be classed as parageosynclinal. In such cases there is a general gentle downwarping of the geosyncline, with compensated accumulation of continental and shallow-water sediments with hardly altered thicknesses in the area of their development, before it is divided into contrasted intrageosynclines and intrageanticlines and before the eugeosynclinal and miogeosynclinal zones are differentiated. Such a regime existed, for example, in the Alps at the end of Permian time, in the Triassic, and at the very beginning of the Jurassic, when continental sand, clay, and conglomerate, lagoon gypsum and dolomite, and shallow-marine sandy-clay sediments and limestone were laid down in the area of the future geosyncline of the Alpine cycle. The geosyncline was broken up into troughs and uplifts early in the Middle Jurassic. In the Caucasus, the Triassic is this same stage of a preceding parageosynclinal regime.

The median mass regime is confined to areas in the interior of geosynclines that can be interpreted either as very broad intrageanticlines or small platforms within the geosyncline and "streamlined" by it. The basis of median masses is "fragments" of the structure produced by previous tectonic cycles. Geosynclines of the Alpine cycle, for example, may contain median masses that are remnants of structures created by folding, metamorphism, and magmatism during preceding Hercynian, Caledonian, or Precambrian cycles. Immediately after their formation these structures may have had a much greater extent, but with deposition and further development of the geosyncline of the Alpine cycle, this old folded, metamorphic and igneous magmatic "shell" wasbroken up and reworked, and only separate fragments survived within the new geosyncline. That a median mass is the remains of the later "breaking up" of ancient structures is underlined by the discordance of its internal structure and that of the surrounding young geosyncline, the boundaries of which cut across the strikes of the older structures within the massif (Fig. 82).

In the development of block-undulatory oscillatory movements, median masses behave as extensive intrageanticlines. The thickness of the sediments being laid down during the geotectonic cycle in which the massif is median is small, and in that resembles an intrageanticline. Folding is weak then and mainly block. Median masses, however, are much fractured by faults and

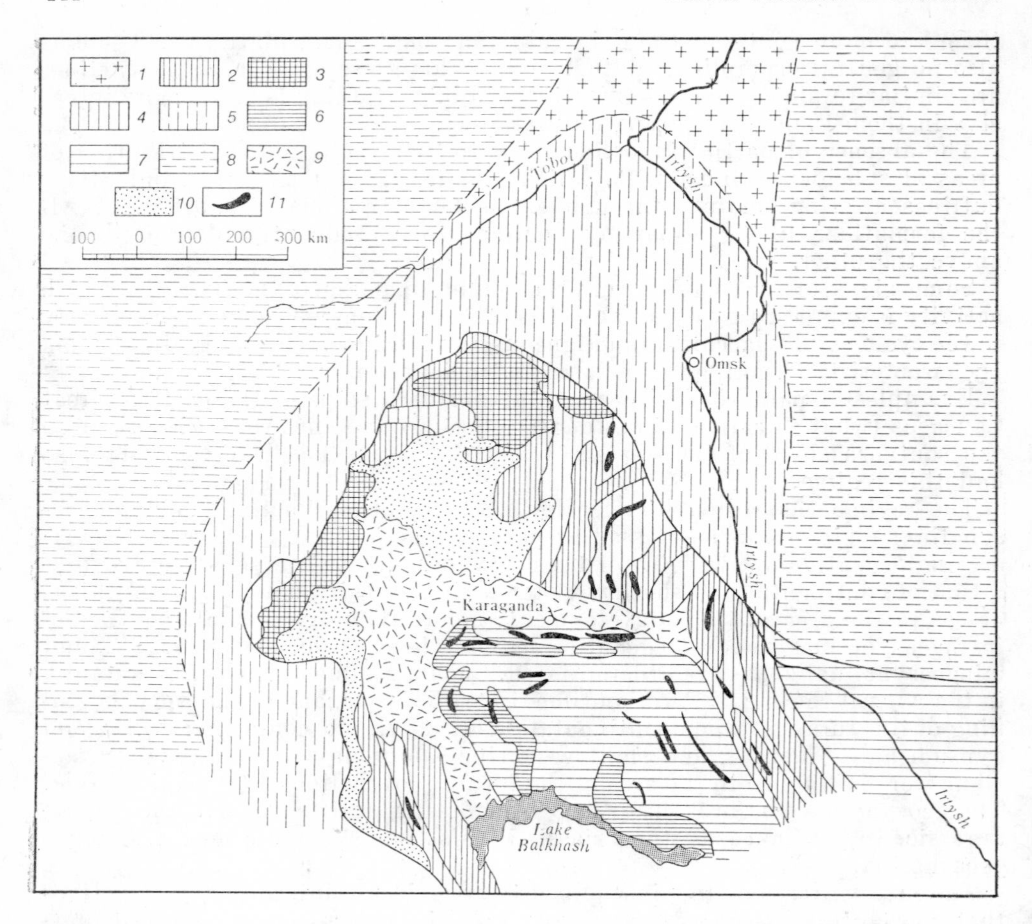

Fig. 82. Median mass with a Caledonian and older structure within a Hercynian geosyncline (Central Kazakhstan) (after Zaitsev):

1—region of Precambrian folding beneath a platformal cover of Palaeozoic and Mesozoic-Cenozoic sediments; regions of Caledonian folding; *2*—anticlinoria; *3*—outcrop of Precambrian rocks; *4*—synclinoria; *5*—sectors overlain by a cover of Palaeozoic and Mesozoic-Cenozoic sediments; regions of Hercynian folding; *6*—anticlinoria; *7*—synclinoria; *8*—sectors over a platformal cover of Mesozoic-Cenozoic sediments; *9*—marginal volcanic belt of the region of Hercynian folding; *10*—basins filled with Upper Palaeozoic sediments; *11*—major anticlinoria. Truncation of the internal structure of the median mass is visible in the area of Karaganda

are distinguished by intensive magmatism, both effusive and intrusive. The composition of the igneous rocks is very diverse; both calc-alkaline and alkaline rocks are found among them. Together with andesites, basalts, and normal granites trachytes, phonolites, and syenites occur here. Acid, intermediate, and basic rocks are generally evenly distributed. Intrusions are confined to faults and form bodies of a fissure type.

Typical median masses are the Macedonia-Rhodope massif in the Alpine geosyncline of the Dinaric Alps and Kraištides and the Armenian massif that embraces southern Armenia and part of Anatolia and lies within the broad, also Alpine, Tauro-Caucasian geosyncline.

Very big median masses, like the Hungarian massif surrounded by the Alpine eugeosynclines and miogeosynclines of the Carpathians, Balkan Moun-

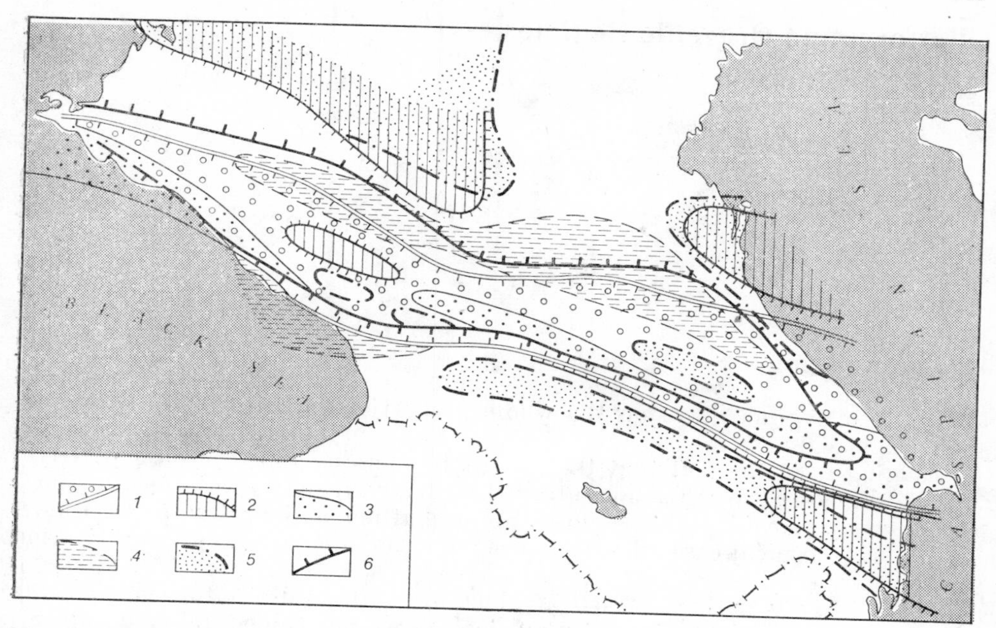

Fig. 83. Scheme of the geosynclinal and orogenic development of the Great Caucasus during the Alpine cycle (after Sholpo, 1978):

1—preinversion (pre-Late Jurassic) intrageosynclines; 2—preinversion intrageanticlines; 3—marginal troughs; 4—parageosynclines; 5—pre-Late Jurassic central uplifts (within the orogenic uplift) and foredeeps and intermontane basins (beyond the orogenic uplift); 6—boundary of Pliocene-Quaternary orogenic uplift

tains, and Dinaric Alps, and the Kolyma massif in the north-east of the USSR, within a Mesozoic geosyncline, have a complex structure. Closed troughs with a considerable infilling of sediments and a development of parageosynclinal or even miogeosynclinal type may be found within them.

The general patterns of development of geosynclinal regimes are shown schematically in Figure 81, which depicts not only geosynclinal regimes, but also the epigeosynclinal orogenic regime that will be discussed later, and the preceding parageosynclinal regime.

In addition, Figures 56 and 83 depict the main stages in the evolution of the Alpine geosyncline of the Caucasus in both section and plane.

The terminology used for geosynclines by different writers varies somewhat. The present writer calls the whole complex area of active endogenous development that includes uplifts and troughs and regions of the manifestation of eugeosynclinal, miogeosynclinal, and parageosynclinal regimes a geosyncline, and terms partial troughs and uplifts intrageosynclines and intrageanticlines (which is the terminology introduced by Tetyaev).

In another terminology (Shatsky's) geosynclinal belts, regions, and systems are distinguished. The first term signifies a group of geosynclines of different age united by their spatial position between two ancient platforms. Geosynclinal regions and systems are parts of belts distinguished by age and position between two median masses or between a platform and a median mass. The term "geosyncline" itself is used to designate partial troughs within a geosyncline (in our sense), i.e., our intrageosyncline; partial uplifts are then called geanticlines.

CHAPTER 16

Platform and Orogenic Regimes

Platform Regimes

Platform regimes have on the whole properties opposite to those of geosynclinal regimes.

Judging from the fact that there is little or no magmatism on a platform in typical cases, we can suppose that the crust in regions of platform regimes is *not very permeable* and as a close approximation can be called *impermeable.*

Again, in typical cases there is no regional metamorphism or granitization on platforms.

There is very little contrast in block-undulatory oscillatory movements of the crust compared with geosynclines. As we mentioned above (Chap. 8), the gradients of the rates of vertical movements are 10 to 30 times lower on platforms than in geosynclines; hence, the transition from the regions of downwarping, called *subgeosynclines*, and that of upwarping or *subgeanticlines*, is smooth and gradual and is complicated only slightly by small scarps of low amplitude associated with faults.

Folding is quite absent over wide areas of platforms. In other areas it is of block type, and in places where there is a relatively big accumulation of sedimentary series, it is of the injection type, chiefly in the form of salt domes.

The platforms now known on the surface of the continents are very stable structures. Once formed they last throughout subsequent time, being "reworked" only slightly by the effect of other regimes.

On the other hand, however, the structure of all to-day's platforms exhibits signs that another regime—a geosynclinal one—existed previously where they are now located. This is indicated by the two-stage structure of all platforms: below a relatively undisturbed cover of rocks corresponding to the time of a platform regime, a more ancient basement is always found, consisting of rocks more or less folded and faulted, metamorphosed, and granitized rocks, i.e., rocks subjected to processes proper to geosynclinal regimes.

We divide platforms into ancient and young ones according to the time when the geosynclinal regime gave way to a platform one. Ancient platforms have a Precambrian basement, young platforms a later one.

The transition from a geosynclinal to a platform regime always happens at the boundary between endogenous cycles. In the terminology, the time when a platform arises on the site of an earlier geosynclinal regime is indicated by the prefix "epi" (i.e., "post"). Epi-caledonian platforms, for example, are ones formed at the boundary between the Caledonian and Hercynian cycles: during the Caledonian cycle there was a geosyncline, while a platform regime has been established since the beginning of the Hercynian cycle.

There is no partial inversion in the development of block-undulatory oscillatory movements on platforms. Regular wave evolution of subgeosynclines, similar to that characteristic of intrageosynclines, is therefore not observed on them. The distribution of regions of downwarping and uplift may in general remain stable on platforms for whole endogenous cycles, being disturbed only by secondary oscillations at the boundaries between these regions.

In other cases, however, the subgeosynclines and subgeanticlines may shift quite considerably and their relative importance change. A pattern then operates in which the most active subgeosynclines are located in close proximity to the geosyncline bordering on the platform that is most active in the given endogenous cycle, and trend parallel to this same geosyncline. The change in the distribution pattern of subgeosynclines and subgeanticlines on the East European platform from cycle to cycle is shown in Figure 84. During the Baikalian and Caledonian cycles, the most active subgeosyncline had a north-eastern strike through the central areas of the platform, parallel to the Proterozoic and Caledonian geosynclines of Scandinavia. In the Hercynian cycle the principal role passed to a subgeosyncline that ran north-south from the Lower Volga to the lower reaches of the Pechora River, parallel to the Uralian geosyncline that was actively developing at the time. Finally, in the Alpine cycle, when the neighbouring active geosyncline proved to be the Caucasian, the Dnieper-Donets subgeosyncline, in the south of the platform, which trends parallel to the Caucasus and the Crimea, became very important.

The **regime of an ancient platform** (or craton) is the most quiescent of platform regimes. The East European platform during the Alpine cycle (i.e., during the Mesozoic and Cenozoic), when sediments not more than several hundred metres thick were laid down and the thickness gradients were measured in fractions of a metre per kilometre, can be taken as the prototype.

An important feature of ancient platforms is the shape of the subgeosynclines and subgeanticlines, which are round or irregular in outline, and hundreds, even thousands of kilometres across. Subgeosynclines usually have more definite, round outlines while subgeanticlines occupy the spaces between them.

The parts of ancient platforms that lack a continental sedimentary cover, and where basement is exposed at the surface, are called *shields*. They are the quietest and most stable parts of ancient platforms. The parts of ancient platforms with a sedimentary cover are usually called *plates*.

The distribution of subgeosynclines and subgeanticlines on ancient platforms, as far as one can judge from the available data, bears no relation to the structure of the basement. Regions of downwarping and uplift prove to be "superimposed" structures, their boundaries cutting the strike of basement structures at arbitrary angles and intersecting its zones of different age. The strike of Precambrian structures on the Baltic and Ukrainian Shields, for example, is cut by the boundaries between the shields and the adjoining syneclises, Moscow, Ukrainian, Black Sea, and Dnieper (Fig. 85).

The **young platform regime** differs, on the whole, from the ancient platform (cratonal) regime in the large-amplitude of the undulatory oscillatory movements and the high gradients of their rates.

An important feature is that many regions of uplift and trough on young platforms are divided by deep faults.

It is often said that the subgeosynclines and subgeanticlines on young

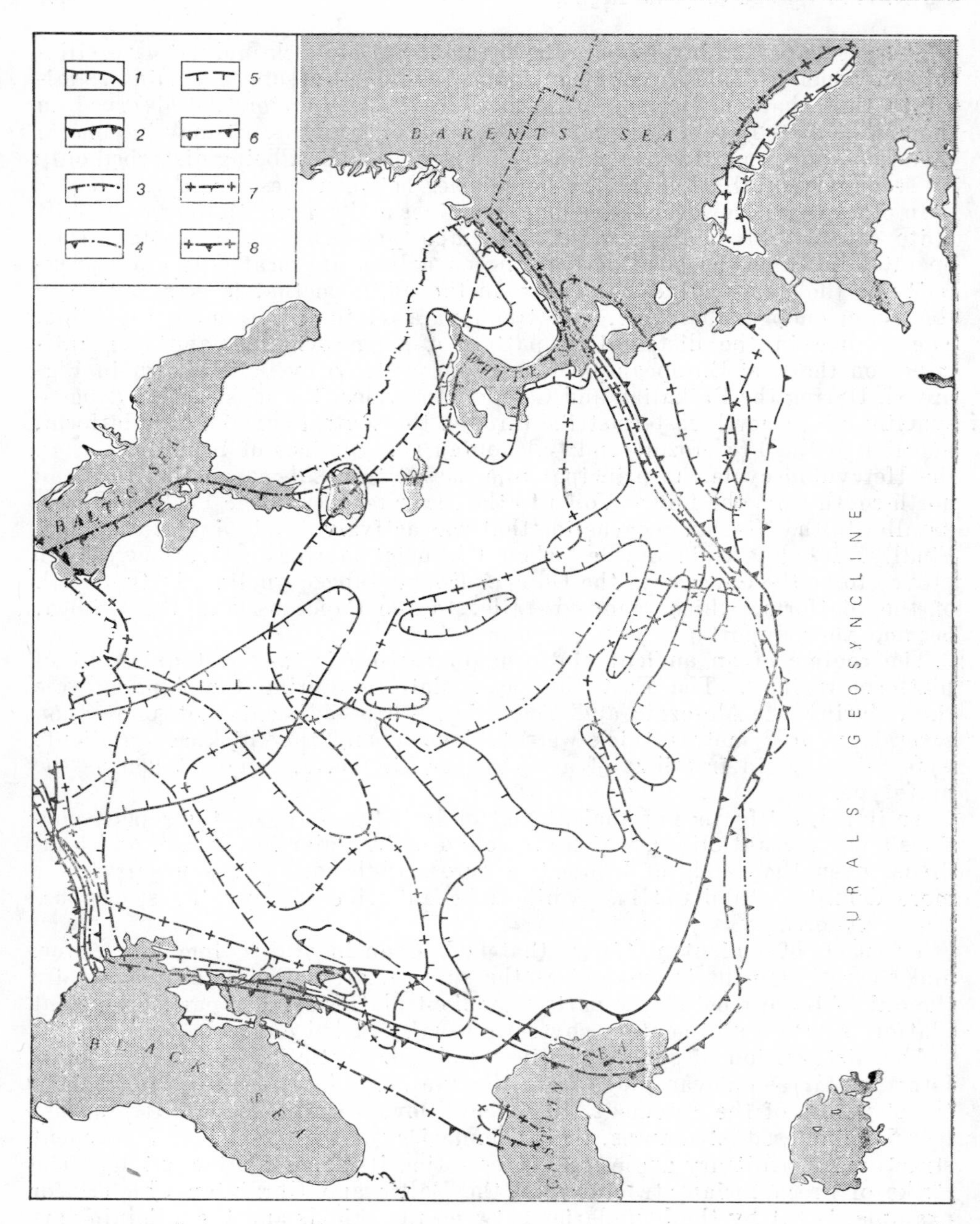

Fig. 84. History of syneclises and anteclises on the East European platform (after Sychyova-Mikhailova):

1—Middle and Late Riphean syneclises; 2—boundary of the Riphean platform; 3—syneclises of the Caledonian cycle (Early Cambrian); 4—boundary of the Caledonian platform; 5—syneclises of the Hercynian cycle (Middle and Late Carboniferous); 6—boundary of the Hercynian platform; 7—syneclises of the Alpine cycle (Early Cretaceous); 8—boundary of the Alpine platform. Additional strokes point inside syneclises

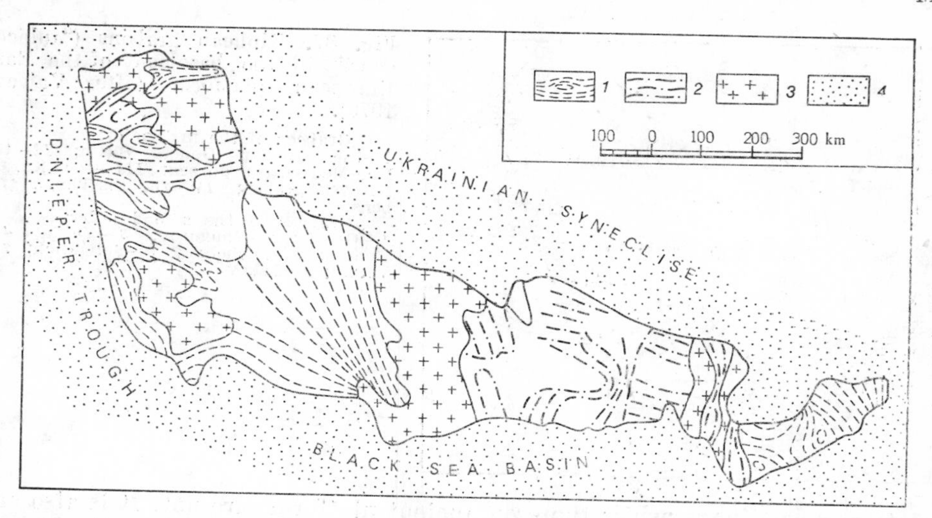

Fig. 85. Strikes of the Precambrian folded structures of the Ukrainian shield (after Geological Map of the USSR):

1—regions of Archaean folding; *2*—regions of Early Proterozoic folding; *3*—granitoids of various age; *4*—Phanerozoic rocks of synеclises. The structures of the shield are cut discordantly by downwarps of the ancient East European platform

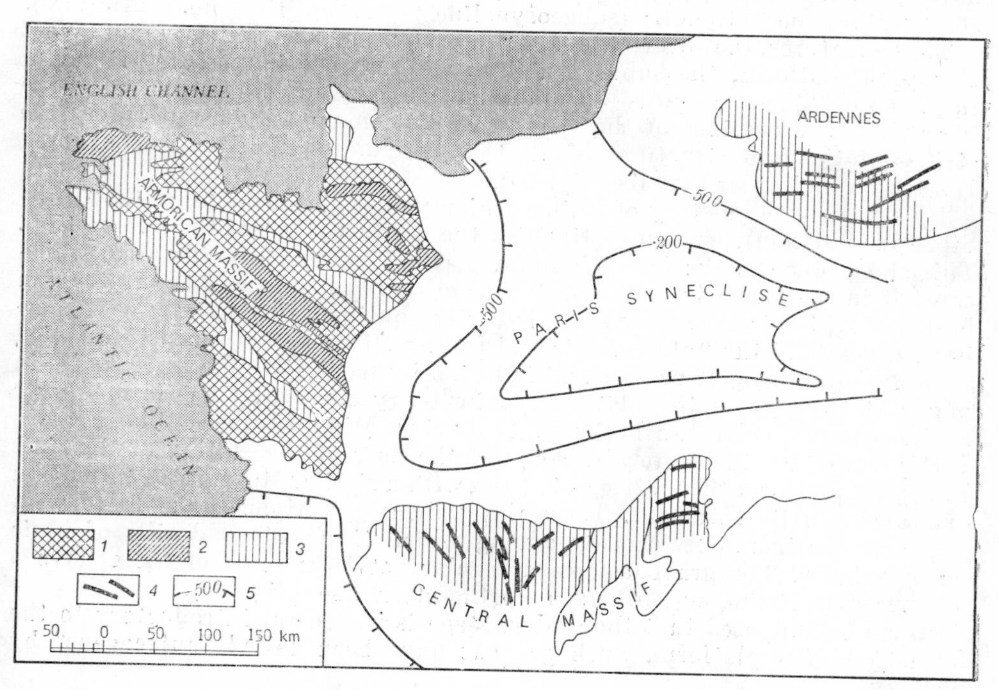

Fig. 86. The unconformably overlying Palaeozoic structures of the Armorican massif, the Ardennes, and the Massif Central of France by sediments of the Paris syneclise. Young Epihercynian platform of the Western Europe (after Beloussov):

1—Precambrian folding; *2*—Caledonian folding; *3*—Hercynian folding; *4*—the strike of folds on the Massif Central and in the Ardennes; *5*—contours drawn of the depth of the folded Hercynian basement in the Paris syneclise (in metres)

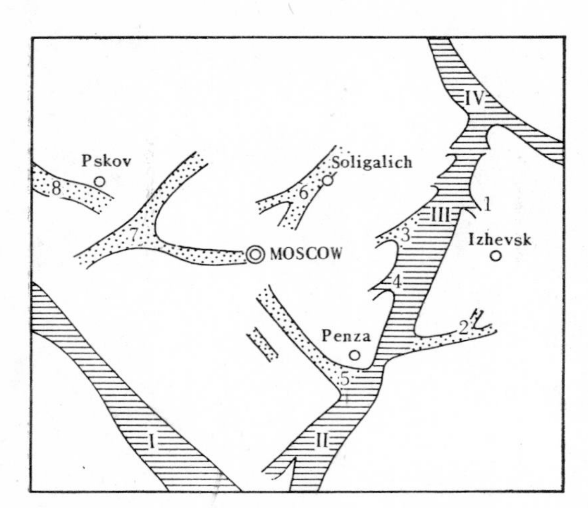

Fig. 87. Riphean grabens ("aulacogens)" in the basement of the East European platform (after Valeev, 1970).

Grabens of the first order:
I—the Greater Donets Basin; *II*—the Don-Medveditsa graben; *III*—the Kazan-Kirovsk graben; *IV*—the Cis-Timan grabens.
Grabens of the second order:
1—Genets; *2*—Shugurov; *3*—Yaran; *4*—Rongin; *5*—Pachelma; *6*—Soligalich; *7*—Valdai; *8*—Pskov

platforms are linear, while those on ancient platforms are not. It is also emphasised, that the location of troughs and uplifts is inherited from the history and structure of the basement. The troughs and uplifts that developed at the end of the preceding, pre-platform cycle (at the orogenic stage) continue to develop in the next cycle, but under platform conditions, as much attenuated "posthumous" troughs (subgeosynclines) and uplifts (subgeanticlines).

These features, however, are not the rule; on the Epihercynian West European platform, the subgeosynclines and subgeanticlines are (1) round in plan and (2) superimposed unconformably on a Hercynian structure of the basement. The outlines of the Paris syneclise, for example, cut across the strikes of the fold structures of all the surrounding uplifted and exposed Hercynian and older massifs (Fig. 86).

By studying the history of Epihercynian young platforms (e.g., the Urals-Siberian platform), we can distinguish the earliest stage of their formation. This stage precedes the formation of a platform cover of a wide area, which consists in the formation of separate grabens on the surface of the Hercynian basement. In this case they are filled with continental, partially coal-bearing sediments of the Late Triassic and Lias. Sedimentation was accompanied with an outpouring of basalts and invasion by basic sills. These rocks lie in bands at the base of the platform sedimentary cover that has formed over a wide area on the Urals-Siberian platform since Middle Jurassic time.

It proves that there are also signs at the base of the section of the East European platform that it began to form with a similar stage of the formation of grabens in its basement. There the grabens are Riphean and filled with Riphean continental sediments among which lie sheets and sills of basic igneous rocks. The grabens form a regular network: some of them have a northwestern strike, and others a northeastern one (Fig. 87).

It can be supposed that the graben stage is in general introductory to the formation of a platform. Such grabens have been called "aulacogens" by Shatsky.

Orogenic Regimes

Orogenic regimes, like geosynclinal ones, are characterised by large amplitudes and contrasts of block-undulatory oscillatory movements; no essential difference can be drawn between these two classes of regime on that basis.

In orogenic regimes, however, uplifts predominate over subsidence, unlike geosynclinal regimes, in which subsidence predominates. The regions of an orogenic regime are primarily mountainous ones and are divided into zones of uplift and subsidence. The uplifted zones are eroded while sediments accumulate in the troughs, forming the molasse formation typical of these regions.

The evolution of a molasse formation clearly reflects a predominance of uplift over subsidence. It begins with the accumulation of fine sandstones and clays, which are marine in many areas, what is called a lower molasse; it is followed by a continental upper molasse, much coarser, with broad development of conglomerates, which is often red. This change in the character of the sediments is a classic example of "overcompensation" of downwarping by accumulation as a consequence of the speedier uplift supplying clastic material, more rapid than the subsidence. The evolution of orogenic regions usually ends with both mountain ranges, and the depressions that become zones of erosion being involved in the general uplift.

The molasse sediments are often intercalated by rocks of a lagoon formation, i.e., various evaporites (rock salt, gypsum, anhydrite, etc.). Coal-bearing series of sediments occur in places, as well.

The uplifts and troughs in an orogenic regime have an arch-block character and are divided by deep slashes.

These deep slashes determine the crust's permeability, which is clearly concentrated in character; that is another major distinction between orogenic and geosynclinal regimes. The deep slashes, which are channels for concentrated penetration, arise during "breaking up" of the crust which has previously been integral, solid, and impervious. With an epigeosynclinal orogenic regime, the crystalline shell that is broken up was the one formed in the geosynclinal inversion stage during regional metamorphism and granitization. With epiplatform orogenesis it is the solid crust of the platform that is broken.

Both extrusive and intrusive magmatism are restricted to deep slashes.

Some folding, also mainly of block type, occurs in the conditions of an orogenic regime. It is associated with the relative rise of small blocks along secondary faults. In addition, injection folds develop where there are plastic rocks (salt or clay). General buckling may occur, however, though on a smaller scale, when pressure is exerted on strata by the weight of relatively uplifted blocks collapsing through gravity.

Epigeosynclinal and *epiplatform* orogenic regimes are distinguished, i.e., regimes developing respectively after geosynclinal and platform regimes.

Until recently the opinion prevailed (many scientists still maintain it) that geosynclinal development regularly and necessarily ends with mountain building, i.e., that an orogenic regime is the inevitable consequence of geosynclinal development and can even be considered the terminal stage of a geosyncline's history.

There are now data, however, that force us to alter our opinion and regard the orogenic regime as an independent one which is not the direct consequence of a geosynclinal regime though it often follows that one.

1. It has become clear that mountains are not formed on the sites of certain geosynclines, in which geosynclinal development is completed with the two stages that were considered above in connection with eugeosynclinal and miogeosynclinal regimes, and are immediately followed by a platform regime without an intermediate orogenic stage. Mountain building did not,

for example, occur before the origin of the Scythian platform after the end of the Hercynian cycle of geosynclinal development in the south of the European part of the USSR. There are no signs of a mountainous relief at the end of Hercynian geosynclinal development over most of the West Siberian depression. Even in those cases when an orogenic regime arises, where a geosynclinal regime had previously existed a more or less prolonged hiatus is observed now and again between the end of the latter and the beginning of the former, during which movements of the crust and all endogenous activity were quiescent.

In the Western Alps, for example, a partial inversion occurred at the end of Late Cretaceous and beginning of the Palaeocene, and flysch accumulated in the Palaeocene and Eocene, while the orogenic regime only began to develop in the Miocene. Before that, during much of the Oligocene, the Alps were flattened hilly country.

An especially long period of intermediate quiescence is present in the history of a number of Cimmerian geosynclines. In the Verkhoyansk-Kolyma geosyncline in the north-east of the USSR, for example, geosynclinal development was completed not later than Early Cretaceous time, while mountain building began only in the Neogene. After geosynclinal development at the end of the Mesozoic, the North American Cordilleras were marshy lowland during the entire Palaeogene. Orogenesis in this region began only late in the Miocene.

2. The uplifts and troughs that are formed on the site of a former geosyncline during orogenic regime do not always succeed and develop those that were in the geosyncline during its inversion stage, or do so only partially. The orogenic rise of the Great Caucasus, for example, only coincides with the central uplift in its south-eastern sector and could be regarded as a further development of the latter. In the north-western half of this same mountain country, the axis of the orogenic uplift is offset from the central uplift: the latter is in Svanetia, i.e., on the southern slope of the present-day orogenic uplift of the Great Caucasus (see Fig. 83).

3. The very existence of an epiplatform orogenic regime contradicts the conception of it as a consequence of geosynclinal development.

The neotectonic events in Central Asia, including the Soviet Tien Shan, are a well-known example of epiplatform orogenesis. In this vast region a young platform regime existed in the Mesozoic and Palaeogene, but from the end of the Oligocene, in the Neogene and Quaternary, an orogenic regime became established.

All this convinces us that an orogenic regime is largely an independent phenomenon. We say largely and not wholly, because the history of the Earth's crust indicates, however, that orogenic regimes were mostly of epidgeosynclinal type. We can say that, although an orogenic regime is independent in origin of the previous geosynclinal development, geosynclines are nonetheless the zones where this regime is most liable to occur. Young platforms are relatively favourable zones for an orogenic regime, while ancient platforms are less often involved. (On ancient platforms, rifting is a sort of "substitute" for epiplatform orogenesis; it can, however, incidentally also occur on young platforms. This regime will be discussed later.)

Like geosynclinal and platform regimes, the manifestation of orogenic regimes consistently fits into the rhythm of endogenous cycles. An orogenic regime always appears at the end of an endogenous cycle.

Orogenic uplifts tend first to be narrow and then to expand. This process is best manifested in the case of epigeosynclinal orogenesis, when uplifts successively incorporate new strips of the former geosyncline and one orogenic uplift may eventually encompass a number of zones of the previous geosyncline. Thus the orogenic uplift of the Great Caucasus Range "has made into a whole" geosynclinal central uplifts, two marginal troughs, a large intrageanticline of the central and north-western part of the Main Range and a parageosyncline of the northern slope. It also has incorporated the margin of the young Scythian platform to the north. Orogenic uplifts also grow in length. Uplifts may consequently become larger than the structures that had dissected the crust in the preceding geosynclinal situation.

The troughs between two uplifts in orogenic regions are called *intermontane*. Those on the margin of an orogenic region are called *foredeeps*. The history of both types consists in their wave-like shift as the orogenic uplifts expand. We make a distinction, we must emphasise, between marginal troughs and foredeeps. Marginal troughs belong to the geosynclinal regime, arise at its inversion stage, and are characterised by the accumulation of a predominantly flysch formation. They are almost completely located within preinversion intrageosynclines, on their periphery and near the central uplift. A foredeep belongs to the orogenic regime and is the site of a molasse formation on the margin of the neighbouring platform. If the axis of an orogenic epigeosynclinal uplift coincides with that of a central uplift, intermontane troughs and foredeeps may coincide in the early stages of their development with marginal troughs; the shifting of the orogenic troughs directly continues the wave-like development of uplifts and troughs that had already begun in the geosyncline in the epoch of partial inversion. The marginal troughs moving towards one another may merge in an intermontane trough located where an intrageanticline was at the beginning of geosynclinal development. This may not be a single trough, but a series of separate basins with residual uplifts of the intrageanticline between them. Orogenic troughs also roll over median masses, these engulfing their margins or may prove to be imposed on their interior as separate basins. The migration of orogenic troughs, by continuing the similar migration that begun during the geosynclinal development, leads to further shifting of the sedimentary lenses: a lens composed of molasse formation proves to be displaced outward, i.e., away from the central uplift, relative to lenses of the upper terrigenous formation.

This type of evolution is shown in the scheme of the development of a geosyncline in Figure 81. An example is the Kura intermontane trough which was mostly occupied at the geosynclinal stage by an intrageanticline up to and including Oligocene time. When an orogenic stage began at the end of the Miocene, this intrageanticline was involved in subsidence that began rapidly from its margin and became an intermontane trough.

Foredeeps may also coincide at first with marginal troughs, but later, simultaneously with rapid expansion of the marginal orogenic uplift adjoining the trough, they migrate as rapidly towards the platform and roll over its margin. A foredeep is asymmetric. The side facing the nearest orogenic uplift is steep while that facing the platform is gently sloping. The folding is also different in character on the different limbs of a foredeep; block and injection folds are more strongly developed and more complex on the inner limb than on the outer.

The folding of the orogenic stage, both in foredeeps and intermontane troughs is the last folding in any endogenous cycle. In the case of the epigeosynclinal type of orogenesis two main stages in the development of folding can consequently be distinguished: geosynclinal, when deep folding and general crumpling occur in the central uplifts; and orogenic folding when block and injection folds are formed in intermontane troughs and foredeeps. In the second case, orogenic folding supplements geosynclinal in the formation of a **folded zone** belonging to the relevant endogenous cycle. This zone consists of folds of different type generated at different stages of the cycle. In the case of epiplatform orogenesis, however, the fold structure is limited to the orogenic folding of intermontane troughs and foredeeps.

As will be seen from the foregoing, two orogenic regimes (epigeosynclinal and epiplatform) have much in common. The differences between them amount chiefly to the composition of the igneous rocks.

An epigeosynclinal orogenic regime is mainly characterised by acid and intermediate magmatism. Among the effusive rocks andesites, dacites, and liparites predominate, while basalts play a subordinate role. As for intrusive rocks, granites and diorites develop, often forming large massifs along deep faults. Examples are Central Kazakhstan and the Altai-Sayan region where, after the end of the Caledonian geosynclinal development (in the Devonian, Carboniferous, and Permian) there was an orogenic regime with intensive outpourings and intrusion of acid igneous rocks. Within the Great Caucasus, the orogenic regime at the end of the Alpine cycle was accompanied mostly by andesite, and to a lesser extent basaltic, outpourings. In Eastern Transbaikalia, where a geosynclinal regime developed during a Mesozoic endogenous cycle and was completed by the end of the Middle Jurassic, the orogenic regime took place in Late Jurassic and Early Cretaceous times and was accompanied with andesite outpourings that were later followed by basalts.

In median masses of the orogenic stage, the volcanism is alkaline in composition. In the median mass of southern Armenia, there was an outpouring of trachyandesites, trachites, phonolites and then (to a less extent) basalts, in the orogenic stage in the Eocene and later.

The signs of magmatism are weaker in intermontane troughs and foredeeps than in uplifts, but outpourings and hypabyssal intrusions of the laccolith and igneous diapir type, mainly alkaline in composition, may be found in them. An example is the alkaline intrusions and extrusions on the northern slope of the Caucasus that occurred at the boundary between the orogenic uplift of the Great Caucasus and the foredeep.

Epiplatform orogenesis is characterised by alkaline magma, mainly basic. In it alkaline basalts, phonolites, and trachites, and hypabyssal intrusions of similar composition are encountered, but there may have been no magmatism.

Examples of epiplatform orogenesis, of the end of the Mesozoic endogenous cycle, are the events that developed in Late Jurassic and Early Cretaceous times in Western Transbaikalia, where intensive differential vertical movements took place, accompanied by faults and outpourings of alkaline lavas along them. An example of epiplatform orogenesis with very weak magmatism is the Tien Shan, where the neotectonic activation of the platform in the Neogene and Quaternary periods was extremely vigorous, but where magmatism was manifested only in a small outpouring of alkaline basalts.

Some Terminological Problems

In the foregoing exposition, in the sections dealing with geosynclines, platforms, and orogenic regions, we spoke of their division into uplifts and troughs in the course of vertical movements of the Earth's crust, called intrageanticlines and intrageosynclines for geosynclines, subgeanticlines and subgeosynclines for platforms, and simply uplifts and troughs for regions of an orogenic regime.

These terms are kinematic or historical, because they reflect the trend of the oscillatory movements of the crust. But they all finally led in our time to the structural forms expressed in the bedding of the strata. Protracted uplift of some sector of the crust produces a convex structure and prolonged downwarping a concave structure. Since wave-undulatory oscillatory movements are manifested unambiguously over vast zones, both convex and concave structures are correspondingly extensive. In geosynclines, the large uplifts and troughs due to oscillatory movements are called *anticlinoria* and *synclinoria*, respectively. Since undulatory oscillatory movements are of different orders, the same is true of anticlinoria and synclinoria, which are all generally crumpled into folds of different type whose distribution patterns were discussed above. It is the combination of anticlinoria and synclinoria with folding of general crumpling, deep, injection, and block folding (depending on the position of one part of the geosyncline or another) that also constitutes the structural picture of the folded zone that takes shape in the site of a geosyncline.

If an orogenic zone is formed at the end of the endogenous cycle on the site of a geosyncline, the structural forms of the geosynclinal origin may be altered. With the development of deep slashes anticlinoria are transformed into horst-anticlinoria and synclinoria into graben-synclinoria. Orogenic folding (block and injection folding and, at places, general buckling) takes place in the latter.

On platforms, undulatory oscillatory movements give rise to anteclises and syneclises, i.e., to gently sloping, broad convex and concave structures.

The considerable complications in the structure of the Earth's crust are linked with the fact that the distribution of rising and sagging areas both in geosynclines and on platforms alters with the passage of time. The structure of each region bears the imprint of the total result of all the vertical movements that occurred here over long geological time, a result that may differ at various levels. Inverted anticlinoria, for example, are produced in which only the upper strata occur as anticlines and the deeper ones as synclines (Fig. 88). Such inverted anticlinoria occur on the sites of central uplifts in geosynclines. In them the synclinal bedding of lower strata reflects the result of the downwarping that occurred before the inversion. This downwarp has survived because the subsequent uplift, which is reflected in the upper strata, was not sufficient to eliminate the downwarp completely. The preinversion suites in inverted anticlinoria form a lens the maximum thickness of which corresponds to the maximum amplitude of the previous downwarping.

In the same way, inverted synclinoria fix the structure of two stages of development: earlier uplift and subsequent downwarping (Fig. 88).

Even more complex relation between surface and deep structures occurs on platforms as a result of the migration of subgeosynclines and subgeanticlines. The topography of the crystalline basement of the East European

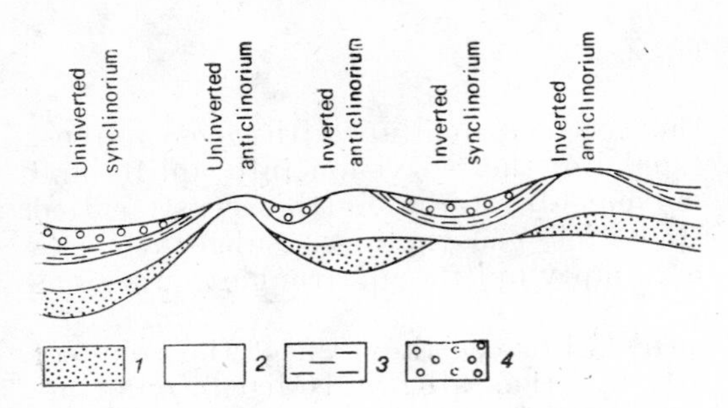

Fig. 88. Scheme of the structure of inverted and non-inverted anticlinoria and synclinoria (after Beloussov):

1, 2—preinversion formations; *3, 4*—post-inversion formations

platform, interestingly, has mostly been affected by vertical movements which were the strongest in the Proterozoic and Early Palaeozoic. On the other hand, younger, Late Palaeozoic and, even more, Meso-Cenozoic movements, which were much less intense, have had almost no effect on the same topography.

In those cases, however, when subgeosynclines and subgeanticlines have remained stable in the same places the syneclises at all levels coincide with subgeosynclines, and anteclises with subgeanticlines. Many parageosynclines were also stable and therefore remain synclinoria to great depths.

CHAPTER 17

Rift, Magmatic Activization, and Continental Margin Regimes

The Rift Regime

The rift regime is similar to the orogenic and is sometimes considered a variety of the latter. A number of specific features, however, lead us to classify it in an autonomous category.

The examples of the rift regime that are well known at present belong exclusively to the Cenozoic, with the most important events in the Neogene and Quaternary, i.e., in the epoch of neotectonic activization. Whether the regime occurred in earlier cycles, and how it was expressed then, remains unclear.

Rifts have attracted great attention of late, following the discovery of a rift belt on the ocean floor. We shall speak about this oceanic rift belt later; here we shall be concerned only with continental rifts.

By rift is meant a tension cleft. In the literal sense every normal or open fault is a rift, but as a geological term "rift" signifies an aggregate of tension structures. The elementary ones are grabens; a rift, consequently, is an aggregate of grabens in a region of crustal tension. We also speak of rift belts and systems, having in mind sets of grabens of different orders.

Rifts are known in various places on the continents, the most typical and best known examples being the Arabian-African rift belt, the Baikal rift, and the Rhine rift valley.

These rifts were formed at approximately the same time (mainly in the Neogene and Quaternary), but on different basements. The Arabian-African belt and the Baikal rift were formed on ancient platforms and the Rhine rift valley on a young epihercynian platform. The basement, it must be emphasized, is preserved in all cases at the bottom of the grabens. It has merely subsided in them and has not been pushed aside, as is supposed for oceanic rifts.

The preliminary stage in the development of a rift regime in all cases was the formation of broad, convex (updomed) structures of anteclise type, but larger in size and amplitude than an ordinary anteclise. We shall call them *arches*. They are irregularly oval in shape, hundreds (and even thousands) of kilometres wide, and several kilometres in amplitude of height.

The arches are subsequently cut by tension faults along which complex grabens are formed, divided into grabens and horsts of the second and even higher orders. The location of these faults in the arches is controlled by two factors: (a) the distribution of the stresses arising as a result of arching through the action of forces directed upward, and (b) inhomogeneities in the structure of the crust. The grabens are generally 30 to 60 km wide, while the amplitude of the faults is several kilometres (5000 to 6000 m as a maximum). The grabens subside gradually over a long period of geological time, in separate pulses. The faults are usually preceded by the formation of flexures.

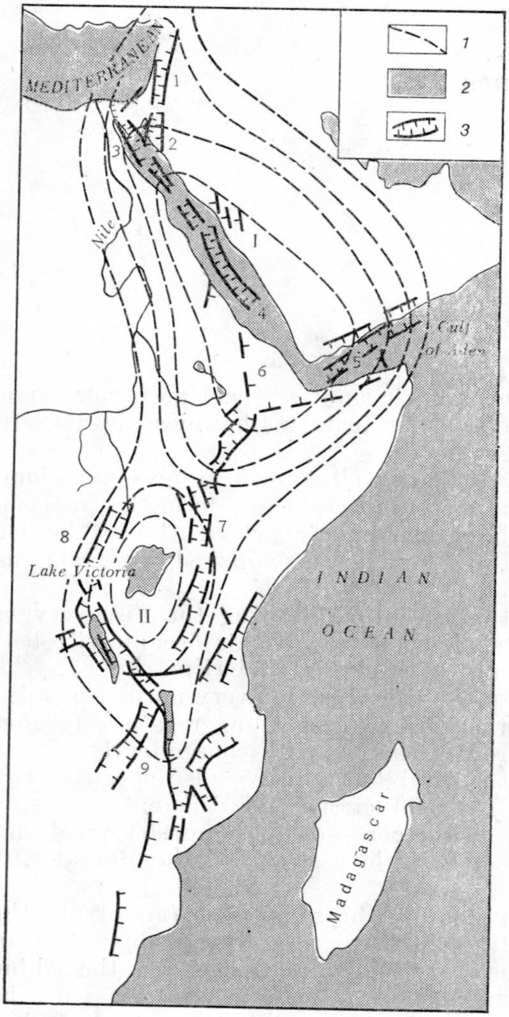

Both the preliminary rise of the arches and subsequent development of the grabens are accompanied with volcanic activity, the volume of which is different, however, in different cases and places. Alkaline rocks that are derivatives of alkaline basalt (phonolites, trachites, etc.) predominate among the lavas. The arches are expressed in the present-day relief as gently sloping uplifts and the grabens as depressions in the form of valleys, in which there are often lakes and river valleys. The grabens are partly filled with volcanic rocks and by lacustrine and fluvial sediments. All rift regions are highly seismic.

Let us glance briefly at the history and [structure of certain rifts.

The Arabian-African rift belt is the biggest structure of this type on the continents (Fig. 89), extending north-south for 6500 km from the Middle East through East Africa to the mouth of the Zambezi River. It consists of two rift systems, the Eritrean and the East African, which are continuations of one another, but are related each to its own arch, the Nubian-Arabian and East African respectively. The former is 2000 km wide and 5 km in vertical amplitude; the latter is 1200 by 2000 km in size and 3 km in amplitude.

Fig. 89. The Arabian-African rift belt (after Beloussov):

1—tentative contours on surface of arches (*I*—Nubian-Arabian, or Eritrean; *II*—East African); *2*—seas and lakes; *3*—normal faults (pointing toward downfaulted block). Figures indicate individual grabens: *1*—Dead Sea; *2*—Aqaba; *3*—Suez; *4*—Red Sea; *5*—Aden; *6*—Ethiopian; *7*—the eastern branch of the East African rift (Gregory rift); *8*—the western branch of the East African rift; *9*—Nyasa

The main structure in the Eritrean system is the Red Sea graben which extends in strip about 60 km wide along the axis of the Red Sea. This graben, in addition, follows the axis of the whole Nubian-Arabian arch. On the northern and southern periclines of this arch, the Red Sea graben fans out into smaller ones, the Suez and Aqaba grabens in the north and the Aden and Ethiopian grabens in the south. In the northern continuation of the Aqaba graben the Dead Sea, Lebanon, and Syrian grabens are distinguished.

This arrangement of grabens on the Nubian-Arabian arch fully accords with a stress field in a sector of the crust that has been upwarped by vertical forces.

The Nubian-Arabian arch was formed in the Mesozoic. Its individual grabens developed at different times from the Oligocene on.

The East African rift system has a more complex structure (Fig. 90). The Lake Rudolph graben is a single one in the north, but it divides in the south into two branches that fringe the central part of the East African arch on the east and west. The eastern arc, called the Gregory rift, crosses Kenya and passes into northern Tanzania, where it fans out into several small grabens and soon terminates. On its continuation the rift is associated with the Kenya arch, which is an uplift of the second order on the flank of the main East African arch.

The western branch runs from Lake Rudolph along the belt of big lakes (Albert, Edward, Kivu, Tanganyika, and Nyasa). The bottoms of some of these lakes are below sea level.

This branch was not preceded by a special uplift of the second order, as happened with the Gregory rift. The marginal uplifts accompanying the graben of the western branch were formed later, simultaneously with the sinking of the graben.

The maximum amplitude of the subsidence of the grabens in this whole system is 3 km.

The East African arch was raised at the end of the Mesozoic and in the Palaeogene. The Gregory rift began to develop in Late Miocene time, but main movements in it occurred in the Pliocene and Pleistocene. The western branch began to form in the Early Miocene, but there too tectonic activity was at a maximum in the Pleistocene.

Volcanic activity of great force accompanied development of the Gregory rift in all its stages.

The first outpourings (phonolites) onto the surface of the East African arch had already occurred in Miocene time before the formation of the grabens; subsequently the area of volcanism shrank and was finally concentrated within the graben itself. The fissure outpourings in the Holocene gave way to eruptions from separate central vent volcanoes.

The western branch is distinguished by much less volcanic activity.

The distribution of the grabens on the East African arch is determined not only by the stresses due to the updoming, but also by inhomogeneities in the crustal structure, i.e., the presence in it of resistant ancient massifs and the much less resistant zones of former Proterozoic geosynclines. The two branches of the East African rift, for instance, follow former geosynclines of Proterozoic age and "flow" around the ancient Victoria massif of Archean age. It can also be seen that grabens intersecting the folds of the Precambrian basement at an angle reflect the "fibre" of the basement in certain details.

The *Baikal rift* embraces not only the basin of Lake Baikal, but also a large number of adjacent depressions, and stretches from the basin of Lake Mopsogol in the south-west across the Tunka, South Baikal, North Baikal, Barguzin, and Upper Angara basins to the Muya, Bauntovo, Chara, Tokko, and other smaller basins in the north-east. Its total length is 2600 km (Fig. 91).

The rift was formed mainly on an arched rise of the surface of the Precambrian basement. Isolated local highs and lows had existed there already in Mesozoic time, but their distribution differed from that of the younger rift

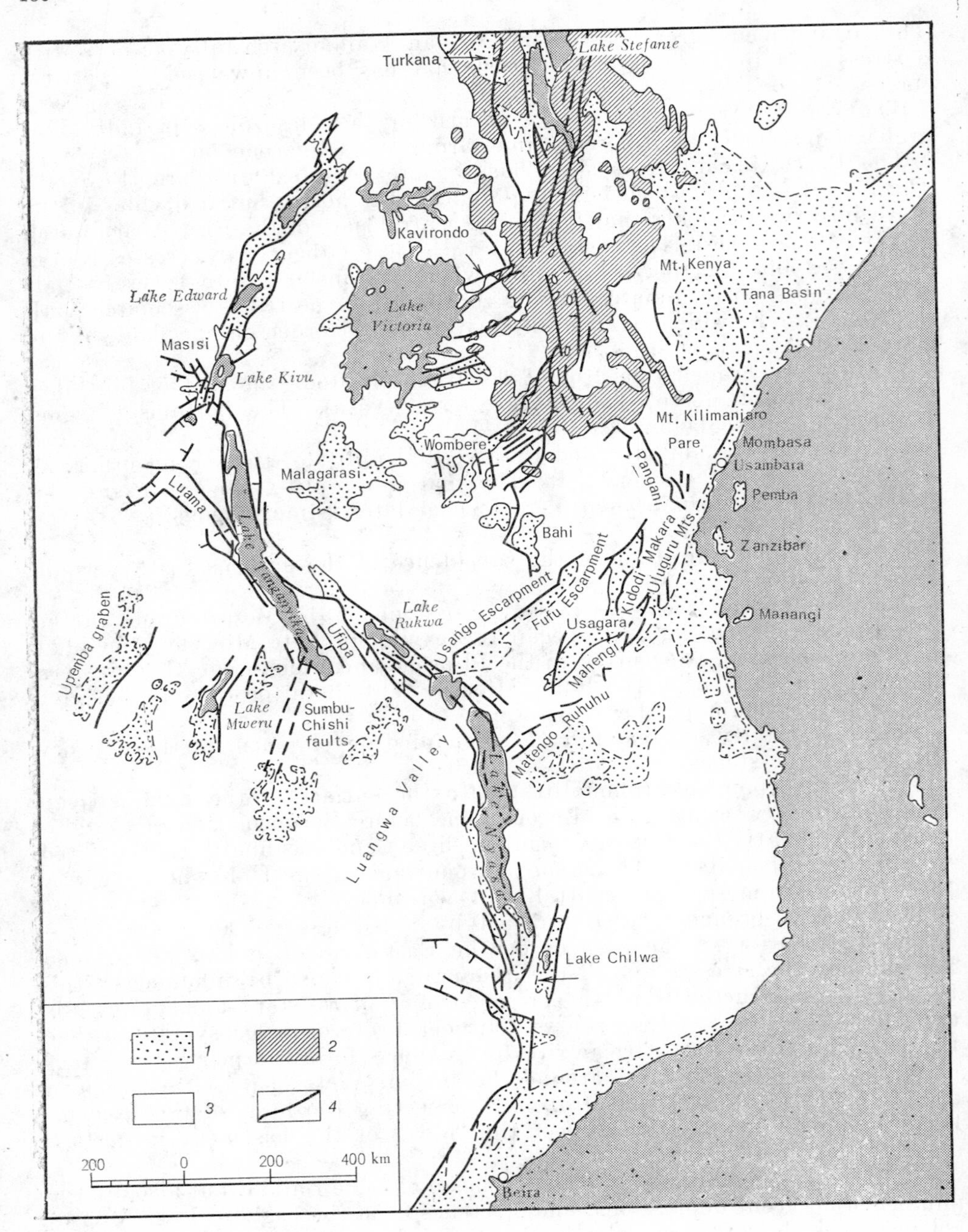

Fig. 90. The African rift system (after Beloussov):

1—Tertiary and Quaternary sedimentary deposits; 2—Tertiary and Quaternary volcanic rocks; 3—Precambrian rocks; 4—faults

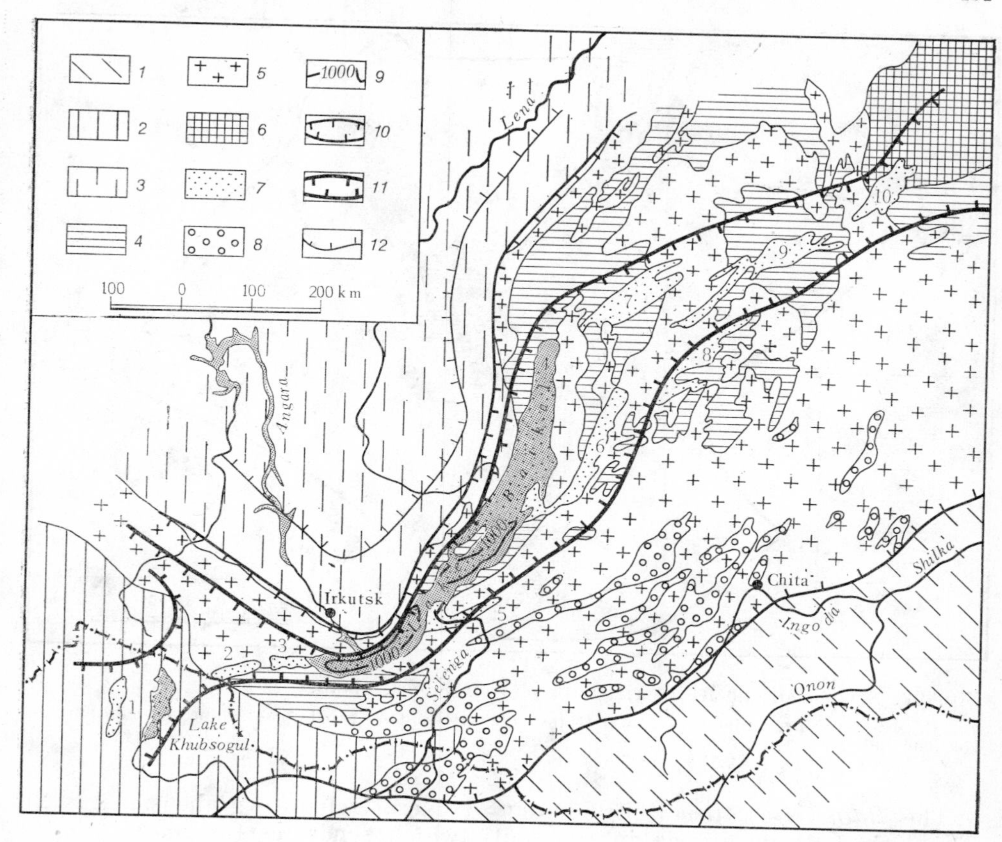

Fig. 91. Scheme of the structure of the Baikal rift and zone of epimesozoic orogeny in western Transbaikalia (after Logachev):

1—zone of Hercynian folding; *2*—zone of Caledonian folding; *3*—Lower Palaeozoic sediments of the Angara-Lena foredeep of the Baikal orogenic zone; *4*—Proterozoic rocks of the Baikal orogenic zone; *5*—Proterozoic granitoids; *6*—zone of Archaean folding; *7*—neotectonic grabens of the Baikal rift; *8*—grabens in the zone of epimesozoic orogenic activization; *9*—1000 m depth contours in Lake Baikal; *10*—boundaries of zones of neotectonic uplift less than 2000 m; *11*—boundaries of the zone of neotectonic uplift less than 3000 m; *12*—boundary of the Irkutsk syneclise. Grabens of the Baikal rift: *1*—Mopsogol; *2*—Tora; *3*—Tunkin; *4*—Baikal; *5*—Lower Selenga; *6*—Barguzin; *7*—Upper Angara; *8*—Bauntovo; *9*—Muya; *10*—Chara

structures. In the Palaeocene and Eocene, the area was levelled, and a mantle of weathering covered the surface. In Oligocene and Miocene times, shallow troughs received fine sediments and coal was formed.

The development of the grabens in their latest form took place in the Late Pliocene, Pleistocene, and Holocene. The whole zone was cut up by a complex network of longitudinal and transverse normal faults. The amplitude of the vertical displacements varies from place to place and is as much as 5000 m. Lake Baikal is 1600 m deep; its bottom is well below sea level.

Simultaneously with sinking of the crust in the grabens on the margin of the rift, the ranges fringing Lake Baikal and other depressions underwent upheaval (e.g., the Khamar-Daban, Primorsky, and Barguzin Mts.). Volcanism was not very intense; only basaltic outpourings of small volume from isolated small volcanic structures are known.

The Baikal rift has no connection with the oceanic rift belt.

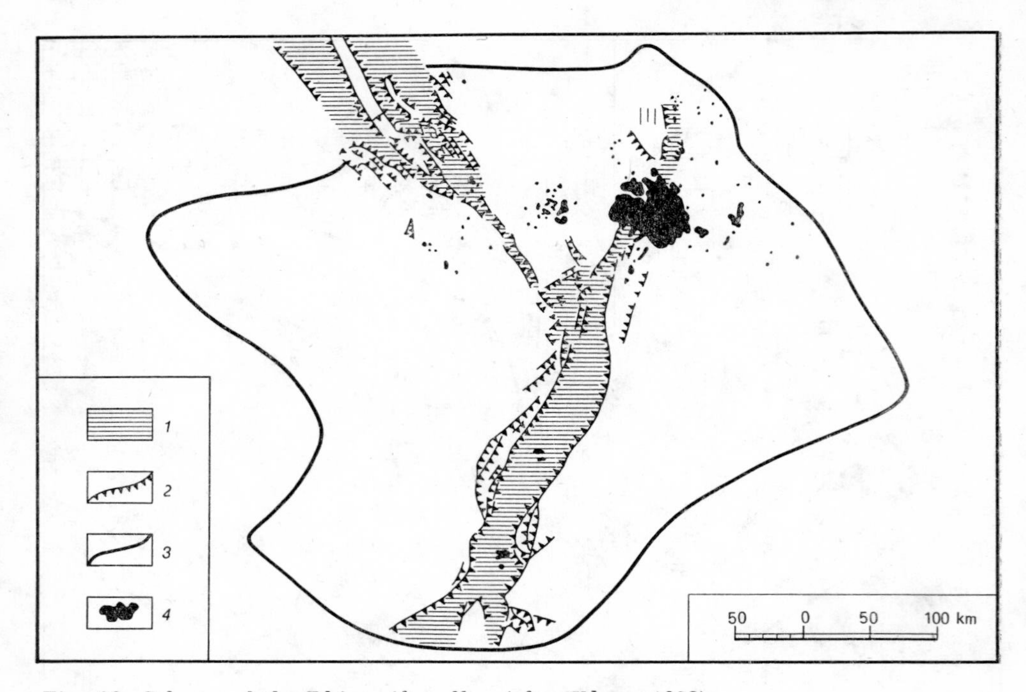

Fig. 92. Scheme of the Rhine rift valley (after Kloos, 1939):
1—grabens (*I*—Upper Rhine; *II*—Lower Rhine; *III*—Hesse); *2*—faults; *3*—Rhine arch contour; *4*—Neogene and Quaternary regions

The *Rhine rift valley* is divided into several separate grabens (Fig. 92). The Upper Rhine graben in the south which separates the massif of the Vosges Mts. in the west and the Black Forest Massif on the east. Its southern end forks near Basel and terminates at the folded zone of the Alps. North of Frankfurt-on-Main, this graben also splits into two branches, one of which (the Hesse graben) continues north and dies out in the area of Göttingen. The other branch (the Lower Rhine graben) trends north-west and terminates in Holland. The Rhine rift is about 600 km long. In the topography it forms a valley about 40 km wide over nearly its whole length.

The Rhine rift is superimposed on a folded Palaeozoic basement and the Meso-Cenozoic sedimentary cover of the epihercynian West European platform. It intersects Palaeozoic folded structures that mostly trend east-west. During the platform stage (in the Mesozoic and Palaeogene) the Rhine arch consisting of the Black Forest, Vosges, Rheinische Schiefergebirge, and Ardennes was formed. The arch is 250 by 350 km in area and has an amplitude of uplift of around 1 km. The Upper Rhine graben coincides with the axis of this arch, and the branching of its ends with periclines. In this case, therefore, as with the Nubian-Arabian arch, the pattern of faults is determined by the stress field associated with upwarping of the crust by vertical forces.

The Rhine rift began to form in the Late Eocene in its southern part, and gradually, during Oligocene, Neogene, and Quaternary times, the settling spread northward. The various stages of its formation were marked by outpourings of alkaline lava. The total amplitude of the subsidence of the grabens attains a maximum of 2500 m. The earthquakes that still occur indicate that the grabens are subsiding even now. Like the Baikal rift, the Rhine rift has no connection with the oceanic rift belt.

It was initially supposed that the formation of graben systems could be explained by stretching of the crust during the forming of the preliminary arches and the subsidence of crustal wedges in the gaps taking shape on the principle of the keystone of an arch, but calculations have shown that such a mechanism is insufficient to account for the observed amplitude of the grabens.

The separation that would have been formed on the East African arch through stretching of the crust alone, as a result of its upwarping, would not have exceeded several hundred metres. At the same time, judging from the average dip of the normal faults (60°-70°) and their vertical amplitude (up to 3000 m), it is readily established that the stretching required to form this rift is as much as several thousand metres.

More precise estimates for the Rhine rift valley have yielded a stretching of 4800 m while that due merely to the upwarping of the arch could not have exceeded a few hundred metres. In all cases the scale of the stretching corresponds approximately to 10% of the width of the tension zone. The subsidence of the grabens also cannot be explained, too, by the ejection of material from the deep interior during volcanic activity, because no relation is observed between the scale of the subsidence of grabens and the volume of volcanic products. The volcanism is much weaker in the Baikal rift, for example, than in the Gregory rift, but the amplitude of the subsidence is much greater.

It would appear that the Earth's crust is influenced by special tensile forces in rifts that ensure the necessary degree of stretching. It is this link between rifts and the dynamic state of tension that makes us classify them in a specific endogenous regime, distinct from the orogenic regime, in the development of which there are no signs of tension. Suggestions as to the nature of the tensile forces will be discussed later.

Regimes of Magmatic Activization of Platforms

Certain regimes were discussed above in connection with which both extrusive and intrusive magmatic activity developed in areas that were previously platforms, but in all these cases the origin of the magmatic processes on former platforms was closely associated with tectonic activization of the latter.

During the history of the Earth's crust, however, "magmatic activization" of platforms has repeatedly occurred, so to speak, in pure form unconnected with major tectonic events on the surface. It has occurred in some one area of either an ancient or a young platform during a relatively short span of geological time. Both before and after there were no manifestations of magmatism in the same area over an incomparably longer time, and a quiescent platform regime developed there.

Two regimes of magmatic activization of platforms can be distinguished.

1. The **plateau basalt, or trap, regime** plays a substantial role in the history of the Earth's crust. It is characterised by outpourings of huge volumes of very uniform tholeiite basalts among which rocks of another composition, products of the differentiation of basalts, play a quite minor role. The outpourings proceed steadily, through fissures, and the basalt sheets spread over vast territories. Their total thickness can attain several

kilometres. Parallel with them sills of the same basic composition (gabbro and diabase) are emplaced. In a region of the outpouring of plateau basalt, the crust is usually rather downwarped, and the effusive series therefore lie in a gentle syneclise. This is the essential difference between this regime and the rift regime in which, as we saw, breaking up of the crust and outpourings are preceded by the formation of arches.

Plateau basalts are known on ancient platforms from Proterozoic time, but massive phenomena of this regime actually first occurred at the end of the Palaeozoic and beginning of the Mesozoic. Interestingly, the outpourings of plateau basalts are almost fully synchronous (naturally in the geological time scale) in various regions of the globe.

On the ancient Siberian platform, plateau basalts and associated basic sills were formed at the end of the Permian and beginning of the Triassic. They cover an area of 1 500 000 million km^2 and are 2000 m thick. These igneous manifestations are mainly restricted to the Tunguska syneclise.

A plateau basalt complex was formed on the ancient platform in South Africa, in the Karroo syneclise, in Late Triassic-Early Jurassic times. Lavas cover an area of 26 000 square kilometres and are up to 1400 m thick. A much greater area (570 000 km^2) in the same region is occupied by dolerite sills of the same age. Some of the intrusives are up to 300 m thick.

In South America, also on the ancient platform, Rhaetian plateau basalts occur in the Parana syneclise.

In addition, plateau basalts of approximately the same age are developed in Tasmania and Antarctica, in both on the ancient platforms.

The next epoch of massive effusion of plateau basalts occurred in the Late Cretaceous and Tertiary Period. During the Late Cretaceous and Palaeogene the Deccan plateau basalts were poured out onto the ancient Indian platform. At present, they cover 500 000 square kilometres, but initially they occupied a larger area: a large part is submerged beneath the waters of the Gulf of Bombay. The Deccan plateau basalts are as thick as 2000 m.

Finally, the plateau basalts of the Columbian Plateau in the North American Cordilleras are of Miocene age. They are restricted to a median mass in a Mesozoic geosyncline.

2. The **regime of central intrusions and explosion pipes** is characterised by intrusives of the most varied composition, from ultrabasic and alkaline to acid, and by kimberlite explosion pipes. The intrusives of this type are nothing but deep parts of the roots of volcanoes.

The Mesozoic (post-Karroo) central intrusions in South Africa are most varied in composition. The numerous large and small intrusives of nepheline pyroxenite, ijolite, nepheline syenite, carbonatite, and olivinite of the Karelia-Kola province belong to the Middle and Upper Palaeozoic.

Alkaline intrusives of Mesozoic age are known on the Aldan Shield.

In South Africa, the Karroo formation is intruded by the Mesozoic Erongo and Branberg granite plutons, the latter being a stock 20 km in diameter.

As for kimberlite pipes, which attract much interest because they bear diamonds, they were first discovered in South Africa, where they are of Late Cretaceous age. Similar pipes were subsequently found in Central and East Africa, Brasil, and, finally in the east of the Siberian Platform.

All these manifestations of intrusive activity are confined to ancient platforms and mark temporary magmatic activization after a very long period of hundreds or even thousands of millions of years of a magmatic development of these platforms.

Regimes of Continental Margins

The margins of continents, including both the coast part of the land and the continental shelf, have special essential features of endogenous development and should therefore be classified in a specific group of regimes that reflect the transition from continent to ocean.

Two regimes of continental margins have been distinguished: the Atlantic and the Pacific.

The **Atlantic regime** is distinguished primarily by unconformable onlap of the margin of the ocean on the truncated pre-Mesozoic structure of neighbouring parts of the continent. This pre-Mesozoic structure may vary greatly. It may be a Hercynian folded zone, or an epi-Caledonian or ancient platform. It may consist of sedimentary, metamorphic, or igneous rocks. It is important that, by Mesozoic time, the Atlantic periphery was characterised everywhere by platform development. This type of continental margin predominates around both the Atlantic and Indian oceans and is the only one around the Arctic Ocean. These margins are accompanied generally with a broad shelf that may be hundreds of kilometres wide in some regions (see the tectonic map of the earth appended to this volume). At present, in a number of areas, both the land and underwater parts of Atlantic margins have been studied geologically and geophysically, and by drilling. The transition zone between the North American continent and the Atlantic Ocean especially has been studied in detail.

Research has shown that no continental margin of Atlantic type contains any indications that an ocean existed in the neighbourhood in pre-Mesozoic time. All the pre-Mesozoic structures of any type and origin are traceable from the land to the continental shelf and are then truncated at its outer edge. The character of the truncation leaves no doubt that these structures originally extended to where the oceanic depths now are. An example of this truncation of pre-Mesozoic structures on the Atlantic coast of Western Europe is given in Figure 93.

The geological section of the coastal part of the land and shelf indicates an abrupt change of conditions in the Mesozoic, when the first signs that an oceanic depression was located nearby are manifested. There are geological indications that the continental margin began to subside in Mesozoic time, the subsidence increasing toward the ocean. This is evidenced by sediments accumulated in the coastal zone and the distribution of their thicknesses. The boreholes drilled on various continental shelves and geophysical profiles have shown that Mesozoic sediments—shallow-water marine and partly continental—lie undisturbed on an eroded pre-Mesozoic basement of the shelf on Atlantic margins. They were laid down in a situation of compensation of the subsidence by accumulation, and their thickness increases toward the ocean, i.e., on the outer edge of the shelf. This means that the outer edge of the shelf subsided faster than the inshore part, and that its downwarping was accompanied with increased tilting of its basement toward the ocean. Passing to the continental slope, sedimentation occurrs in a situation of uncompensated deep-water accumulation and the thickness of the sediments rapidly diminishes. A profile through the Atlantic margin of North America is depicted in Figure 94.

Subsidence of the shelf began at different times in different areas. Judging by the age of the sediments at the base of the sedimentary cover of shelves, the earliest commencement of subsidence occurred in Early Jurassic time,

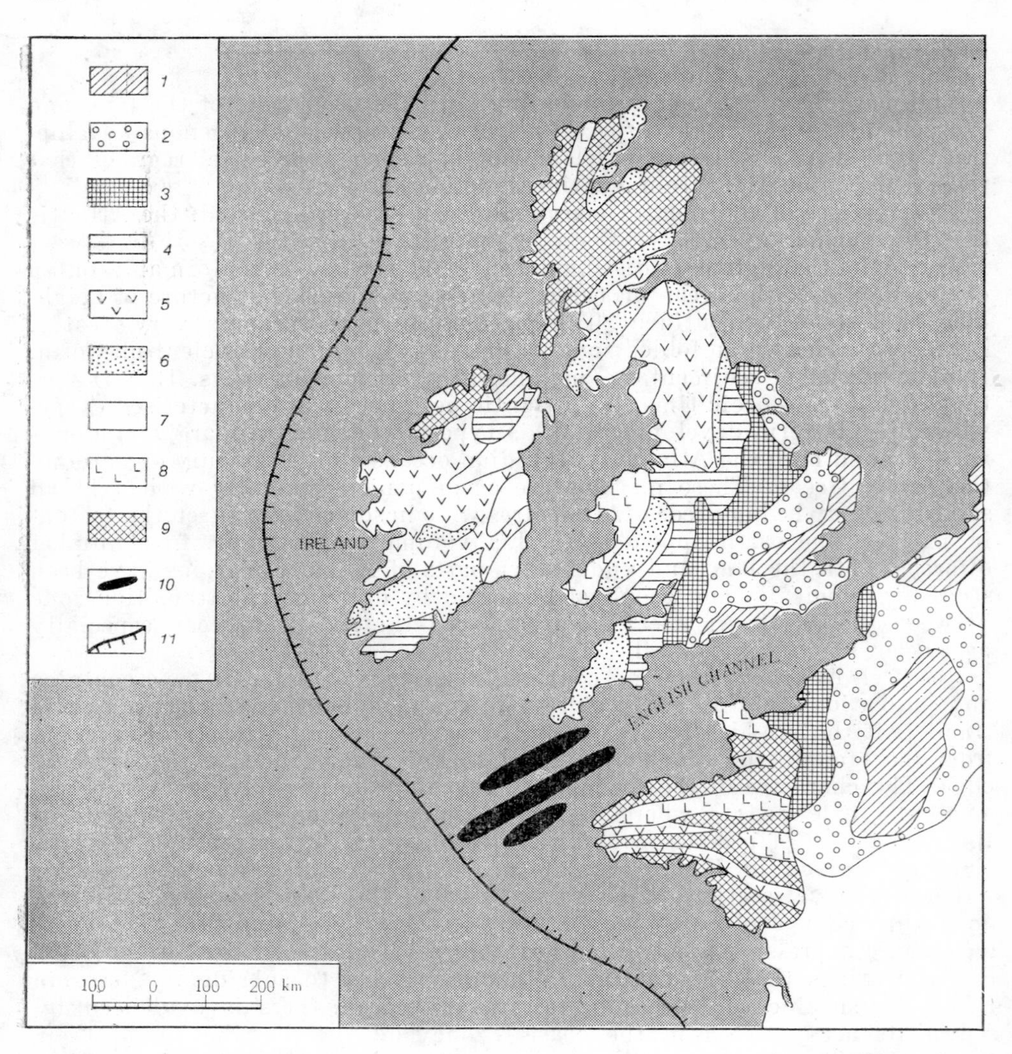

Fig. 93. The truncation of pre-Mesozoic structures of the margin of the Atlantic Ocean in Western Europe (after Beloussov):

1—Tertiary; *2*—Cretaceous; *3*—Jurassic; *4*—Triassic; *5*—Carboniferous; *6*—Carboniferous and Devonian not separated; *7*—Ordovician and Silurian not separated; *8*—Silurian, Ordovician, and Cambrian not separated; *9*—Precambrian; *10*—strike of magnetic anomalies in the English Channel; *11*—edge of the continental shelf

but subsidence and tilting of the shelf began much more frequently in the Early Cretaceous, and all the shelves on peripheries of this type have been involved in subsidence since the beginning of the Late Cretaceous. The amplitude of subsidence of the shelf's basement on its outer edge is generally 2000-3000 m.

Below the shelf, on the continental slope, scarps have been found in places at depths of 1000 to 2000 metres, so-called marginal plateaux. These plateaux are hundreds of kilometres wide; the best studied one is the marginal Blake Plateau, off the coast of Florida (see Fig. 95). It is 900 km long and 300 km

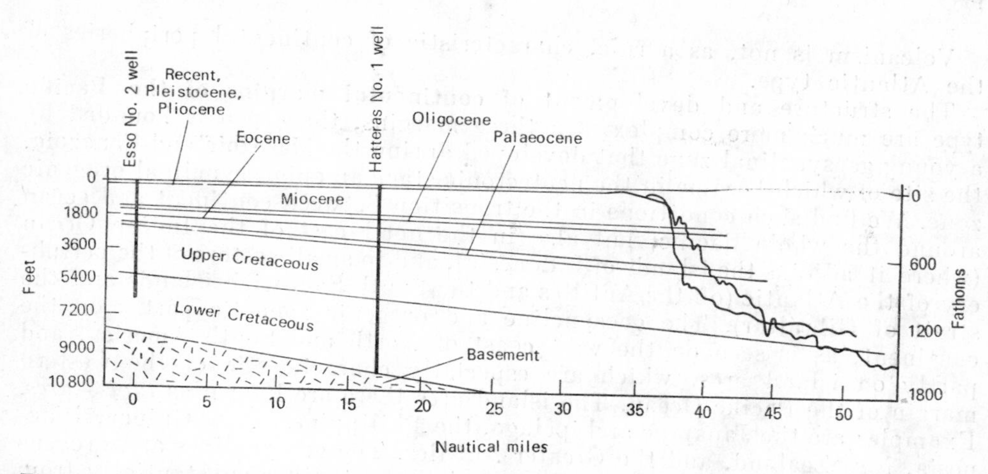

Fig. 94. Section through the shelf of the Atlantic coast of North America near Cape Hatteras (after Heezen *et al.*, 1959)

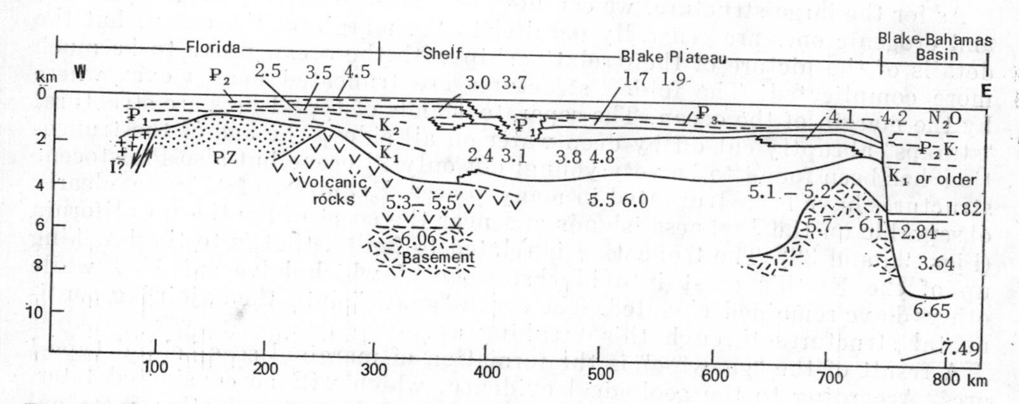

Fig. 95. Section through the continental shelf, marginal Blake Plateau, and continental slope off the coast of Florida (after Sheridan *et al.*, 1966). The figures are p-wave velocities, km/s

wide and varies in depth from 600 m in the north to 1200 m in the south. On the opposite coast of the Atlantic, off the Iberian Peninsula, there is the marginal Iberian Plateau 250 km wide, at depths between 1800 and 2700 m. Other marginal plateaux have been found in both the South and North Atlantic.

Drilling on the Blake Plateau has indicated that it was part of the shelf until Eocene time. Evidence of this is the shallow-water Early and Late Cretaceous sediments, laid down on its eastern bench. The shelf was therefore much broader than at present. The Eocene and younger sediments are already deep-water ones which means that the marginal plateau broke away from the shelf and sank at the end of the Cretaceous and beginning of the Eocene. The Iberian Plateau also subsided, apparently, at the same time. The subsidence of marginal plateaux is a process that disturbs the steady increase in the tilting of the shelf toward the ocean.

Volcanism is not, as a rule, characteristic of continental peripheries of the Atlantic type.

The structure and development of **continental margins of the Pacific type** are much more complex. On these margins, the ocean is bounded by a young geosynclinal zone that developed during the Mesozoic and Cenozoic, the site of which became in the neotectonic stage an epigeosynclinal orogenic zone. We find such conditions in the transition between continent and ocean around the whole Pacific, but also in the north-east of the Indian Ocean (where it adjoins the islands of Indonesia), and in small sectors of the periphery of the Atlantic (off the Antilles and the South Sandwich islands and the strait of Gibraltar). The geosyncline and orogenic zone lie partly on the continent, as is seen on the west coast of North and South America, and partly on island arcs, which are especially characteristic of the western margin of the Pacific Ocean. The island arcs there are known as *type-I arcs.* Examples are the Japanese archipelago, the Philippines, New Guinea, Indonesia, New Zealand, and the Greater Antilles. The geosynclines and orogens of type-I island arcs, it must be stressed, do not differ substantially from those located wholly on the continents. They develop along the same lines, and the structures formed in them are the same.

As for the large structure, we can note that Mesozoic-Cenozoic geosynclinal and orogenic ones are generally parallel to the margin of the ocean, but the details of the picture of their relationship with the ocean prove to be much more complicated. The folded structures are truncated nearly everywhere by the margin of the ocean. The separate islands of type-I arcs are structural "stumps" abruptly cut off by deep water on all sides. The age of the truncation has been found to be very young; not only Pliocene but also Pleistocene structures have been truncated in many places. Examples of this are clearly observable on the Japanese islands and near the coast of Southern California (Figs. 96 and 38). The truncation is related to faults, that is to the breaking up of the Earth's crust into blocks, some of which have subsided while others have remained elevated. The ocean "gnaws" into the adjoining continental structures through this breaking up.

A result of the "gnawing" is the formation of marginal seas behind type-I arcs. According to the geological evidence, which will be considered later, the Bering Sea, the Sea of Okhotsk, the Sea of Japan, and other marginal seas on the western periphery of the Pacific are very young. Their formation occurred predominantly in the Neogene and even Quaternary. It is through their formation that type-I island arcs are isolated from the continent. In some cases, however, the process of separation has not yet been completed; Kamchatka, for example, with its young geosyncline and orogen, is an "unfinished" type-I island arc that still retains its connection with the continent.

More pronounced indications of deep break-up of the crust are also observable on peripheries of Pacific type.

The geological data indicate that very big faults have been formed around the Pacific since the Late Jurassic, and have become channels for the intensive volcanic activity. In Late Jurassic time the Okhotsk-Chukotka fault was formed, which was a zone of the outpouring of a huge volume of andesite and andesite-basalt lava at the end of the Jurassic and during the Cretaceous. In Late Cretaceous time outpourings of andesite also occurred along the western coast of the Sea of Japan, in southern Korea, and in the south-east of China. The lines of the faults associated with these outpourings can

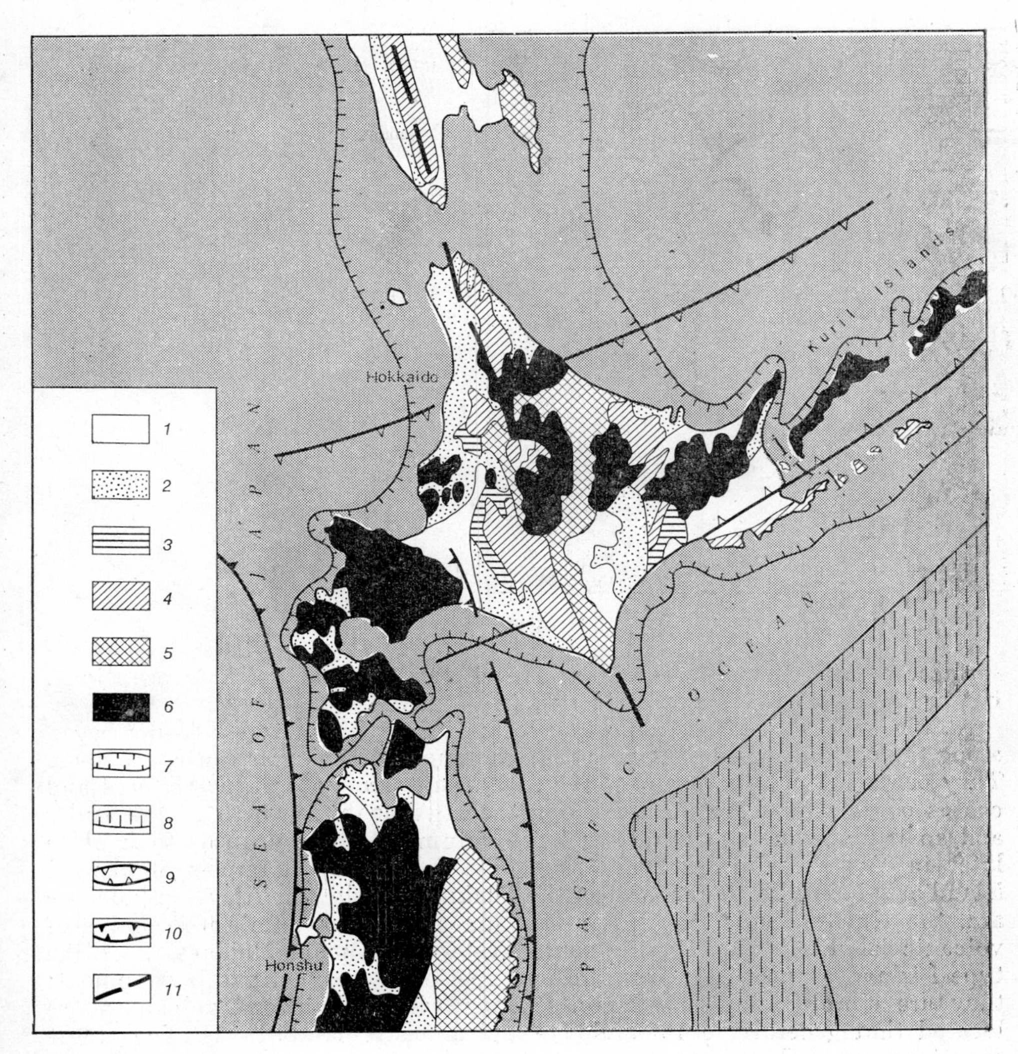

Fig. 96. Scheme of the structure of Hokkaido (after Beloussov):

1—Quaternary; *2*—Neogene; *3*—Palaeogene; *4*—Cretaceous and Jurassic; *5*—Upper Palaeozoic; *6*—Quaternary and Neogene volcanic outpourings; *7*—boundary of the shelf; *8*—deep trench; *9*—boundary of Kurile island arc (type II); *10*—boundaries of Japanese island arc (type I); *11*—axis of Sakhalin-Hokkaido folded zone

be considered the first tracing of the boundary along which marginal seas will later be formed.

On the other side of the Pacific, in the Andes and Cordilleras, the Late Jurassic and Cretaceous were the time of the formation of a huge granodiorite batholith that stretches along these mobile zones for thousands of kilometres. This batholith can be explained by a deep break-up of the tectonosphere, that involved the upper layers of the mantle and deep layers of the crust, thereby creating conduits for the rise of hot magmatic asthenoliths from the mantle into the crust. This break-up did not, however, reach the earth's

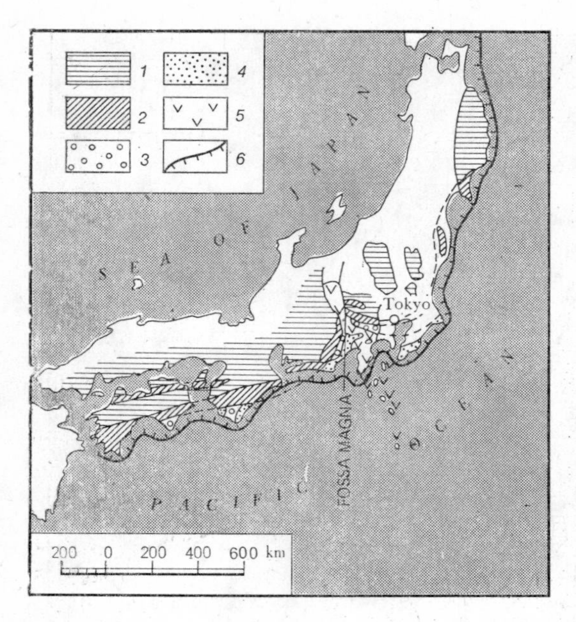

Fig. 97. Scheme of the structure of part of the Japanese archipelago (after Beloussov):

1—Palaeozoic; *2*—Mesozoic; *3*—Palaeogene; *4*—Neogene; *5*—Neogene and Quaternary volcanic rocks; *6*—edge of the shelf

surface, so that the process did not reach the stage of volcanic eruptions, but remained at the stage of intrusion.

At the end of the Cretaceous the formation of big new break-ups began along the periphery of the Pacific that were now closer to its central regions. They occur mainly in the form of gently bent, known as type-II, arcs and chains of volcanoes are formed along them. A further growth of the faults and an increase in the number and the volume of the volcanoes took place later in Miocene and Pliocene times. We find many examples of type-II island arcs in the north and west of the Pacific, namely, the Aleutian volcanic arc, the Kuril Islands, the Bonin-Mariana arc, and a number of chains of volcanic islands located in the south-western Pacific in Melanesia. Unlike type-I island arcs, type-II arcs did not experience geosynclinal development: they are simply deep-seated faults along which chains of volcanoes are located that generally erupt andesite and andesite-basalt lavas. The oldest eruptions are of Eocene age; many of the volcanoes are still active.

A necessary element of a type-II island arc is a deep trench bounding it on its outer side. These trenches are the deepest zones in the ocean. They are about 100 km wide, the slope adjoining them is steep while the opposite side is gentler. The trenches are generally more than 7000 m deep and in one of them, the Mariana Trench, a record depth of 11 023 m has been registered.

The relationship between the two types of arc presents special interest. Type-II volcanic arcs are now and again superimposed on type-I arcs, so that chains of volcanoes erupting andesite and andesite-basalt lava arise on them. In this case, the volcanic arcs may be superimposed on the geosynclinal and orogenic structures of the type-I arcs at various secant angles. Examples are the volcanic zone on Honshu in Japan, known as the Fossa Magna, which lies precisely on the continuation of the type-II Bonin-Mariana arc, which approaches the middle of the island, near the Iju Peninsula at almost a right angle (Fig. 97). There is no doubt that the Fossa Magna, which intersects the folded structures on Honshu at a right angle, is an extension of the Bonin-Mariana arc. Structurally, it is a narrow, deep trough filled with Miocene sediments (clays and volcanic rocks). Downwarping ended in Middle Miocene time, when Miocene sediments were buckled into folds during inversion of the trough. Large andesite volcanoes were formed here in Quaternary time (Fuji, Hakone, and others).

Another example of a type-II arc superimposed on a type-I arc is to be found on the Hokkaido, where a chain of volcanoes intersects folded and orogenic structures at an acute angle (see Fig. 96). In other cases the volcanic arc may be superimposed along the strike of geosynclinal and orogenic structures, as can be seen on Kamchatka, where a chain of andesite volcanoes follows the trend of the structural zones of a young geosyncline and folds. In these cases a deep trench lies along a type-I arc.

Parallel with the formation of type-II island arcs multiple break-up of the crust around the Pacific Ocean is manifested in another way. The huge volumes of volcanic outpourings of the Cascade complex in the North American Cordilleras, which on the whole spanned Late Eocene-Quaternary time have been noted. Since the end of Miocene time the folds of Southern California have been strongly faulted. Extremely vigorous andesite volcanism goes hand in hand with the history of the Andes in Eocene, Miocene, and Quaternary times.

A similar complex of phenomena, including forceful break-up of the crust, vertical block movements of the crust, and intense volcanism also characterised the Mesozoic-Cenozoic evolution of other continental peripheries of Pacific type.

The regime of Pacific continental margins therefore differs substantially from that of the Atlantic type. On Atlantic margins, there was a progressive subsidence of the coastal zone during the Mesozoic and Cenozoic, with the basement beneath the shelf gradually tilting toward the ocean. On Pacific margins, however, the subsidence of the land, although it is the main background of development, was not quiet and gradual, but occurred through faulting and the subsidence of separate blocks of the crust along the faults so formed. While the Atlantic regime is practically devoid of volcanism, the Pacific-type regime, on the contrary, is characterised by extremely vigorous, mainly extrusive magmatism as a necessary component. The composition of the magmas is specific: andesites and andesite-basalts are decidedly predominant in outpourings, and granodiorites in intrusives.

A quite common misunderstanding must be pointed out here that is due to no distinction being made between the island arcs of the two types, in particular between the geosynclinal and orogenic structures of the former and the purely volcanic structures of the latter. It is thus supposed that the andesite volcanoes of Japan, the North American Cordillera, and the South American Andes belong to a cycle of orogenic epigeosynclinal development and that they are directly linked with the long previous history of these arcs.

It follows from all the geological data, however, that the andesite volcanism is a process superimposed on the geosynclinal and orogenic development of type-I arcs, and superimposed not earlier than Eocene time, and mainly from the Miocene epoch. This process of superimposition belongs to the Pacific-type marginal regime, whereas the preceding geosynclinal and

Table 3. Endogenous Regimes

Class of regimes	Regime	Stage of development	Contrast and Intensity (rate and amplitude) of oscillatory movements	Relationship between subsidence and uplift	Predominating sedimentary formations	Character and degree of the crust's penetrability
Geosynclinal	Eugeosynclinal	Ophiolitic	High	Considerable preponderance of subsidence; uncompensated downwarping	Lower terrigenous and deep-water siliceous (jasper)	Diffuse, high
		Inversion	Moderate	Equilibrium between subsidence and uplift; slight preponderance of uplift at the end of the stage	Upper terrigenous (flysch, caustobiolithic)	Concentrated, low
	Miogeosynclinal	Compensated downwarping	Moderate	Preponderance of subsidence	Lower terrigenous (slate followed by shallow-water limestone)	Diffuse, low
		Inversion	Low and moderate	Equilibrium between uplift and subsidence; slight predominance of uplift at the end of the stage	Upper terrigenous (flysch, caustobiolithic)	Concentrated, low
	Parageosynclinal		Low and moderate	Preponderance of subsidence; slight preponderance of uplift at the end of the cycle	Sandy-clay (lower terrigenous) followed by limestone, which frequently predominates; changes upward into upper terrigenous including flysch and caustobiolithic	Concentrated, low

Characteristic magmatism	Regional metamorphism and granitization	Type of folding	Characteristic type of fault	Principal minerals	Examples
Basic underwater outpourings and sills (stocks, laccoliths); ultrabasites (ophiolites); at the end of the stage—intermediate and acid outpourings and plagiogranite intrusions	Propylitization	Folding not characteristic; block folds on intrageanticlines	Deep slash faults	Sulphide ores of copper, zinc, and lead; igneous deposits of chromite, platinum and titamagnetite	The Sayans in the Early Palaeozoic; the eastern slope of the Urals in the Early and Middle Palaeozoic; the Alps in the Jurassic and Early Cretaceous; the Sierra Nevada in the Triassic
Granite batholiths	Regional metamorphism and granitization (batholiths)	General crumpling and deep folding	Thrust faults, nappes of Helvetian and Pennine types	Skarn deposits of tungsten; hydrothermal deposits of gold, copper, molybdenum, lead, and zinc and pegmatite; deposits of tin and tungsten	The Sayans, Northern Tien Shan in the Silurian; the eastern slope of the Urals in the Carboniferous; the Alps in the Palaeogene; the Sierra Nevada in the Late Jurassic
Few or no manifestations of basic outpourings and sills	Absent	Absent or weak block-type folding	Deep slash faults	Disseminated ores of vanadium, molybdenum, and radioactive elements; phosphorites	The Great Caucasus in the Mesozoic; the western slope of the Urals in the Early and Middle Palaeozoic; the Late Balkhash zone in Central Kazakhstan in the Middle Palaeozoic
Granite; small intrusions, normal and alkaline	Moderate greenschist metamorphism and granitization	Block and injection and to a less extent general crumpling	Thrust faults, reverse faults, and nappes of Helvetian and Pennine type	The same as in the inversion stage of eugeosynclines	The Great Caucasus in the Oligocene; the western slope of the Urals in the Middle Carboniferous; the Verkhoyansk-Kolyma zone in the Middle Jurassic
Limited magmatism manifested as basic dykes, sills, alkaline laccoliths, and igneous diapirs	Absent	Block and injection	Reverse and slash faults	Coal, oil, rock salt and potassium salts; hydrothermal deposits of mercury, antimony, tin, gold, lead, and zinc deposits of radioactive and rare earth elements	The Donets Basin; Limestone Daghestan; the Pyrenees in the Alpine cycle

Class of regimes	Regime	Stage of development	Contrast and Intensity (rate and amplitude) of oscillatory movements	Relationship between subsidence and uplift	Predominating sedimentary formations	Character and degree of the crust's penetrability
	Median mass		Low	Equal development of slight uplift and subsidence	Sandy-clay and limestone	Concentrated, high
Orogenic	Epigeosynclinal		High	Considerable preponderance of uplift	Molasse and lagoonal (including caustobiolithic)	Concentrated, high
	Epiplatform (tectonic activization of platforms)		High and moderate	Considerable preponderance of uplift	Molasse and lagoonal (including caustobiolithic)	Concentrated, high to moderate
Rift			Moderate and high	Subsidence of grabens on a broad gently sloping uplift	Molasse and lagoonal	Concentrated, high
Platform	Ancient platform		Very low	On the whole equilibrium, with a slight preponderance of subsidence at the beginning of the cycle and of uplift at the end	General ascending sequence of formations: (a) lower terrigenous (including bituminous); (b) limestone; (c) upper terrigenous (including caustobiolithic); (d) molasse and lagoonal	Impermeable

TABLE 3 (continued)

Characteristic magmatism	Regional metamorphism and granitization	Type of folding	Characteristic type of fault	Principal minerals	Examples
Various extrusive rocks, basic and acid, and also alkaline; fissure intrusives, basic, acid, and alkaline	Absent or weak	Block	Slash and reverse faults	Hydrothermal, metasomatic, and skarn deposits of base, rare, and precious metals	The Armenian massif in Transcaucasia; the Hungarian massif; the Macedonia-Rhodope massif
Preponderance of outpourings of intermediate and acid lavas; basic, intermediate, and acid fissure intrusives	Absent	Block and injection; local manifestations of general crumpling	Deep slash faults, normal, reverse, and thrust faults	Hydrothermal deposits of base, rare, radioactive, and precious metals; skarn lead-zinc deposits; sedimentary deposits of iron, copper, vanadium, and uranium; rock salt and potassium salts, gypsum, anhydrite; bauxites, coal, gas, and oil	The Alps, Caucasus, Apennines, and Dinaric Alps in the neotectonic stage
Basic and alkaline outpourings; fissure intrusives of the same composition	Absent	Block and injection; local manifestations of general crumpling	Deep-slash, faults, normal, reverse, and thrust faults	Hydrothermal deposits of tin, tungsten, lead, zinc, mercury, and antimony; bauxites; placers of gold, monazite, etc.; rock salt and potassium salts; coal, gas, and oil	The Tien Shan, Altai, Sayans, Kunlun Shan, the mountains of Scandinavia and Scotland in the neotectonic stage
Basic and alkaline outpourings	Absent	Block	Slash, normal, and reverse faults (main form of structure—grabens)	Thermal waters; diatomite, coal	The Arabian-African rift belt; Baikal and Rhine rifts; rift of the Great Basin
Weak or no manifestations of basic and alkaline outpourings and minor intrusions	Absent	Block and injection	Slash, normal and reverse faults	Coal, gas, and oil; rock and potassium salts; sedimentary deposits of copper, iron, and manganese	The East European, Canadian, Siberian, and African platforms

Class of regimes	Regime	Stage of development	Contrast and Intensity (rate and amplitude) of oscillatory movements	Relationship between subsidence and uplift	Predominating sedimentary formations	Character and degree of the crust's penetrability
	Young platform		Low	On the whole equilibrium, with a preponderance of subsidence at the beginning of the cycle and of uplift at the end	General sequence as with ancient platforms	Concentrated, very low
Magmatic activization	Plateau basalt		Very low	Slight preponderance of subsidence		Concentrated, moderate or high
	Central intrusions and explosion pipes		Very low	Indefinite		Concentrated, moderate or high
Continental margins	Atlantic		Low	Subsidence predominates	Sandy-clay, lagoonal, caustobiolithic	Impermeable
	Pacific		Very high	Subsidence predominates	Sandy-clay	Concentrated, high

Note. The present writer used Academician Smirnov's suggestions in filling the twelfth

TABLE 3 (concluded)

Characteristic magmatism	Regional metamorphism and granitization	Type of folding	Characteristic type of fault	Principal minerals	Examples
Slight manifestations of basic outpourings, fissure intrusions, and laccoliths of alkaline composition	Absent	Block and injection	Slash, normal and reverse faults	Coal, gas, and oil; rock and potassium salts; sedimentary deposits of copper, iron, and manganese	The Urals-Siberian epihercynian platform; the West European epihercynian and epicaledonian platform
Plateau basalts	Absent	Absent	Normal and slash faults	Igneous deposits of nickel and copper; skarn iron deposits; bauxites of the mantle weathering	The Deccan Plateau in India; the Tunguska and Parana syneclises
Intrusions—(fissure-type, laccoliths, and igneous diapirs—of various composition (mostly alkaline); kimberlite pipes	Absent	Block	Normal and slash faults	Diamonds; skarn and hydrothermal deposits of iron, rare earths and apatite	The Kola Peninsula on the Baltic Shield; the Anabar Shield; South Africa
Absent	Absent	Block and injection	Weak development of normal and slash faults	Gas, oil, and rock salt; placers of rutile, ilmenite, zircon, and monazite	The eastern margin of North and South America; the northern margin of Eurasia
Vigorous terrestrial outpourings of andesite and andesite-basalt; granodiorite and granite intrusions	Slight	Block, general crumpling	Strong development of normal and slash faults; deep thrust faults	Volcanogenic and hydrothermal deposits of gold, silver, zinc, tin, tungsten, copper, mercury, sulphur, etc.; placers of rutile, ilmenite, zircon, and monazite	The western margins of North and South America; the zones of the island arcs of East and South-East Asia

graph of the table.

orogenic regimes of type-I arcs belong to normal geosynclinal and orogenic regimes quite similar to those in any other part of continents.

We have come to the conclusion that both marginal regimes consist essentially in subsidence of the Earth's crust. In the one case, however, this happens through steady bending of the crust, with a very minor role for faulting; in the other case it is fully related to deep break-up of the crust and block collapse, and is accompanied by very vigorous magmatic activity.

The data on the features of all the continental endogenous regimes listed above are summarized in Table 3.

CHAPTER 18

Principal Patterns of the Endogenous Development of Continents

General Evolution of Endogenous Regimes

The history of endogenous regimes can be reconstructed by studying rocks of different age, beginning with the oldest, and signs of the tectonic, magmatic, and metamorphic processes that have affected them. The most ancient rocks are found on crystalline shields like the Baltic, Canadian, Anabar, Guyanan, Brazilian, and African.

The data on the geological structure of shields is still, unfortunately, extremely incomplete. The absence of identifiable fossils in the ancient rocks, the very complex deformation, the quite ubiquitous metamorphism and granitisation greatly complicate the solution of both stratigraphic and tectonic problems. Opinions on the age of the rocks are based mainly on absolute determinations, which, as we know, cannot always be interpreted unequivocally.

In spite of these difficulties, the progress in study of the geology of shields has been great in recent decades and has made it possible to divide the Precambrian, albeit roughly, into main stages differing in character of endogenous development.

A feature of the structure of shields is that the crust underwent multiple metamorphism, granitization, igneous intrusion, partial fusion, and complex deformation within them during the Precambrian. All the processes listed were largely repeated in one and the same region, that is, similar processes reworked the same crustal material many times. Hence, we may speak of repeated endogenous pulses.

Beyond the regions of such repeated superimposition of similar processes, however, areas survived each time to which the recurrent metamorphic, magmatic, and tectonic processes did not penetrate, or did so only slightly. These regions that were not involved in a routine pulse of the endogenous processes and retained the results of a similar previous pulse intact in its structure are obviously of the greatest interest for reconstructing the history of development and distribution of endogenous regimes, because they help us trace what was introduced new into the crust by each new endogenous pulse. It proves that the deeper we descend in geological history, the smaller is the area of the segments that have retained a structure intact that is older than that around it. As we approach the end of the Precambrian, the sectors preserving relatively older structures become bigger and bigger.

For examples let us turn to four shields: the Baltic, Ukrainian, Canadian, and African. Schemes of their structure are given in Figures 98, 99, and 100.

The most ancient rocks on all the four shields vary in age from 3000 to 3500 million years, and in general are the oldest crustal rocks so far known. On the Baltic Shield they are exposed as blocks of biotite-hypersthene and biotite-amphibolite gneisses, and of amphibolites (tens of kilometres wide),

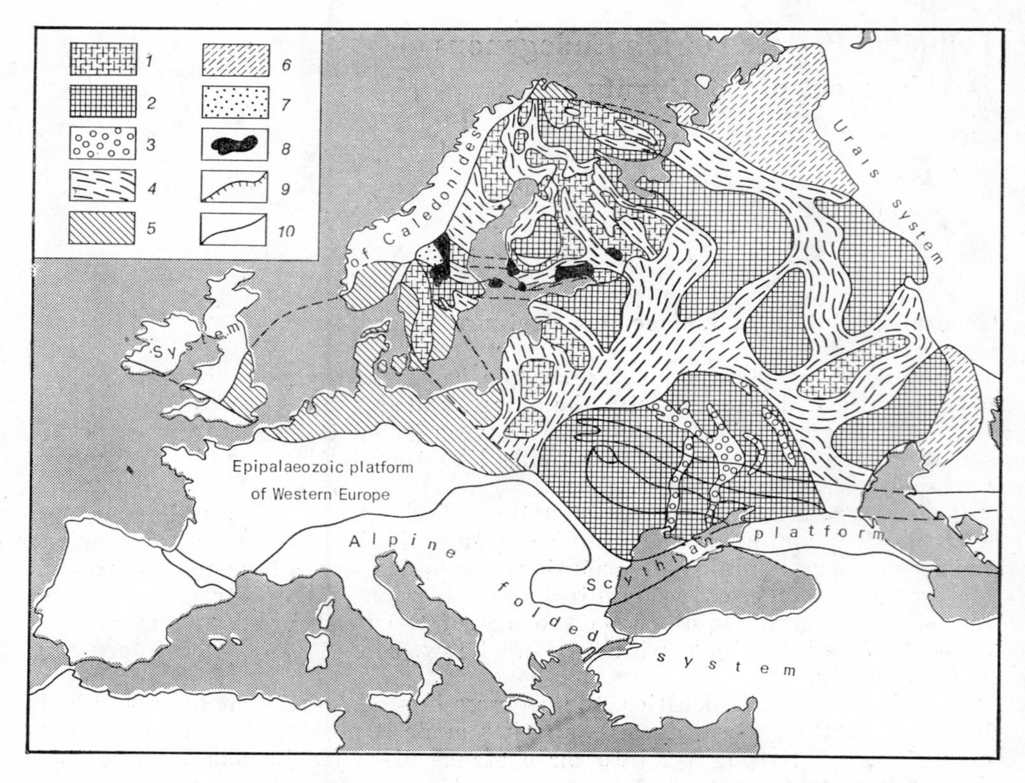

Fig. 98. Scheme of the structure of the folded basement of the East European platform (after Muratov, 1966). Beyond the Baltic and Ukrainian shields, beneath the sedimentary cover, the contours are extrapolated from the geophysical data and rare boreholes:

1—gneiss massifs and relicts of the protoplatform; *2*—similar, but smaller massifs reworked by Karelian and Sveccofennian granitoids; *3*—troughs of the Saxaganide system; *4*—geosynclinal troughs of Karelides and Sveccofennides; *5*—Gothlandian system; *6*—system of Timanides (Baikalides); *7*—areas underlain by Jotnian sandstone; *8*—Rapakivi granites; *9*—contour of the sedimentary cover; *10*—boundary of the East European platform

i.e., mainly metamorphosed basic volcanic rocks. Blocks of basic metarocks, 3200 million years old, have also been found on the Ukrainian Shield.

On the Canadian Shield, the most ancient rocks form two massifs, the Lake Superior and the Yellowknife, and consist of gneisses, and also of gabbro and peridotites of the Keewatin suite, more than 3000 million years old. In Africa, rocks of the same age, consisting of various gneisses, and also of quartzites and basic igneous rocks converted into amphibolite, are exposed within two massifs, the North African and South African. They have been studied best in Southern Rhodesia, where they are part of the Sebakwa and Bulawayo systems.

On all shields, the most ancient rocks have been strongly granitised. The age of the granites and migmatites is 2700 million years to 2800 million years.

As a whole, the characteristic ubiquitous features of these rock series are the following: strong regional metamorphism to the granulite facies in places (although the amphibolite facies predominates), granitization, and complex deformation. The deformation is represented by two basic types: (1) domes

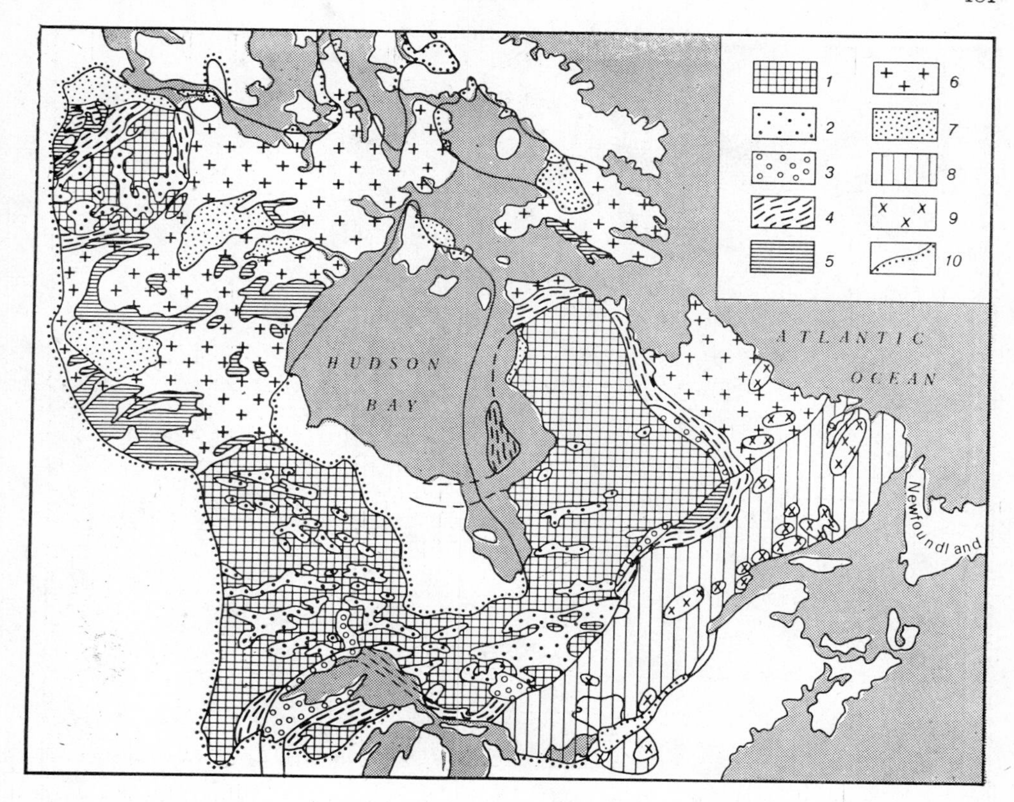

Fig. 99. Scheme of the structure of the central Canadian Shield (after Muratov, 1966):
1—protoplatform—Archaean granites; *2*—areas and strips of folded Archaean rocks of the Keewatin series, the Timiskaming and other troughs; *3*—the oldest sedimentary cover of the protoplatform, corresponding to the Huronian; *4*—folded Huronian (Lower Proterozoic) rocks; *5*—gneisses and other rocks involved in Hudsonian folding; *6*—Hudsonian granites and gneisses with inclusions of more ancient rocks; *7*—sedimentary cover of the Middle to Late Proterozoic (post-Hudsonian); *8*—gneisses and other metamorphic rocks, and granites of the Grenville system; *9*—anorthoclase intrusives (anorthosites); *10*—boundary of the sedimentary cover of the North American platform

and swells formed by gneisses, granite gneisses, and migmatites; and (2) very complex disharmonic folding, a feature of which is the superimposition of folds of different orders and strikes. The folding is developed in strips between the domes and swells. An absence of lineation is characteristic of the whole structure: the domes are round or irregular in shape, the swells bend along their strike and often form complex loops. Even less can be said of any definite strike of the folds, for which vertical hinges are normal. An example of such a structure was shown in Figure 67. This folding has to be classed as deep-seated; it occurs in a situation of low viscosity of the rocks because of the effect of volumetric forces causing a continuous flow of the material.

Since strong deformations, metamorphism, and granitization are observed in these rocks wherever they have survived, it can be supposed that the same features of composition and structure characterised them where they were subsequently reworked by tectonic, magmatic, and metamorphic processes. This suggestion is based, of course, on observations of the very

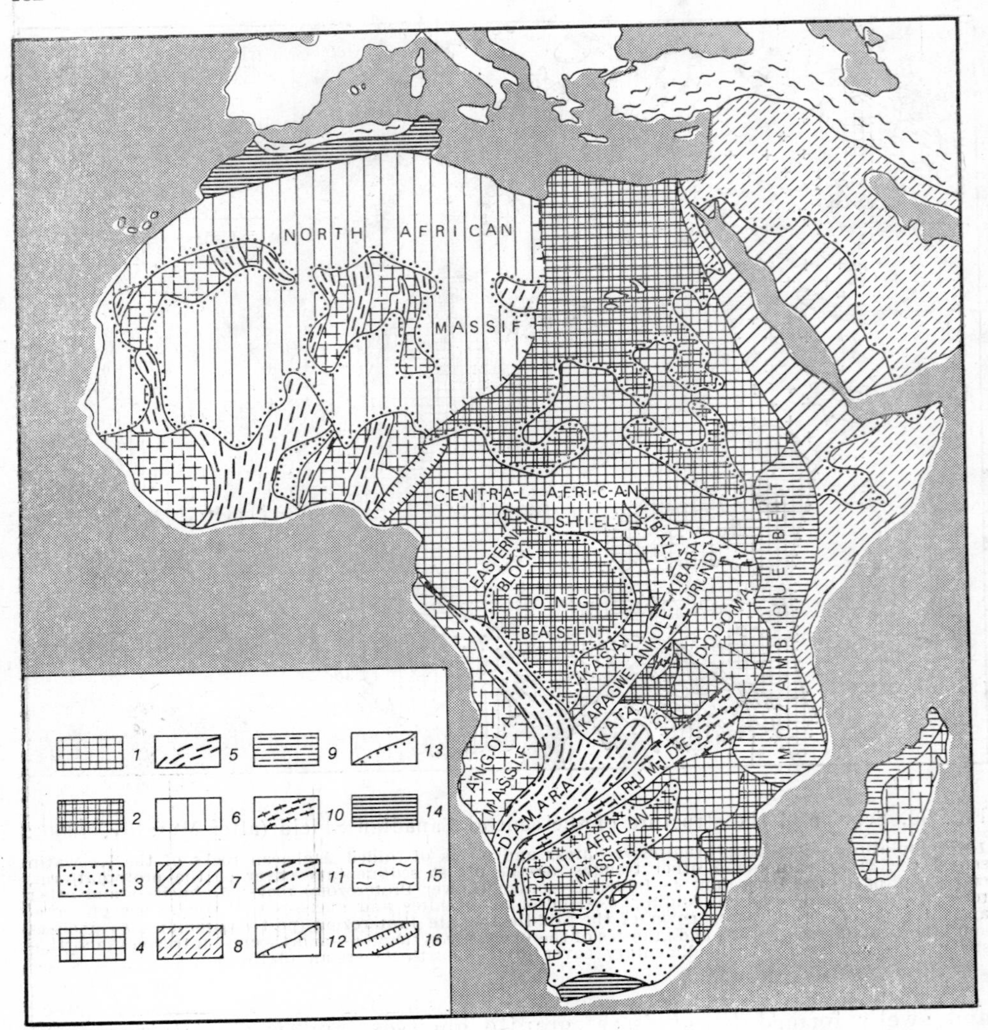

Fig. 100. Scheme of the structure of the folded basement of the African platform (after Muratov, 1966):

1—the protoplatform of South Africa and the eastern block of North Africa, and separate massifs older than 2500 million years; *2*—same beneath sedimentary cover; *3*—Lower Proterozoic cover on the protoplatform of South Africa; *4*—ancient massif of the western block of North Africa, and the Kibali-Lukoshi, Mayumba, Angola, and other massifs; *5*—graben-like basins filled with Birrimian and Faruzian formations in the western block of North Africa (1820 to 1730 m.y.); *6*—same and ancient massifs of the western block of North Africa beneath sedimentary cover; *7*—the late Precambrian folded system of the Red Sea coast; *8*—same beneath sedimentary cover; *9*—the Mozambique Belt; *10*—the folded systems of the Irumides and Karagwe-Ankole-Kibara-Urundi (900-1200 m.y.); *11*—the Katanga and Damara systems; *12*—major Precambrian faults; *13*—contours of the chief areas of the sedimentary cover; *14*—Palaeozoic folded structures of the Cape system and Mediterranean belt of North Africa; *15*—the Alpine system of North Africa; *16*—the Mesozoic Benue graben

limited areas of Archaean outcrops, but since they occur on all shields and are the same everywhere, it seems plausible.

It can therefore be concluded that between 3500 and 2800 to 2700 million years ago, i.e., through Archaean time, conditions seemingly prevailed everywhere on continents that meet the criteria of eugeosynclines. Metamorphosed basic igneous rocks, which may represent the ophiolitic stage of the development of geosynclines are in fact widely developed among Archaean rocks; they experienced strong regional metamorphism, granitization, and complex folding, which can be regarded as the mark of the inversion stage. The occurrence everywhere of manifestations of eugeosynclines permits us to call this stage in the crust's development, the oldest known to us, *pangeosynclinal* or *permobile*.

The absence of lineation in the disposition of structures mentioned above, which is characteristic of this stage, must be ascribed to the complete absence, or very minor role of deep-seated faults, a circumstance that accords with the character of the deformations, which indicates a very plastic state of the rocks. The main tectonic form is a deep-seated diapir in the form of a granite-gneiss dome or swell, but folding in the metamorphic schists bordering them is mainly associated, it would seem, with the bursting pressure of the diapirs on the country rock.

The granitisation, which is confined to the epoch 2700 to 2800 million years ago, led to the formation of a crystalline shell in the crust that seems to have overlain vast areas, extending far beyond the limits of the sectors where it has survived. The shell, however, was later broken up tectonically and troughs were formed on the site of the breaking-up which became filled with sedimentary and volcanic rocks subsequently transformed into metamorphic rocks, mainly of the amphibolite facies, i.e., various crystalline schists and gneisses, jaspilites are also characteristic. These rocks are not only metamorphosed, but are also granitized and complexly deformed. Their age varies between 2600 and 1900 million years. This applies to the White Sea gneiss suite of Karelia and the Kola Peninsula and the Krivoi Rog series of the Ukrainian Shield. On the Canadian Shield the Timiskaming series (around 2500 million years old) belong to this stage. On the African Shield bands of conglomerates, quartzites, and phyllites are known about 2700 million years old. Metamorphic rocks, 2100 to 1800 million years old, are also developed there (the Ubendi-Ruzizi system).

All these rocks belonging to the Early Proterozoic, form bands or strips bordered by the Archaean blocks about which we spoke above. The blocks are sometimes so small that they look like "islands" among the bands of Early Proterozoic rocks. In other cases, on the contrary, the bands of Early Proterozoic rocks are only a score or so kilometres wide; the outcrops of the Krivoi Rog series on the Ukrainian Shield, for example, form very narrow strips. The ones composed of rocks of the same age on the African Shield are much broader. These strips have straight or slightly curved, mutually parallel boundaries truncating the internal structure of the neighbouring Archaean blocks discordantly over considerable distances. There are signs of lineation not only on the boundaries of the bands of Early Proterozoic rocks, but also in their folded structure: many swarms of folds in them are characterised by lineation, although gneiss and granite-gneiss domes are also common. There are grounds, consequently, for believing that the role of the deep faults that control the lineation of the structures, became greater in the Early Proterozoic than in Archaean time.

Since the rocks composing the strips described above are largely volcanic and igneous, metamorphosed, granitized, and strongly deformed, we are justified in viewing them as geosynclines. These are Early Proterozoic eugeosynclines, similar in many ways to the Archaean ones.

The geosynclines, however, are now opposed by the surviving Archaean blocks that were not subjected to the same strong endogenous effects in the Early Proterozoic. In many places sedimentary and volcanic series, slightly metamorphosed and deformed, and resembling the cover of younger platforms, were laid down on them during the Early Proterozoic. These sectors, consequently, play the role of platforms in relation to the Early Proterozoic geosynclines.

The most ancient platforms, however, for all that, differ substantially from younger "authentic" ones. (1) They are small and resemble median masses rather than platforms. (2) The cover of the Early Proterozoic platforms, an example of which is the sedimentary and volcanic Early Proterozoic series (2700-1600 million years old), occurring on the South African massif and several kilometres thick, did not remain unmetamorphosed, though its metamorphism is generally within the greenschist facies. Local granitisation is also observed in the cover. (3) Large sills are very typical of the cover of the Early Proterozoic platforms, basic in their initial composition but stratified as a result of differentiation into ultrabasic to acid rocks. In the cover of the South African Archaean massif, for example, there is the huge Bushveld lopolite, up to 9000 m thick, and on the Canadian shield there is the Sudbury lopolite, famous for its sulphide copper-nickel deposits.

Because of these features of the cover's structure, which distinguish it from younger platforms, these oldest platforms known to us can be called *protoplatforms*, i.e., primitive platforms.

Protoplatforms were not such stable elements of the crust, as we shall see later, as the younger "ancient" platforms became. They were not yet "nuclei of consolidation", from which platform regimes could spread areally at the expense of geosynclinal ones. The Early Proterozoic protoplatforms were subjected to a "breaking-up" and part of their area was involved in a geosynclinal regime that "reworked" their basement. The geosynclines of that time accordingly also did not form permanent, spatially stable geosynclinal belts within which further geosynclinal development would be concentrated. Geosynclines arose here and there and in the next endogenous cycle were superimposed on the areas that had been platforms in the preceding cycle. The geosynclines of that time therefore also differed from later ones and so can be called *protogeosynclines*. We conclude that the permobile stage in the development of the Earth's crust that took place in Archaean time was followed in the Early Proterozoic by a *stage of protogeosynclines and protoplatforms*. As all of them were unstable, this stage can also be called "unstable protogeosynclinal-protoplatform" or, more briefly, "unstable *protogeosynclinal*".

The duration of the unstable stage varies on the various shields. On the northern ones—the Baltic and Canadian—it was completed by the beginning of the Late Proterozoic. On the Baltic Shield, troughs were formed 1900-1750 million years ago, in which sedimentary and volcanic series of the Karelian formation accumulated. Its rocks were subsequently transformed into crystalline schists and amphibolites, folded (the Karelian epoch of folding), and penetrated by granites 1850-1750 million years old. The youngest granites (about 1650 million years old) represent a special variety, the rapakivi.

Analogous sedimentary and volcanic rocks of the Huronian series accumulated simultaneously on the Canadian Shield in troughs, formed on the protoplatform, which were buckled, metamorphosed, and intruded by granites 1850 to 1650 million years ago. This is the Churchillian folding

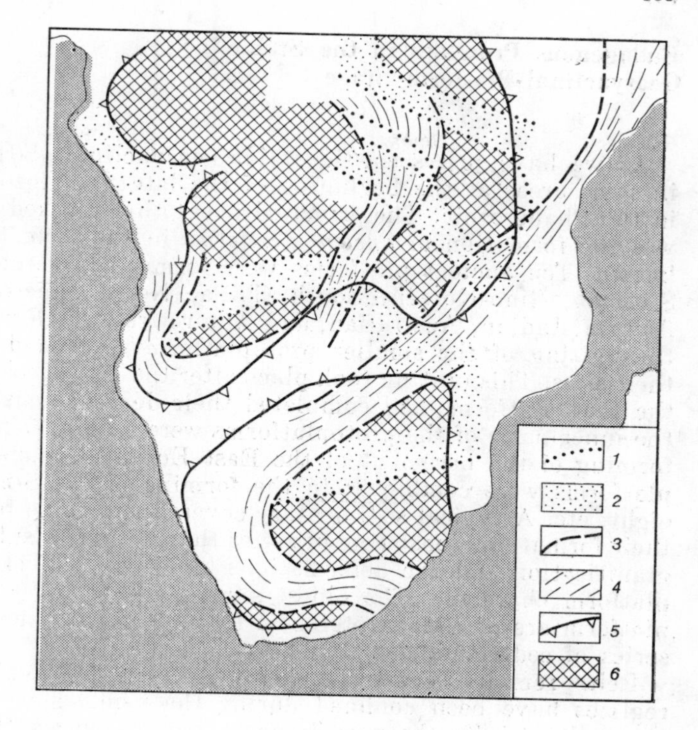

Fig. 101. Precambrian zones of South Africa (after Vail, 1968):

1, 2—boundaries and regions of geosynclines with an age of metamorphism around 2000 m.y.; *3, 4*—boundaries and regions of geosynclines with an age of metamorphism around 1000 m.y.; *5*—boundaries of geosynclines with an age of metamorphism around 550 m.y.; *6*—protoplatforms

epoch, which coincides in time with the Karelian. The Karelian endogenous pulse proved to be the final one of the unstable protogeosynclinal stage, not simply on the Baltic and Canadian Shields, but also over the whole area of the East European and North American platforms. This lasted around a thousand million years on them and was completed by the beginning of Late Proterozoic. After the Karelian pulse, the last "stable geosynclinal-platform" stage began on these platforms, which has lasted more than 1500 million years to the present and is characterised by the "normal" platform, geosynclinal, and orogenic regimes described above.

This last stage began with the uniting of the former protoplatforms into a monolithic "ancient" platform, the inner sectors of which no longer have broken up and were the nuclei from which the platform regime spread later areally. All the features of this stage will be considered in the next section.

On southern platforms the unstable protogeosynclinal stage lasted longer, to the end of the Proterozoic, i.e., until 550 to 600 million years ago. From Figure 101 it will be seen that instability of disposition characterised geosynclines on the South African massif with an age of folding and metamorphism of 2000, 1000, and 550 million years. Geosynclines changed their positions from one cycle to another and in so doing crossed the platforms formed earlier, partially reworking them. Only by the beginning of the Palaeozoic did a stable "ancient" platform form there, like that formed a whole thousand million years earlier in the north.

Endogenous Processes in the Stable Geosynclinal-Platform Stage

As we have just seen, the stable geosynclinal-platform stage commenced in some areas at the beginning of the late Proterozoic and in others early in the Phanerozoic. The principal event that marked the onset of this stage was the formation of *ancient platforms* in the form in which they are now known. Ten ancient platforms were formed altogether: the East European, Siberian, Sino-Korean, South China, North American, South American, African, Indian, Australian, and Antarctica, which were all formed through the uniting of the smaller protoplatforms that had previously existed on their sites. This uniting took place after the protogeosynclines that separated the protoplatforms had completed their development and consolidated, and the inner parts of the protoplatforms were no longer being broken up. In the forming of the basement of the East European ancient platform, 14 protoplatforms were combined; in the forming of the Siberian ancient platform eight, etc. A typical sedimentary cover began to be formed from the time of their formation, consisting of series that no longer suffered metamorphism or granitisation and the deformations of which were slight and had a typical platform character. The oldest strata of this cover on the East European platform are of Late Proterozoic (Riphean) age (the Jotnian and Ovruchan series of sediments).

Belts remain between ancient platforms to which alone geosynclinal regimes have been confined during the whole subsequent time. These are *geosynclinal belts*, the most important of which are the Pacific and Mediterranean belts. The former almost encircles the whole Pacific Ocean, while the latter runs east-west from the Mediterranean through Central Asia to the Pacific Ocean. A fragment of this belt can also be seen in Central America. Other, less important belts are the Urals-Mongolian which separates the East European platform from the Siberian and the latter from the Sino-Korean; Atlantic, represented by the Caledonides of Norway, Scotland, Wales, Ireland, and Greenland, and the Caledonides and Hercynides of the Appalachian Mountains; belt of the Franklin geosyncline in the north of Canada; and the Argentine belt, which may once have been connected with the Cape Belt in the south of Africa (Fig. 102).

The lineation of tectonic zones and separate structures that occurred in the previous stage, but did not become universal then, becomes the general rule for the stable geosynclinal-platform stage. This indicates that faults were much more important for it; they determined the boundaries between tectonic zones and "dictated" the strike of individual major structures. We have already spoken about this role of deep faults above (Chap. 9). Even where some one major structure is bent into an arc (e.g., the mountain arcs of the Carpathians and Western Alps), more detailed examination shows that the arc consists of a combination of many straight segments each of which is possibly associated with its own deep fault.

The evolution of the regimes during the stable geosynclinal-platform stage was largely determined by the expansion of their area and contraction of the area of geosynclinal belts. This tendency predominated, although, no doubt, there was reverse capture of areas previously occupied by platforms by geosynclinal regimes. This secondary process was of local and temporary importance and did not disturb the general line of development.

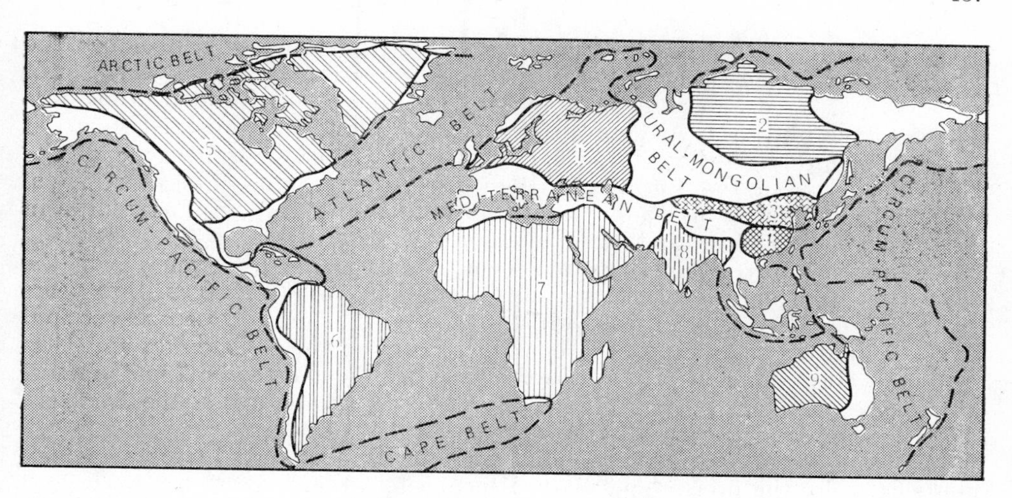

Fig. 102. Ancient platforms and geosynclinal folded belts (after Muratov, 1966):
1—the East European platform; *2*—Siberian; *3*—Sino-Korean; *4*—South China; *5*—North American; *6*—South American (Brazilian); *7*—African; *8*—Indian; *9*—Australian

The platforms expanded mainly through "accretion" of ever younger platforms along their peripheries. In the sectors where that happened the geosynclinal regime was followed by a platform one. All platforms, both ancient and young, therefore, consist of two parts: a lower one composed of thick, much metamorphosed and dislocated sedimentary-volcanic and igneous rocks and corresponding to a geosynclinal stage; and an upper one that is a platform cover that lies quiescent, unmetamorphosed, and almost devoid of igneous rocks. The later a given area was involved in the growth of the platform, the younger both the geosynclinal and platform stages are.

We have already spoken several times above of endogenous cycles. One of their manifestations was that platform growth was not continuous, but broken, and in spurts, so to speak, from one cycle to another. When one cycle ended, part of the geosynclinal belt became a platform; at the end of the next cycle new sectors were added to the platform that had been part of this geosynclinal belt before, and so on.

The ancient East European platform, for example, which was formed about 1650 million years ago after the Karelian endogenous pulse, expanded approximately 1300 million years ago through accretion in the south-west (in southern Sweden, southern Norway, Denmark, and adjoining areas of continental Western Europe) of a large area where the Gothlandian system of sedimentary and volcanic rocks had previously accumulated and undergone deformation and metamorphism. Later still (about 600 million years ago) a zone was added to this platform, but this time along its north-eastern margin in the Timan region, that had been occupied by a geosyncline where rocks of the Hyperborean system had accumulated and suffered folding and metamorphism in the Baikalian epoch of diastrophism, at the boundary of the Proterozoic and Palaeozoic. In the Palaeozoic, there were two more "spurts" in growth of the East European platform: one at the end of the Caledonian cycle (about 400 million years ago), when the Caledonian geosyncline of Scandinavia, Scotland, Wales, and Ireland, which had completed its development, was added to the platform, and the other after the Hercynian cycle (about 250 million years ago) when the whole zone between the East European and Siberian platforms was converted into a platform and, in addition, a considerable area of Western and South-West Europe was added to the platform. As a result the geosynclinal conditions had survived to the beginning of the Alpine endogenous cycle only in the zone of the Mediterranean, in the Betic Cordillera, Atlas Mountains, Pyrenees, Alps, Apennines, Carpathians, and Dinaric Alps, on the Balkan Peninsula, in Anatolia, the Crimea, and the Caucasus.

The ancient North American platform also expanded in "spurts", on the transition from one endogenous cycle to another. About 1000 million years ago the Grenville tectonic zone, which had earlier belonged to a geosynclinal belt, was united with it. At the boundary between the Caledonian and Hercynian cycles, the platform's growth involved the Northern Appalachians, and at the boundary between the Hercynian and Alpine cycles the Southern Appalachians and the area adjoining the Gulf of Mexico.

The time of the formation of platforms, as we know, is indicated terminologically by the prefix "epi" (i.e., after), before the name of the cycle that was the last geosynclinal one in the area. In this way, as we pass from younger to more ancient platforms, epi-Cimmerian, epi-Hercynian, epi-Caledonian, and epi-Baikalian platforms were formed. For the earlier young platforms, formed during the Late Proterozoic, these terms are seldom used, though one could well speak of epi-Gothlandian or epi-Grenville platforms. Terms of the same type are also used for ancient platforms, which, as we have seen, may be epi-Karelian (the East European and North American platforms) and epi-Baikalian (the African platform).

Ultimately, huge platform regions are formed consisting of platforms of various age united with more ancient "nuclei", mainly on the inside and younger platforms on the periphery. Because platforms once formed persist stably in this stage of the crust's development, those that were formed earlier exist longer and were therefore able to survive several endogenous cycles. Their tectonic evolution, expressed in the oscillatory movements and dislocations characteristic of platforms, of course, continued all the time. The structure of the epi-Baikalian platform, for example, therefore experienced changes during the Caledonian, Hercynian, and Alpine cycles.

In this connection, it is often insufficient, in order to denote a particular structure of one platform or another, to call it epi-Karelian or epi-Caledonian; it is also necessary to assign the structure interesting us to a definite cycle. We can therefore speak, for example, of the Caledonian structure of ancient epi-Karelian platforms, or of the Alpine structure of an epi-Caledonian platform, and so on.

Platforms grew in such a way that the Circum-Pacific and Mediterranean geosynclinal belts existed longest, up to and including the Cenozoic, though they became much narrower. The other belts listed above were transformed into platforms earlier, by the beginning of Mesozoic time. It is important to underline that the growth of platforms did not occur at random, but in a definite direction, so that by the final cycle the geosynclines had not disintegrated into randomly dispersed patches, but survived as continuous zones along these two main geosynclinal belts.

The geosynclinal regime sometimes gave way to a platform one through transitional states. In Central Kazakhstan a Caledonian eugeosyncline was transformed, in the Balkhash zone, into a Hercynian miogeosyncline, and only after the beginning of the Alpine cycle did a young platform regime come into existence there. Eastern Transbaikalia was a eugeosyncline in the Hercynian cycle and became a miogeosyncline (and in places a parageosyncline) in the Mesozoic cycle. In Late Cretaceous time, an epi-Mesozoic young platform was formed there.

Study of the internal structure of platforms of different age indicates that platform growth does not consist in a parallel articulation of new consolidated structures to older ones.

As a rule, the strikes of the basement structures of an older platform are truncated discordantly by the margin of an adjoining younger one and by the strike of its basement structures. It will be seen from Figure 98 above that the edge of the epi-Gothlandian platform truncates the Karelian trends in the south-west of the East European platform, in Scandinavia the epi-Caledonian platform abruptly truncates the strikes of all the older structures, and similarly the boundary of the epi-Hercynian platform of the Urals discordantly truncates older strikes. On the North American platform, the epi-Grenville platform also adjoins earlier structures discordantly.

These phenomena should probably be understood in such a way that each time after a recurrent endogenous cycle the geosyncline gave way to a platform, for a short period of time, over a broader area than that in which the platform regime subsequently survived: some of this area soon after, already at the beginning of the next cycle, was again involved in geosynclinal development. It may once more be a matter of partial break-up of the crystalline shell that was formed during the geosynclinal process and which would have constituted the basement of a future platform, but in fact survived only in part. With this break-up secant contacts naturally arose between structures of different age, but its scale was such that the platforms grew larger and larger at the expense of the geosynclinal belt.

Let us recall that the structure of the cover on ancient platforms and the location of subgeosynclines and subgeanticlines on them proves to be superimposed discordantly on the structure of the basement (see Fig. 85). We find much evidence that substantial structural reconstruction takes place in the crust during the transition from one endogenous cycle to another.

During the stable geosynclinal-platform stage an orogenic regime repeatedly manifested itself, the traces of which are not very clear for earlier stages. The signs for the younger orogenic regime belonging to Neogene and Quaternary time are the forms of the relief, while the signs for older regime (Mesozoic and Palaeozoic) are formations of molasse type derived from the destruction of older mountain ranges.

Orogenic regimes, both epigeosynclinal and epiplatform, always arose at the end of an endogenous cycle, but the onset of such a regime after a geosynclinal one was not the rule: some geosynclines ended their development and were transferred into platforms, without an orogenic stage, by gradual dying out ("degeneration") of all endogenous processes.

When a geosynclinal regime was followed by an orogenic one the orogenic uplifts and troughs (or horst-anticlinoria and graben-synclinoria) generally proved larger than such intrageosynclinal zones as intrageanticlines and intrageosynclines, central uplifts, and marginal troughs. Several intrageosynclinal zones could be "unified" by one uplift or trough. The orogenic process, moreover, also spread to the margin of the neighbouring platforms: foredeeps "rolled" over them, and they could be involved in an uplift within a certain zone. The correlation of the intrageosynclinal and orogenic zones of the Great Caucasus was depicted in Figure 83. As we see, the orogenic uplift of the Great Caucasus Range encompassed not only the former central uplift, but also the adjoining intrageanticlines and parageosynclines, while the foredeep was superimposed on the margin of the Scythian platform, and the intermontane trough occupies the site of the former Kura-Rioni intrageanticline.

As we have already said, orogenic structures could be superimposed discordantly on the previous tectonic plan.

When an orogenic regime originated at the end of a cycle, evolution was possible at the beginning of the next cycle towards either a platform regime or recurrence of a geosynclinal one.

The change of an orogenic regime into a platform could proceed gradually, through the smoothing out of smaller contrasts and amplitudes of vertical movement of the crust. Signs of this are the gradual reduction in the coarseness of molasse, and the appearance of coal-bearing sediments in it, which is evidence of a marshy terrain. In the Mongolia-Okhotsk zone, for example, strong orogeny developed at the end of the Hercynian cycle. Intensive vertical movements continued in Early Mesozoic time, when coarse sediments accumulated in the basins. Subsequent gradual quiescence of the vertical movements, however, led to a gentle relief in Late Cretaceous time, to swamping and the establishment of a typical platform regime. In the Urals, however, the orogenic regime at the end of the Hercynian cycle was followed, right at the very beginning of the Alpine cycle, by platform conditions. Finally, in the Altai, the orogenic conditions of the end of the Caledonian cycle gave way, at the beginning of the Hercynian cycle, to geosynclinal development.

With a succession of regimes at the boundary of two cycles, both inheritance of the strike of the disposition of the main structural forms and discordant superimposition of completely new strikes on the older ones can be observed. In the Tien Shan and Appalachians, the Caledonian and Hercynian geosynclinal structural zones are parallel, but in Figure 103 a case of sharply discordant superimposition of Hercynian geosynclinal trends on Caledonian in the Sayans and Altai is depicted.

In exactly the same way unconformity of the strike of structures of a previous geosyncline and those of the subsequent one may occur during the renewal of geosynclinal conditions. In the Western Alps, for example, the Alpine trends do not coincide with the Hercynian.

In these cases, too, as in those discussed above in connection with the growth of platforms, the impression is created that the new structures engendered by a new regime are largely formed during the break-up of structures created by previous regimes. One can imagine, for example, that the forming of orogenic structures, or the dismembering of the crust into zones of intensive subsidence and uplift in connection with the formation of a new geosyncline on the site of an older one, takes place during the destruction of the crystalline shell created in the preceding epoch of regional metamorphism. The strike of the structures may then be both inherited and new.

We will conclude this section by considering a feature of the crust's development that may be called the conservativeness of tectonic movements and of the structures formed by them. When vertical movement of the crust, opposite to the previous direction, begins somewhere during its development, the new movement overcomes, so to speak, the resistance of the old one. It is superimposed on the older movement, added to it, and over a certain period the older movement "shines" through the younger.

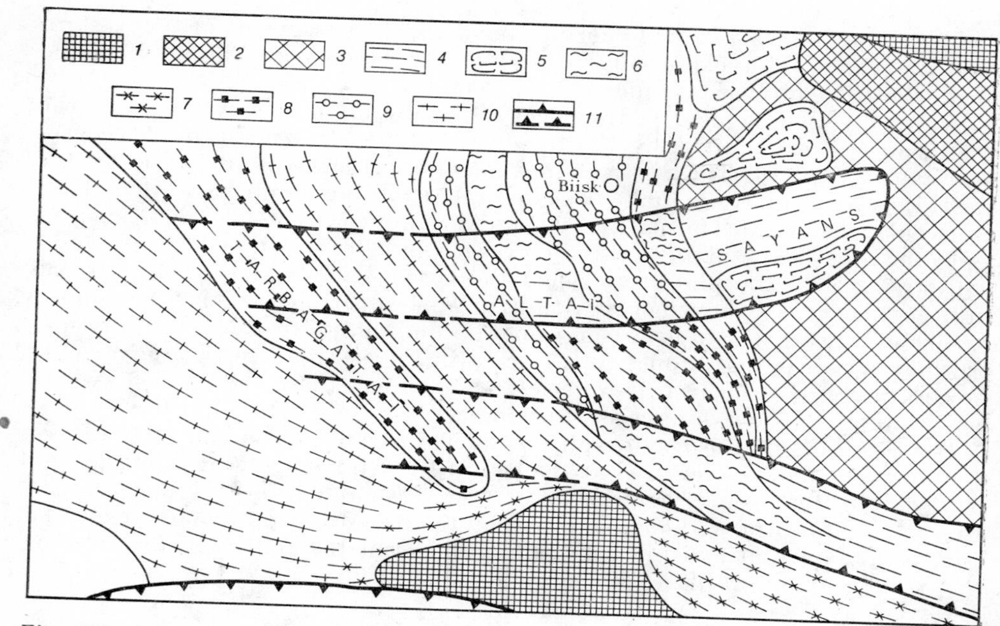

Fig. 103. Intersection of the Hercynian and Caledonian folded zones in the Altai-Sayan region (after Rudich, 1972):

1—pre-Baikalian platforms; *2*—Baikalian folded zones; *3*—Early Caledonian (Salairian) folded zones; *4-6*—Late Caledonian folded zones; *7-10*—Hercynian folded zones; *11*—boundaries of Late Caledonian folded zones (established and inferred)

Foredeeps are known to vary in both width and depth. They are divided into deep ovals separated by less downwarped bridges. Foredeeps may wedge out in places, and an orogenic uplift adjoin a platform directly, without an intervening foredeep.

N.S. Shatsky has shown that such changes in the shape of a foredeep are associated with movement of those parts of the platform on which it is superimposed. Where a sinking sector of a platform (subgeosyncline) adjoins an orogenic zone, the foredeep is broad and deep, but where it is superimposed on a rising sector of the platform, it becomes narrow and shallow or even wedges out. The Ciscaucasian Neogene and Quaternary foredeep is divided into two strongly sagged sectors, the Terek and Kuban with a raised bridge between them that lies on a southern continuation of the Stavropol subgeanticline, whereas the sectors of sagging are superimposed on the Caspian and Azov-Kuban subgeosynclines.

The geological map clearly shows the difference in structure of the eastern and western halves of the Great Caucasus. The eastern half has very thick Lower to Middle Jurassic sandy-shaly sediments (totally as much as 6000 m). In the western half of the range, the Lower to Middle Jurassic sediments are much less thick, not more than 3000 m. In the east of the range, there are no exposures older than the Lower Jurassic, while in the west a broad axial strip of the range is occupied by Palaeozoic metamorphic and igneous rocks. There is no doubt that downwarping in Jurassic time was much greater in the east of the Great Caucasus than in the west. The eastern part, however, lies on an extension of the Caspian subgeosyncline, which sagged extremely deeply throughout Mesozoic-Cenozoic time, while the segment of the western part, in which old rocks occur, is largely located on a continuation of the north-south-trending Stavropol subgeanticline.

In this case, consequently, the effect of the direction of movements on the platform was not limited by the foredeep but penetrated far into the geosyncline beyond it. The vertical movements of the crust observed on the East European platform had already been evolving for a very long time apparently before the formation of the Alpine geosyncline of the Caucasus and (and this is the main point) involved a broader territory than the present-day area of the platform. When the Great Caucasus geosyncline came into existence, the movements within it were superimposed on these much earlier downward and upward movements, and added to them. As a result, a deeper geosynclinal trough was formed in the east, where two movements in one direction, downward, were added together, than in the west, where geosynclinal downwarping was superimposed on an earlier uplift.

The Scandinavian orogenic uplift, which occurred at the end of the Caledonian cycle and was renewed at the end of the Alpine cycle, was not accompanied by a foredeep either earlier or now. It borders directly on the East European platform without an intervening trough. This anomaly may be due to the fact that a rising part of the platform, the Baltic Shield, adjoins the Scandinavian uplift.

The Rhythm of Endogenous Processes

We have spoken several times already, above, of endogenous cycles and endogenous rhythm. In this section we intend to generalize the data on the rhythmicity in the development of endogenous processes and consider its manifestations more fully. It is a matter that has long been discussed in geological literature; tectonic cycles are usually meant, but a wider sense has been given to the term, since metamorphic and magmatic processes are also kept in view together with tectonic movements. The term "endogenous cycles" therefore seems to be more correct.

By endogenous cyclicity is meant the recurrence of a certain group of endogenous processes (tectonic, magmatic, and metamorphic) that are combined in a regular fashion at different stages of the cycle. The sequence of the cycles creates what can be called endogenous rhythmicity, with the proviso, however, that the terms "cycle" and "rhythm", and also the frequently used term "periodicity", are used conventionally. Recurrence does not literally mean recurrence. A certain fundamental, but fairly rough scheme of the sequence and combination of processes recurs, but the separate cycles may differ greatly from one another in many of their essential features. Not only are cycles belonging to different times distinguished, but also simultaneous ones manifested in different places. Rhythmicity and periodicity do not mean that there is some precisely established period of recurrence of the processes and that the course of endogenous processes can be depicted by a geometrically regular curve. The length of the separate cycles is undoubtedly different, although the principal Phanerozoic cycles seem to be similar in their duration, in any case to within approximately 30 or 40 million years, which is around 15-20% of the average length of a cycle. It should also be remembered that the boundaries between cycles are not quite the same age in different regions.

Nevertheless, in spite of all these reservations and conditions, recurrence in the endogenous development of the Earth's crust is a remarkable phenomenon that puts its stamp on the crust's whole structure and the distribution of various rocks in the geological section, and determines the most important patterns in the laws of distribution of minerals.

The character and degree of preservation of the geological evidence makes it possible to record the whole complex of endogenous processes in principle only for the stable geosynclinal-platform stage, and that is now feasible in practice only for Phanerozoic time. Older cycles cannot be studied as thoroughly because the stratigraphy is undeveloped, but mainly because of the widespread metamorphism, which largely conceals original composition of the rock.

In general, we derive our main information on endogenous rhythm for the Precambrian from data on the absolute age of regional metamorphism and granitization. When all the absolute dates obtained for the Precambrian are gathered together, it turns out that they are concentrated mainly around certain epochs, while there are many fewer intermediate dates. These epochs of concentration of absolute dates are as follows: 3000-2800, 2600-2500, 2200-2000, 1800-1600, 1200, 1000-900, 600-500 million years ago. The last named epoch makes the boundary between the Proterozoic and Phanerozoic.

As absolute age is mainly determined from minerals arising from metamorphic processes and granitization, the dates given indicate that epochs of predominant occurrence of regional metamorphism and granitization recurred every 300 to 600 million years in the Precambrian. Since later endogenous cycles indicate that regional metamorphism and granitisation occupy a definite place in them, confined to the middle of the cycle, we are justified in seeing evidence in the separate epochs for several endogenous cycles in Precambrian time. Some of them could be characterised more fully, since stages of the predominant occurrence of basic volcanism, for instance, have been established for them in which examples of the ophiolitic stage of the eugeosynclinal regime can be seen. There are too few data, however, for wholly reliable results to be obtained, therefore let us turn at once to the Phanerozoic cycles for which a fuller picture can be obtained.

Study of Phanerozoic history shows that the basis of endogenous cycles is the rhythmicity of the largest general oscillations of the Earth's crust. A cycle begins with a state in which uplift prevails, while subsidence plays a minor role. Then a period sets in during which uplift gradually loses its significance and subsidence begins to predominate. In the middle of this cycle the predominance of the latter is at its maximum. A reverse process of growth of uplift relative to subsidence then sets in, and at the end of the cycle we are back at the initial state. The cycle is then divided into two parts of opposite tendency, one is characterised by an increase in subsidence, the other by an increase in uplift. The end of one cycle and the beginning of the next one merge into an epoch of maximum predominance of uplift.

Major transgressions and regressions are associated with this periodic development of general oscillations. The facies of sedimentary rocks change together with them, whereas their general sequence in time proves to be contrary at the beginning and end of a cycle: at the beginning evolution proceeds from continental and in-shore sediments rich in coarse clastic material to finer and finer ones and finally to carbonate sediments; and at the end of the cycle the sequence is reversed.

The time of the transition from an increase in subsidence to an increase

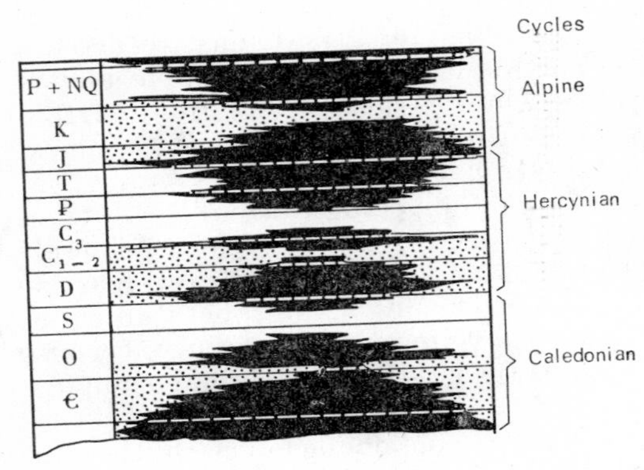

Fig. 104. Periodicity of transgressions and regressions on the North American platform: solid black represents regressions and white and dotted areas transgressions (after Sloss, 1969). Within all the cycles there are expressions of transgressions and regressions of the second order (division into cycles after Beloussov)

in uplift, when there is equilibrium of the two, is a time of *general inversion* of the regime of the crust's oscillatory movements.

The rhythm of general oscillatory movements is manifested in regions of any regime, but is expressed in its clearest form on platforms where it is not obscured by other phenomena. The change in the area of sedimentation as a percentage of the total area of the East European platform in the Phanerozoic was shown in Figure 29. The curve indicates a division of the Phanerozoic into three major cycles, each around 200 million years long. A cycle begins with a very small area of sedimentation, i.e., a predominance of uplifts (the state inherited from the end of the previous cycle). There is then a considerable expansion of the area of sedimentation (increasing subsidence). After a certain period of upward and downward oscillation, the curve begins to sag, which suggests an increase in uplift.

In Figure 104 the periodic character of transgressions and regressions on the North American platform is represented. In it, too, the area of sedimentation during Phanerozoic time is depicted, but in a qualitative manner and different form. Second-order cycles are outlined on the background of the main cycles.

On both platforms, three major cycles of general oscillation have been established for Phanerozoic time, viz., the Caledonian, Hercynian, and Alpine. Within them secondary cycles are distinguished, but we shall ignore them here.

How far did the general inversions in the different cycles occur synchronously on these two platforms? In the Caledonian cycle general inversion took place on the East European platform in Ordovician time, while on the North American platform, if we combine the first two partial cycles, it must be assigned to the boundary between the Ordovician and Silurian. In the Hercynian cycle, the turning point for both platforms is the Late Carboniferous. In the Alpine cycle general inversion occurred on the East European platform at the end of the Jurassic and on the North American platform in the middle of the Cretaceous. As to the boundary, that between the Caledonian and Hercynian cycles is synchronous (early Devonian) on both platforms, while that between the Hercynian and Alpine cycles is rather later (Early Jurassic) on the North American platform than on the East European (Middle Triassic). These relationships well illustrate the synchronism of the endogenous cycles manifested in regions of the globe remote far from one another. The lack of synchronism in this case does not exceed 30 million years, which is 15% of the total length of the cycle.

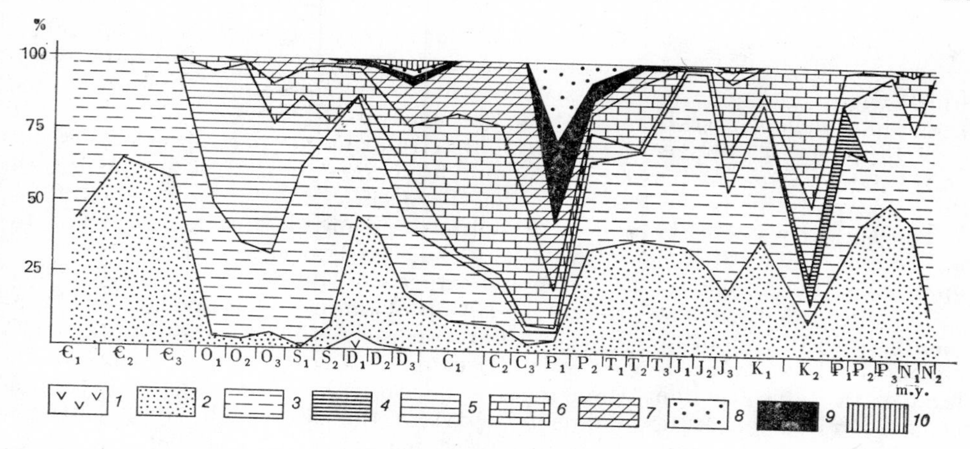

Fig. 105. Periodicity of sedimentary formations on the East European platform. Changes in their relative percentage volumes in the Phanerozoic (after Ronov *et al*, 1969):
1—conglomerate; *2*—sandstone; *3*—clay; *4*—siliceous rocks; *5*—marl; *6*—limestone and chalk; *7*—dolomite; *8*—gypsum; *9*—salt; *10*—extrusives and tuffs

In Figure 105 the change in facies of the sedimentary series on the East European platform during the Phanerozoic time is depicted. In this change we see a periodicity due to the same cycles of general oscillatory movement.

The rhythm of general oscillations is also reflected in the development of geosynclines.

The ophiolitic stage of a eugeosyncline and its deep-water sediments, great contrast of vertical movement with a predominance of subsidence, basic and ultrabasic volcanism, and the stage of compensated downwarping of a Miogeosyncline correspond to the first part of the cycle. The inversion stage is characterised by a balancing out of uplift and subsidence as well as by regional metamorphism, granitisation and folding, and this feature would seem to indicate that the time of partial inversion was also one of general inversion, and that the formation of central uplifts coincided with the turning point in the trend of the general oscillatory movements. In some cases that is actually so, but other cases are known in which the balancing of uplift and subsidence in the stage of partial inversion proved to be only a temporary halt in the course of increasing subsidence, and the increase continued for a certain time after the partial inversion occurred. The central uplifts, once formed, are then involved in this general subsidence, their role as sources of terrigenous material diminishes, and they may be transformed from zones of absolute uplift to zones of relative uplift.

The first case, coincidence of partial and general inversion in time, is seen in the Western Alps, where partial inversion took place in the inner Pennine zone on the boundary between the Late Cretaceous and Palaeogene and since when uplift has in general been of increasing importance. The Eocene flysch was followed by a lower molasse of Oligocene age and by an upper molasse of Neogene age.

An example of the second case is to be found in the Caucasus. In the Somkheto-Karabakh intrageosyncline in Transcaucasia, the ophiolitic stage spanned the Jurassic and Neocomian. At the end of the Neocomian stage, a partial inversion occurred, but during the Late Cretaceous the central uplift was involved in the continuing general increase of subsidence, sank below sea level, and was overlain by limestone while remaining, however, a zone of slight relative uplift. Only in Palaeogene time did the central uplift again emerge from the sea. In this case, consequently, the general inversion began later than that of the partial.

Still longer lag in general inversion in relation to partial is observable in the Great Caucasus, where the partial inversion took place in the Middle Jurassic, and the general only in the Oligocene.

The end of a cycle, both in geosynclines and on platforms, is marked by an increase in uplift. Here, too, there may be two cases. If there was no epigeosynclinal orogeny, the inversion stage continued to the close of the cycle and ended with a gradual cessation of undulatory oscillatory movements, overall smoothing out of their contrasts, and a slight uplift that converts a locality previously very dissected into low-lying, slightly rolling land with separate depression basins, mostly lacustrine. The geosyncline progressively degenerates and is followed in the next cycle by a platform which slowly and slightly rises in accordance with the trend of the crust's general oscillations. An example of such development can be seen in the Eastern Transbaikalia where, after the Mesozoic miogeosynclinal development, the platform regime set in from the beginning of the Late Cretaceous with weak undulatory oscillatory movements and slight general uplift.

When epigeosynclinal orogeny is manifested, however, it occupies the end of the cycle. Immediately after equalization of the volume of uplift and downwarping and smoothing out of the contrasts of vertical movements that characterises the inversion stage, an increase of uplift begins relative to subsidence, at first weakly but later more strongly; at the same time the contrast in manifestations of block undulatory oscillatory movements increases. The geosynclinal regime gives way to an orogenic one. In the neotectonic epoch in the Great Caucasus a slight increase in uplift relative to subsidence began in the Oligocene, which can be considered the end but of the inversion stage in the geosyncline's development. Uplift intensified in the Late Miocene and became extremely intensive in the Pliocene and Quaternary. We can consider this uplift as orogenic from the Late Miocene. It was accompanied with the same accelerated development of troughs into which clastic material carried down from the uplifts accumulated. The change in character of the sediments toward ever coarser material and the replacement of marine sediments by continental indicate a considerable preponderance of uplift.

An increase in contrast, rate, and amplitude of block-undulatory oscillatory movements and analogous acceleration of uplift accompanied by basic and alkali volcanism is observed at the end of the cycle on platforms, where epiplatform orogeny is manifested (e.g., in the Tien Shan).

Data on the change of regimes for several geosynclines, platforms, and orogenic zones in the Phanerozoic are given in Figure 106. The figure shows the following: epochs when subsidence prevailed over uplift, epochs of a balance between them, epochs of moderate subsidence and uplift, and stages of a great preponderance of uplift. In addition, geosynclinal, platform, orogenic, and rift regimes are distinguished

Examination of this figure allows us to draw a number of conclusions.

First of all it can be established that the Palaeozoic is divisible into cycles usually of one kind in the most varied regions of the globe. Caledonian and Hercynian cycles are distinguished corresponding to the Lower and Middle Palaeozoic in the one case and the Upper Palaeozoic in the other.

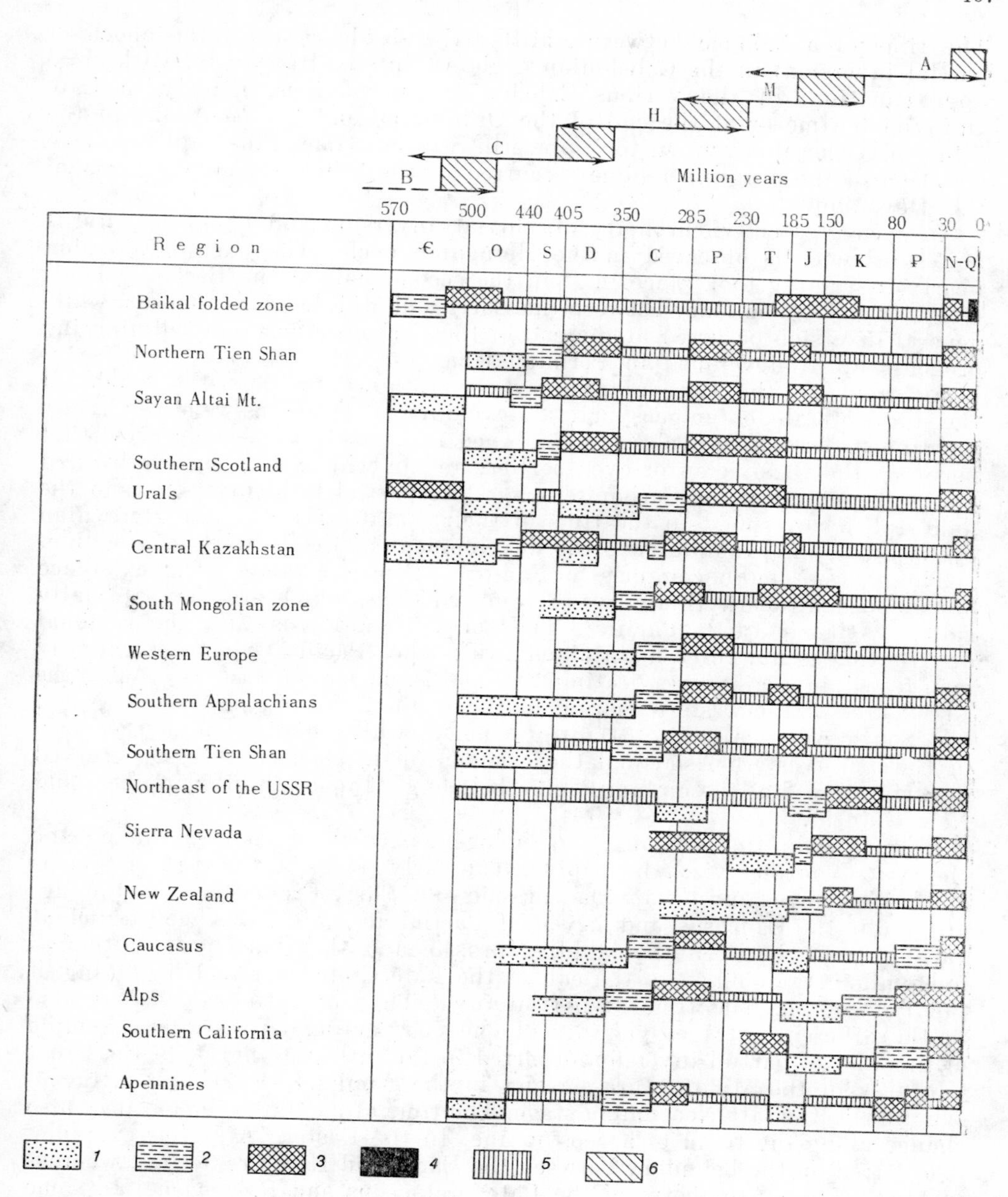

Fig. 106. Succession and rhythmicity of endogenous regimes (after Beloussov): *1*—the ophiolitic stage of eugeosynclinal regime and, in general, a time of considerable predominance of subsidence; *2*—inversion stage of eugeosynclinal and miogeosynclinal regimes; *3*—orogenic regime; *4*—rift regime; *5*—moderate subsidence and uplift (platform, parageosyncli al, and miogeosynclinal regimes); *6*—"excited" regimes. Cycles: B—Baikalian; C—Caledonian; H—Hercynian; M—Mesozoic; A—Alpine

The time of a balance between uplift and subsidence (i.e., the epoch of partial inversion) in the Caledonian cycle set in simultaneously, with close approximation, in the various Caledonian geosynclines, namely in Late Ordovician time or at the end of the Ordovician and the beginning of the Silurian. General inversion took place in Silurian time. Final uplift (manifestation of the orogenic regime) occurred in the Silurian and beginning of the Devonian.

An even greater synchroneity of partial inversion and balance of uplift and subsidence is observed in the Hercynian cycle. The partial inversion nearly invariably took place in it in the second half of the Early Carboniferous or on the boundary between the Early and Middle Carboniferous, while general inversion occurred at the end of the Carboniferous. The culminating orogenic uplift developed in Permian time.

Far from all the Hercynian geosynclines, however, passed through all the stages of Caledonian geosynclinal development. Often, as we have already mentioned, these two cycles are combined in a single Palaeozoic cycle that began in the Ordovician or even earlier, and in which subsidence prevailed to the beginning of the Carboniferous, while partial local inversion and the final uplift correspond to the time of their manifestation in the Hercynian cycle. Such, for example, was the development of the Uralian geosyncline.

Things were more complex for Mesozoic-Cenozoic times. The existence of two schemes of rhythmicity is recognised then, which are conventionally called "Atlantic" and "Pacific". The conventionality is that the regional distribution is not fully maintained; while the "Pacific" scheme is actually manifested in regions gravitating to the Pacific Ocean (Eastern Asia, the island arcs of the western margin of the Pacific, and in the west of North and South America), the "Atlantic" scheme, while manifested in the main in Europe, Western Asia, and the east of North America, is also observed in certain areas of the circum-Pacific belt (e.g., Japan, the Philippines, and New Guinea).

In the "Atlantic" scheme, one endogenous cycle developed during the Mesozoic-Cenozoic, viz., the Alpine. The early stages of geosynclinal development occurred mostly in the Jurassic and Early Cretaceous (in the figure 106 in the Caucasus and beyond). Partial inversion was manifested at various times between the Middle Jurassic and the Palaeogene, although predominantly confined to the end of the Late Cretaceous and beginning of the Palaeogene. The time of partial inversion proves to be different also in the different intrageosynclines of one geosyncline. We saw an example of this above: partial inversion occurred in the intrageosyncline of the Great Caucasus in the Middle Jurassic and in the Somkheto-Karabakh intrageosyncline in the late Neocomian stage. The time of partial inversion may also change along the trend of a geosyncline. In the Eastern Alps, for example, it occurred on the boundary between the Early and Late Cretaceous, whereas in the Western Alps between the Late Cretaceous and Palaeogene. In some cases, the general inversion coincides with the partial, but more often it has to be assigned to the Miocene.

The final uplift of all Alpine geosynclines occurred simultaneously in the Neogene (mostly in the Pliocene) and Quaternary, and is expressed almost everywhere in the form of epigeosynclinal orogeny.

In the "Pacific" scheme, two cycles are distinguished in the Mesozoic-Cenozoic instead of a single Alpine cycle. One of them is called the Mesozoic or Cimmerian, and the other the Alpine or Pacific. The early stage of geosynclinal development in the Mesozoic cycle spans the late Carboniferous, Permian, Triassic, and Early and Middle Jurassic (the north-east of the USSR, the Sierra Nevada, and New Zealand). At the end of the Middle Jurassic or during the Late Jurassic, partial inversion occurred followed in Cretaceous time by general inversion and slight uplift of a platform type (epi-Mesozoic platform), but the orogenic regime, if any, arose in Neogene and Quaternary times, i.e., in the same epoch of "neotectonic activization" as in the regions of the other rhythm. This synchroneity of the development of the last orogenic regime in the history of the crust is an extremely interesting phenomenon.

The Alpine cycle of the "Pacific" scheme does not differ essentially from that of the same name of the "Atlantic" scheme.

The Mesozoic and Alpine cycles in regions of the "Pacific" scheme are manifested differently in different regions. The Alpine cycle usually developed next to the region of a just completed Mesozoic cycle. The downwarping of the crust at the start of the Alpine cycle on Kamchatka, for example, took place in Cretaceuos time just after a slight uplift at the end of the Mesozoic cycle of development in the neighbouring Verkhoyansk-Kolyma regions. The Alpine geosyncline of Southern California began to develop immediately after the final uplift on the site of the neighbouring Mesozoic geosyncline of the Sierra Nevada.

The diagram of Figure 106 clearly demonstrates the link of the succession of different regimes with the general endogenous rhythmicity. Certain regimes and all their intrinsic features of tectonic, igneous, and metamorphic processes occupy certain places in the cycles and the change from one regime to another is also predominantly confined to certain stages of the cycle. Several examples were given in the previous sections of this chapter.

General rhythmicity affects the development of any regime, geosynclinal, platform, or orogenic. The development of the platforms in Mesozoic-Cenozoic times, for example, may obey both the Atlantic and Pacific schemes of rhythmicity: the East European platform followed the "Atlantic" scheme and its single Alpine cycle in Mesozoic-Cenozoic times, whereas elements of the "Pacific" scheme and its Cimmerian cycle were clearly felt in the development of the Siberian platform and the western part of the North American platform. These elements are mainly expressed in the final uplift of these platform regions having developed before that on the East European platform (already in Mesozoic time).

In general, however, it can be noted, when general oscillations on a platform are compared with those in neighbouring geosynclines, that the movements on platforms, both up and downward, lag rather behind those in geosynclines. The Uralian geosyncline downwarped as early as Silurian time, whereas the East European platform was uplifted at the time and only began to sink in the Middle Devonian. When an uplift developed in Late Carboniferous and Permian times on the site of the Uralian geosyncline, broad regions of downwarping still existed on the East European platform. In the Alpine cycle, the same platform was involved very gradually in subsidence. In the Early Jurassic, when major downwarps had already occurred in the Caucasian geosyncline, only the southernmost margin of the platform began to subside, whereas much of the platform was involved in subsidence only in Late Jurassic time. At the end of the Alpine cycle—in the Palaeogene and Neogene—there were still zones of downwarping on the platform, while the Caucasian geosyncline had already been involved in its final uplift.

14-0601

We have more than once mentioned that orogenic regimes, both epigeosynclinal and epiplatform, were restricted to the final stage of the endogenous cycles. This pattern is seen from Figure 106.

The rift regime is confined to the end of the Alpine cycle and occupies the same place in this cycle as the orogenic regime in other regions.

It is generally accepted that the ophiolitic stage in the development of eugeosynclines occurs at the beginning of the endogenous cycle, while the orogenic regime is characteristic of the end of the cycle. The diagram in Fig. 106 shows that these two regimes often coincide in time: when the orogenic regime is developing in one zone, the ophiolitic stage of the eugeosynclinal regime has already begun in another, i.e., the end of one cycle is superimposed, so to speak, on the beginning of the next.

We may note this phenomenon here, and shall return to special consideration of it in the last chapter, when we will clarify the reason for distinguishing the "excited" regimes mentioned in the legend of Figure 106.

CHAPTER 19

Principles of the Zoning of Continents by Endogenous Regimes

The distinguishing and comprehensive description of types of endogenous regime helps us obtain an idea of the structure and character of the geological development of some one area of a continent with the maximum economy of words and time. It is sufficient, in fact, to say that a given region was characterised during some particular period of geological time by an orogenic regime for a picture to arise of the contrasted division of that region into uplifts (of a horst-anticlinorium type) and subsidences (of a graben-synclinorium type), with a predominance of uplift over subsidence, of sediments of a molasse and lagoon formation of block and injection folding (with a possible local manifestation of general crumpling along deep faults), of the absence of metamorphism, and of magmatism of the mainly extrusive character, the chemical composition of which depends on whether we are dealing with epigeosynclinal or epiplatform orogenesis.

If another zone is referred to, however, as a geosynclinal central uplift, much more complicated ideas are evoked. We shall think of the general structure of the central uplift as of an inverted anticlinorium complicated by intensive general crumpling. Helvetian-type nappes are possible on the flanks of the uplift. If the central uplift was formed in a eugeosynclinal zone, it should be made up, largely, of rocks of the ophiolitic stage: deep-sea siliceous sediments, jaspers, limestones, and ophiolites proper (i.e., basic and ultrabasic igneous rocks). In the central uplift we shall expect by rights minor intrusions of a gabbro-plagiogranite formation, and large granite batholiths and fissure intrusives of varied composition. There should be developed regional metamorphism, with which deep diapirism may be associated, in particular, in the form of Pennine-type nappes.

The importance of the classification is also that it allows us, having established a regime's type by *certain* features, to predict the probability of finding *others* characteristic of the regime. A coarse, clastic red molasse formation, disturbed by local block folds, may prove a sufficient indication to class this zone as orogenic and to forecast a possibility of finding rocks of a lagoon formation within it not yet encountered, injection folds, deep-seated faults dividing uplifts and troughs, volcanic outpourings, and fissure intrusives.

This prognostic value of the classification of endogenous regimes can be deepened and broadened if, in addition to the general characteristic of each regime, the patterns of their historical development and restriction to certain stages of endogenous cycles are also employed. If, for example, signs of an ophiolitic formation are found and it is established that it is of Devonian age, then, basing on the general patterns of development of endogenous processes, we can suppose with grounds that we have here a geosyncline of the Hercynian cycle. That will suggest that partial inversion and consequently also major folding, regional metamorphism, and granitization occurred here, probably, during the Early Carboniferous, that flysch accumulated in the Middle and Late Carboniferous, and that, if the region was involved in epigeosynclinal orogenesis, it happened in Permian time, when orogenic uplifts and troughs were formed, sediments of molasse and lagoon formations were deposited, outpourings of lavas occurred, and fissure intrusives were formed. All these assumptions would have to be checked, of course, but they are useful for elaborating the lines of further investigations.

The great practical importance of classifying endogenous regimes is that by basing ourselves on it, we can, albeit in general form, predict the distribution of minerals, the formation of which is closely associated with the composition of the magma, the character of the magmatic activity, the conditions of the formation of sedimentary series, the character of the tectonic structures, and with metamorphic processes. Depending on the genetic type of a mineral, its concentrations are controlled by one or more of these factors, but they all depend on the endogenous regime, and endogenous regimes may therefore also be considered metallogenous regimes (understanding by metallogenesis the formation not only of metals, but of any mineral).

The practical importance of classifying geological regimes has long been understood in the USSR. The work of most value in this field is that of S.S. Smirnov, Yu.A. Bilibin, and V.I. Smirnov.

Tectonic Zoning

Work on classifying endogenous regimes at first took the form of tectonic zoning, i.e., the distinguishing of zones of different tectonic regimes. Very soon, however, in addition to the purely tectonic features of the development of an area, phenomena of magmatism, and also of metamorphism, began to be included in the description. Tectonic zoning was converted, essentially, into zoning by types of endogenous development, though it continued to be called tectonic.

The most important element in zoning is the age of the regime concerned, which is usually assigned to a definite endogenous (or, as is often said, tectonic) cycle.

The first tectonic schemes of the USSR were those of A.D. Arkhangelsky, M.M. Tetyaev, and D.V. Nalivkin, published in 1933. Arkhangelsky's scheme, which was repeatedly revised and improved in the years following (together with N.S. Shatsky), was of outstanding value.

The schemes were compiled primarily on the basis of distinguishing regions with different ages of the last, "main" folding. Regions or zones of Precambrian, Caledonian, Hercynian (Variscan), Mesozoic, and Alpine (Tertiary) folding were delineated and in addition several other features of the structure (depth to the basement on ancient platforms; outcrops of older folded complexes surrounded by younger ones; the "reworking" of older structures by younger movements, and so on) were distinguished on them.

In 1956 on the initiative of Soviet scientists (N.S. Shatsky and A.A. Bogdanov), work began on compiling international tectonic maps, for which the appropriate international commission was set up. The work was concentrated at first on Europe, and an International Tectonic Map of Europe, edited by Shatsky, was published in 1964 (1 : 250 000).

Compared with the first tectonic schemes of the USSR, the list of conventional symbols used on this map expanded considerably. Zones of the Svecofennian, Karelian, Gothic and Dalslandian, Baikalian and Cadomian, Caledonian, Variscan (Hercynian), and Alpine folding are distinguished. On platforms, both ancient and young, the age and depth of the basement are shown. Eugeosynclines and miogeosynclines, and marginal troughs (by which are meant our *foredeeps*) are delineated. In folded regions the rock section is divided into structural stages, three usually being distinguished, corresponding in general to the following stages of orogenesis: ophiolitic, inversion, and epigeosynclinal.

In addition, the most important structural forms (anticlinoria, synclinoria, major anticlines and synclines, and large dislocations on platforms) are shown.

Tectonic maps of the other continents were in preparation at the time of writing.

Tectonic maps of this type, it must be noted, have drawbacks as well as very great positive qualities.

The age of the last, "main" folding determines the endogenous cycle to which a given folded zone belongs, but this criterion is inadequate, of course, for establishing whether the zone belongs to a definite regime. The problem is resolved to some extent by distinguishing eugeosynclines and miogeosynclines and foredeeps, but the orogenic regime is not distinguished on the maps, the type of folding remains unclarified, zones of regional metamorphism are not indicated, and magmatism is quite inadequately reflected. These drawbacks are associated with the fact that, from the very beginning, only the tectonic criteria of zoning were borne in mind and at the time (more than 40 years ago) exaggerated importance was attached to folding at the expense, for example, of oscillatory movements. But the breakdown of folding into types had not yet been worked out and the time of its occurrence, it seemed, was the main criterion that should be used in zoning. When, however, it became clear that it was only one of the many characteristics of geological development, attempts were made to broaden the content of the maps by formal addition to new symbols in the legend without revising the fundamentals of zoning.

An interesting step towards improving the principles of tectonic zoning has been made by T.N. Spizharsky who proposed the idea of zoning by tectonic regimes rather than age of folding, which brought us much nearer already to the idea of zoning by endogenous regimes.

Starting from the conception that the crust differs in composition and structure in geosynclines from that of platforms, T.N. Spizharsky divided continental tectonic regimes into regimes of a geosynclinal crust and regimes of a continental crust. In considering the former, he distinguished two stages in the evolution of a geosyncline: (a) demission (predominance of subsidence) and (b) inversion. The main criterion for the dividing of geosynclinal regimes was taken to be magmatism, i.e. its composition. Four kinds of geosynclinal regime were distinguished: (a) femic, i.e., characterised by very intensive manifestations of primary basic and ultrabasic magmatism; (b) salic-femic, distinguished from the first by the greater role of acid magmatism; (c) femic-salic, with a preponderance of acid (granite) magmatism over basic; and (d) salic, in which the stage of basic and ultrabasic magmatism is wholly or almost wholly lacking. Ural Hercynian geosyncline is considered an example of a femic regime; a salic-femic development is encountered in the Caledonian cycle in the Sayans and Altai; a femic-salic regime is characteristic of the Tien Shan in the Hercynian cycle; and a salic regime is observed in the Verkhoyansk Mts., in the Kimmerian cycle. In this sequence the transition from more mobile geosynclinal regimes to more and more quiet ones is traceable: a femic regime is a typical eugeosynclinal one, while a salic regime is typically miogeosynclinal.

T.N. Spizharsky, in addition, identified regimes of median masses and a so-called impositive one, to which he assigned the Okhotsk-Chukota Cretaceous andesite belt, in particular. The regimes of the continental crust he divided into an omogeosynclinal one which coincides almost fully in its properties with our parageosynclinal regime and regimes of "completed folding", which he divided into orogenic and koilogenic regimes, meaning by the latter young platforms in which subsidence predominates and a sedimentary cover is formed, whereas the orogenic regime is characterised by a predominance of uplifts and erosion.

An example of a koilogenic regime is the epi-Hercynian West Siberian platform.

The platform regime in Spizharsky's scheme embraces only ancient platforms; their development is subdivided into the following stages: aulacogen, Kalumogen (moderate subsidence) and emission (moderate uplift).

For all the interest that this classification of regimes arouses, it is difficult to employ. Its drawbacks, in particular, are the complicated new terms and the vagueness of the criteria for distinguishing various types of geosynclinal regime differing in the character of their magmatism.

Zoning by Types of Endogenous Regime

Our proposal is to take the classification of endogenous regimes presented in the previous chapters as the basis for zoning. The properties of these regimes were set out above and summarized in Table 3.

The detailed character, accuracy, and fullness of zoning by types of endogenous regime is affected by its scale and the geological period for which it is carried out. While very small-scale maps clearly can show all geosynclinal regimes, on the one hand, only in a generalized way together, and all platform regimes, on the other hand, larger-scale maps can distinguish eugeosynclines, and miogeosynclines, and ancient and young platforms. With zoning on an even larger scale central uplifts, marginal troughs, and stable geanticlines can be shown in geosynclines, and ancient and young platforms, shields and sediment-covered platforms, and subgeosynclines and subgeanticlines distinguished among platforms. A further increase in scale allows us to show the following structural elements: anticlinoria and synclinoria of various orders, types of folding or even individual folds, faults of various types, degree of metamorphism, and intrusives of various composition.

The geological time for which a zoning scheme is compiled affects the content of the map (1) through the periodicity of the endogenous processes, and (2) through the accessibility of formations for observation. In the periodicity of endogenous processes, zoning can best be related to an endogenous cycle as a whole, or to some certain stage of it. When the Hercynian cycle, for example, is being zoned, the map would show geosynclines, platforms, orogenic zones, and rifts developing during the Hercynian cycle, i.e., in the middle and late Palaeozoic. On a schematic diagram of the final stage of this cycle, geosynclines will not figure, and attention will be focused instead on epigeosynclinal and epiplatform orogenesis.

The accessibility of formations to observation, as will be readily realised, affects zoning in this respect, that it is easy to zone for the complex of rocks that is more exposed, i.e., that occurs on the surface. In general, therefore, zoning is easiest for endogenous regimes that developed during the youngest cycle, i.e., the Alpine. In these cases certain difficulties may arise only where the structures that originated at the end of the cycle have substantially altered structures belonging to earlier stages of the same cycle. In a case of epigeosynclinal orogenesis, for example, the orogenic structures are superimposed on the geosynclinal ones, and these can only be identified by "removing" all the distortions introduced by the orogenesis. A similar problem arises when geosynclinal or platform structures are overlain by rift structures.

The compiling of zoning schemes for older cycles calls for "removal" of the results of subsequent endogenous processes. The influence of processes superimposed in subsequent cycles is minimal, or so small as to be ignored where the structures of the cycles being studied were affected by a later platform regime. In Central Kazakhstan, for example, the structures created by the geosynclinal regime of the Hercynian cycle have remained practically unaltered because a platform regime was active there in the subsequent (Alpine) cycle. In the Donets Basin, the degree of preservation of the Hercynian structures is similar. The subsequent changes in structures relating to older cycles may be very slight if an orogenic regime of a later cycle, with low-intensity tectonic movements, was superimposed on them. The Hercynian structure of the Urals, for example, has survived well, although some faults and gentle uplifts and troughs in the Urals are connected with weak orogenic movements of the Alpine cycle. In other cases the superimposition of subsequent processes can have caused such distortion of the structures of the cycle being studied that it is impossible to establish their original shape. The possibility of distortion increases, obviously, the earlier the endogenous cycles we penetrate. The epiplatform orogenesis at the end of the Alpine cycle in the Tien Shan allows us, without special difficulty, to reconstruct the structures not only of the young platform that existed there in the earlier stages of this cycle, but also the regimes of the Hercynian and, for the Northern Tien Shan, Caledonian cycles. At the same time, the conditions that existed in the Great Caucasus or the Alps in the Hercynian cycle are reconstructed with great difficulty and incompletely, because they were strongly distorted by the intensive geosynclinal processes of the Alpine cycle.

It will be clear from the foregoing why zoning schemes are usually compiled for the youngest regimes; they are most complete and best substantiated. The zoning of earlier regimes, however, though incomplete, is of great practical as well as theoretical interest, because certain groups of minerals may be associated with more ancient regimes. In the Tien Shan, for example, telethermal and epithermal deposits of mercury and certain other elements are associated with the orogenic regime of the Hercynian cycle. In connection with the geosynclinal regimes of the Hercynian and Caledonian cycles, complex ore deposits were formed and a number of deposits associated with the basic and ultrabasic igneous rocks of the early stage of geosynclinal development.

In the Caucasus, certain groups of deposits, connected with the Caledonian, Hercynian, and Alpine regimes have been distinguished. The distribution patterns of each of these groups can apparently be elucidated only by zoning the regimes for each particular cycle and gradual "exposure" of the older structures beneath the younger ones.

PART II

Tectonics of Oceans

CHAPTER 20

Relief, Sediments, and Rocks of the Ocean Floor

Oceans and seas cover 70% of the Earth's surface. If we subtract the area of the continental shelf and epicontinental seas, which are submerged parts of the continents, the share of the oceans proper and deep seas is 50%. General ideas of the structure and evolution of the Earth's crust quite obviously cannot be obtained without taking its oceanic half into account. It is only two decades or so ago that scientists began to unfold the mysteries of the structure of the ocean floor. This became possible because of the development of a number of new methods of research, mainly geophysical, and because of the introduction into common practice of new technical means, above all, specially equipped research ships.

The progress that has been made in study of the structure of the ocean floor is remarkable, but one must stress that we are still at the beginning, and that our knowledge of ocean depths should not be over-rated. The continents, which have been studied by geologists in much better conditions for 200 years, have not yet yielded up many of the mysteries of their structure. Even greater enigmas must be concealed in structures that have been studied for only twenty or thirty years and which lie, in addition, under a layer of water several kilometres deep. At this time, however, the opinion has spread in certain scientific circles that we have as good knowledge of the ocean floor, if not better, than of the continents and that "continental" geology should be revised in the light of the new data on the structure of oceans. It is categorically impossible to agree with that. General conceptions of the Earth's crust should be based on a reasoned, balanced synthesis of "continental" and "oceanic" views.

The following methods are now employed to study the structure of the ocean floor:

1. bathymetry,
2. dredging of rocks from the floor,
3. geophysical methods.

1. **Bathymetry of the ocean floor.** This technique of surveying the relief of the ocean floor is carried out by means of echo sounding, i.e., the sending of sound signals from a ship to the floor and reception of their reflections.

Study of the topographic forms of the ocean floor, it must be stressed, is more important for problems connected with endogenous dynamics than study of landforms. On land, the manifestations of tectonic and volcanic processes in the relief are usually much distorted by erosion, which destroys elevations, and sedimentation, which fills depressions. Because the processes of destruction and accumulation proceed incomparably slower on the ocean floor than on land, its topography reflects tectonic and volcanic processes to a much greater degree.

2. **The dredging of rocks from the floor.** This is done by means of dredges for hard rocks, and for loose sediments by special pipes, which in falling pierce the bottom to a depth of 10 m or more.

In recent years, offshore drilling has not only used in shallow water, but also become widely used in the deep sea. The American research ship *Glomar Challenger* of the USA has made a great many drilling profiles across all the oceans. The record depth of water in which drilling has been carried out, is 6000 m and the maximum depth of drilling 1200 m. The section of the sedimentary series lying on the floor has been studied in many areas of the ocean by means of underwater drilling, but the hard basalt underlying them has only been penetrated to a slight degree. At present a project of drilling is being carried out through the oceanic sedimentary layer with penetration in the basaltic layer.

3. **Geophysical methods.** Seismic methods in various modifications are the geophysical techniques of greatest importance. To study the structure of the sedimentary series of the ocean floor and the relief of the underlying hard basement, ultrasonic profiling based on the reflection of ultrasonic waves from various reflecting surfaces within the sedimentary formations is employed. To penetrate to greater depths, where hard rocks rather than loose sediments occur, deep seismic sounding is used, based on the recording of elastic waves caused artificially by explosions in the water. This method helps us obtain an idea of the elastic properties of the medium down to a tens of kilometres below the floor. Deep seismic sounding is supplemented by placing geophones on the floor that record weak natural earth tremors and help us judge the elastic properties of rocks at different depths.

Gravimetric measurements are important for elucidating the picture of the distribution of masses beneath the ocean floor and their degree of equilibrium. They help distinguish submarine elevations of volcanic origin from topographic forms with another history. A combination of gravimetric and seismic methods makes it possible to extend the results of the latter far beyond the limits of individual seismic profiles.

Magnetic surveys help us to study magnetic anomalies on the ocean floor. As we shall see, much importance is now attached to them, and palaeomagnetic research is playing a special role.

Geothermal research indicates the considerable variations in terrestrial heat flow in various regions of the oceans, which is important for developing ideas of deep-seated processes.

The geological and geophysical data obtained directly in the oceans are supplemented by the findings of observations on islands and the adjoining continents made by the normal land methods.

The Relief of the Ocean Floor

The largest geomorphological elements, distinguished in the topography of the ocean floor are the underwater margins of the continents and the bed of the world ocean (see the appended tectonic map of the world[4]).

The underwater continental margins are divided, in turn, into the continental shelf, the continental slope, the continental rise, and deep trenches.

---

[4] We also recommend the reader to use the maps of the oceans published by the Main Board of Geodesy and Cartography of the USSR: the Pacific Ocean, 1 : 25 000 000, 1968; the Atlantic Ocean, 1 : 20 000 000, 1964; and the Indian Ocean, 1 : 5 000 000, 1964.

The *continental shelf* is a shallow-water zone adjoining the land, characterised by a depth of less than 200 m almost everywhere and a smooth relief with a general gentle gradient (not more than 1 : 1000) toward the ocean. The shelf varies in width from a few kilometres to 800 km. An especially wide shelf fringes the Arctic Ocean. The continental shelf also encompasses a vast area in Indonesia. The Baltic and North Seas, Kara, Laptev, East Siberian and Chukchi Seas, the Yellow Sea, Java, Arafura, and Timor Seas are fully within the shelf and hence are typical epicontinental seas, i.e., lie on the submerged parts of the continents. Considerable parts of the Bering, East China, South China, Andaman Seas and Baffin Bay lie within the shelf.

Many irregularities have survived on the surface of the shelf since the time it was above sea level, as during the epoch of Quaternary glaciation, when considerable masses of atmospheric water were bound in the continental ice sheets, and the level of the world ocean was 100 to 150 m lower than now. The submarine valleys on the bottom of the North Sea, mentioned above, are examples of subaerial erosion forms. A deep submarine valley has been dug in the shelf off the estuary of the Hudson River on the Atlantic margin of North America; similar submarine valleys have been found off the estuaries of many other rivers.

Other irregularities of the surface of the shelf are connected with an uneven accumulation of sediments. On the whole, however, the shelf has an extremely gentle topography, which is the consequence of the rewashing of sediments by waves and levelling of their surface at the level of the wave base.

The *continental slope* connects the shelf with the areas of greater depth. Its slope varies between 3° and 10°. Where there are no deep trenches, i.e., on Altantic-type continental margins, the continental slope goes down to depths of 1500 to 3500 m. Where there are trenches, on Pacific-type margins, however, the slope reaches much greater depths, and is steeper (up to 10°). In some regions terraces have been found on the continental slope at depths of several hundred to several thousand metres. An even more important element of the relief of the continental slope is the marginal plateau discussed earlier (see Chap. 17). These are benches with levelled surfaces, similar to the shelf, but lying at depths between 1000 and 2000 m. The most typical example of a marginal plateau, as we have said, is the Blake Plateau off Florida.

The *continental rise* is observed where there is no deep sea trench. It is a gradual transition from the relatively steep continental slope to the practically horizontal surface of the abyssal plains. Its gradient is between 1 : 100 and 1 : 700.

An element common to both the continental slope and the continental rise is the submarine canyon. These are valleys, several hundred metres deep, running down the slope, and 1000 to 15 000 m wide. They have steep, often precipitous sides and an uneven bottom divided into benches. Canyons generally begin at the top of the continental slope, but, frequently, though far from always, they lie on a continuation of submarine valleys located on the shelf. They are down the slope to various depths and some go beyond the limits of both the slope and the continental rise and are traceable on the oceanic plains.

Unlike the submarine valleys on the surface of the shelf, continental canyons are of submarine, rather than subaerial origin and are formed by what are known as turbidity currents, that is flows of water whose density is higher than that of clean water because of an accumulation of suspended mud particles. These turbidity currents develop predominantly on the edge of the shelf, which consists, on its surface, of loose silt and which is stirred up, especially by earthquakes or catastrophic run-off from the adjoining land; the resulting suspension carries a certain volume of water down the slope. Turbidity currents move extremely fast and have great erosional energy. A submarine canyon, once formed, acts as a channel for subsequent turbidity currents, which deepen and lengthen it over a long period.

Deep trenches are confined to Pacific-type continental margins. In them the greatest depths have been observed, with a record one of 11 023 m. Trenches are generally 2000 to 3000 m deeper than the adjoining abyssal plains. On the continent side they border immediately on the continental slope.

A trench is generally 100 to 150 km wide. Its cross-section is asymmetric, the landward slope (10°-15°) being steeper than the seaward one (2°-3°). Both slopes are complicated by scarps, some parts of which are very steep (up to 45°). The bottom of a trench is level, nearly flat, but the flat strip is narrow, only a few kilometres wide, and the transition from the steep slope to the flat bottom is very sharp. Trenches are divided into elongated ovals along their trend by transverse rises. The individual ovals are hundreds, even thousands of kilometres long.

The floor of the ocean includes, as geomorphological elements, deep ocean basins and various rises, expressed in the relief both as long ridges and as separate, rounded or irregular swells.

Deep ocean basins are the largest features on the bed of the ocean, their floors lie at depths of 5000 to 6000 m. Abyssal plains and abyssal hills are developed on the floor of deep ocean basins. In the Atlantic Ocean plains predominate over hills, which are found only in areas furthest from the continents. In the Pacific Ocean, on the contrary, abyssal hills strongly predominate, occupying 80 to 85% of its area.

The abyssal hills on the floor of deep basins are dome-shaped rises 50 to 100 mm high and 1 to 10 km wide at their base. The most typical hills are 300 m high and about 6 km in diameter. Abyssal plains are ideally levelled expanses with gradients of less than 1 : 1000. In the Atlantic Ocean they occupy parts of the basins close to the continents. In the Pacific Ocean only the Gulf of Alaska is a coastal plain, the other abyssal plains are in the central part of the ocean around groups of volcanic islands (the Hawaiian, Marshall, and Gilbert Islands, and others).

An important feature of the topography of the deep basins in the eastern part of the Pacific Ocean is east-west rises of vast length. The Mendocino Escarpment begins near California and runs westward along 40° North latitude for 3000 km. The Murray Escarpment starts near Los Angeles and continues westward to the Hawaiian Islands, a distance of 4000 km. The Clarion Escarpment, which lies on the latitude of Mexico City, is 2500 km long. Finally the Clipperton Escarpment stretches from the Isthmus of Panama west-south-westward for more than 3000 km. We have listed here only the most important of these escarpments.

The escarpments divide the ocean floor into areas of different depths. The difference in depth on either side of a rise may be 1000 to 1500 m as is observed along the Mendocino and Murray Escarpments, or be limited to a few hundred metres (the Clipperton Escarpment). The zone of a rise itself has a complex relief, of which two types can be distinguished: (1) an asymmetric ridge sloping gently toward a shallower area of the floor and steeply toward

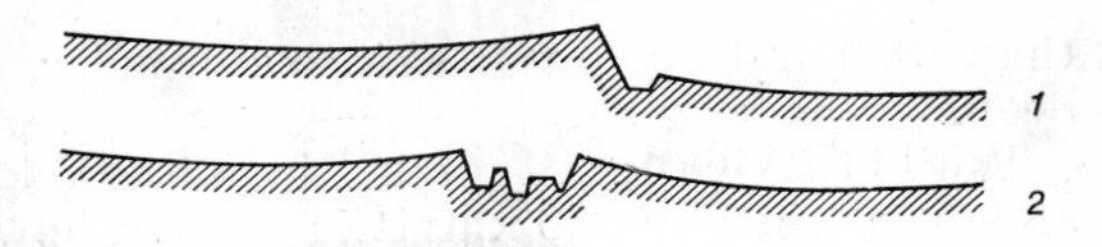

Fig. 107. Schematic profiles across fault zones in the Eastern Pacific, showing two types of relief:

1—a unilateral horst; *2*—two facing unilateral horsts with a graben in between (after Menard, 1966)

a deeper area accompanied with a steep-sided rift; and (2) two asymmetric ridges facing each other along their steep slopes and divided by a rift (Fig. 107). An example of the first type is the Mendocino Escarpment which is 1500 m high. A submarine range runs along the edge of the rise; it is 1500 metres above the elevated side of the rise and 3000 metres above the sunken side. An example of the second type is the Clipperton Escarpment. Here a low shell stretches along the rise with an axial trough, 15 to 40 km wide, and more than 1200 m deep. North of the swell, the floor is 300 to 400 m lower than on the south. The width of the swell is about 100 km. These rises are usually called faults in the literature (the Mendocino Fault, Murray Fault, etc.). Geologically, as we shall see, these faults can be classed in the type of structure as "aseismic" ridges described below.

Mid-ocean ridges are the biggest positive elements in the relief of the ocean floor, and form a single system more than 60 000 kilometres long, extending from the Atlantic Ocean into the Indian Ocean, and thence into the Pacific.

The ridge most thoroughly studied is the Mid-Atlantic. Its crest has been traced almost exactly along the axis of the ocean. A characteristic depression—a "rift valley", shaped like a kind of cleft—stretches along the crest and is broken up into a number of separate troughs or basins by transverse walls or bridges. The bottom of the rift valley is 4000 m deep, on average, while the neighbouring peaks of the ridge rise to 2000 m below sea level, and are observed in places even above sea level, forming islands like St. Paul, Ascension, and Bouvet Island. The valley is between 25 and 50 km wide. Its bottom is rugged and there are almost no level areas on it.

Belts with an extremely complicated relief, with uplifts and troughs up to 1000 m of relative amplitude border the rift valley on both sides. The relief becomes smoother lower down the slope of the ridge; several scarps, hundreds of metres high, divide the slope into steps.

In the area of Bouvet Island in the far south of the Atlantic, the Mid-Atlantic Ridge turns abruptly north-east and passes medially between the southern tip of Africa and Antarctica. There it is called the African-Antarctic Ridge. Continuing further it passes into the Western Indian Ridge, and near Rodriguez Island it forks, the Western Indian Ridge continuing northward (now called the Arabian-Indian or Carlsberg Ridge). To the north the Carlsberg Ridge turns north-eastward towards the entry to the Gulf of Aden where it ceases to be a geomorphological element. From Rodriguez Island, another ridge, the Central Indian, trends south-eastward toward Amsterdam Island (Nouvelle Amsterdam) and St. Paul Island. Further on it turns east-south-east and passes between Australia and Antarctica (the Australia-Antarctic Ridge) (see the appended schematic tectonic map).

The crestal zone of the Western Indian and Carlsberg Ridges, like that of the Mid-Atlantic Ridge, is complicated by a rift valley. At the entry to the Gulf of Aden, interestingly, the ridge ceases to be a positive form of relief, but the rift valley continues as a chain of basins. From there it can be traced into the Red Sea, where an axial depression is known 60 km wide and more than 2000 m deep, "inserts" into a broader, shallow zone not more than 600 m deep.

While the Western Indian Ridge is 300 km wide, the Central Indian Ridge attains a width of 900 km; it is gentler, however, and has no large rift valley. Instead, there are small, isolated deeps. The Australian-Antarctic Ridge is even gentler.

The continuation of the ridge into the Pacific forms the Southern Pacific and East Pacific Ridges, which are often called rises rather than ridges because of their very gentle relief. The East Pacific Rise is a swell 2000 to 3000 m high and 2000 to 4000 km wide. From 55° South latitude to the equator it runs in a north-south direction, then its trend becomes north-westerly and it passes into the entry to the Gulf of California in the same way as the Carlsberg Ridge runs into the Gulf of Aden. No longitudinal rift valley has been observed at the top of this rise.

Mid-oceanic ridges are cut by numerous transverse fissures. The largest on the Mid-Atlantic Ridge are observed in the equatorial belt, where it makes a sharp S-bend. The biggest at 11° North latitude has been named Vema, it varies in width between 8 and 20 km and is a little less than 5200 m deep. The mountains bordering it on the north and south rise to 3000 m above it. The bottom of Vema is surprisingly flat. It forms a kind of corridor connecting the basins on both sides of the ridge. It has been noted that the axis of the Mid-Oceanic Ridge is offset, as if along a transform of strike-slip fault, where it is cut by transverse fracture zones. The offset along the Vema is left-lateral and has an amplitude of 460 km. The appended schematic tectonic map of the world gives an idea of the transverse offsets of other ridges as well as of the Mid-Atlantic Ridge. Mid-ocean ridges are characterised throughout their length by high seismicity.

Among the positive topographical features of the ocean floor there are also "*aseismic*" *ridges*, so called because, unlike the mid-ocean ridges, earthquakes are not registered on them now. An example is the Walvis Ridge in the Atlantic, which has been traced for 2000 km in the south-eastern part of the ocean. The crest of the ridge lies at depths between 1500 and 2000 m. The ridge is between 200 and 250 km wide and is asymmetrical with a flat top; the eastern flank sinks steeply to a depth of 4500 m while the western flank slopes gently.

A whole number of "aseismic" ridges are known in the Indian Ocean. One of them, for example, is the Mascarene Ridge, which bends in an arc, convex on the eastern side. The Seychelles lie on this ridge, which is more than 2000 km long and between 200 and 450 km wide. The arch is flat and slopes steeply (up to 15°). The ridge rises 4000 m above the surrounding basins. Another of these ridges is the Maldive, which runs north from the Chagos Archipelago for 2500 km. The Eastern Indian Ridge, which runs for 4250 km almost due north-south along 90° East longitude from the Andaman Islands in the north to 30° South latitude, is an interesting feature. It is often called the "Ninety-east Ridge". It is about 200 km wide and rises between 1000 and 3500 m above the surrounding floor. Its slopes are as steep as 25°. The southern end of the ridge is joined by the east-west "aseismic" Western Australian Ridge.

In the Pacific Ocean typical "aseismic" ridges occur only between the Eastern Pacific Rise and the continent of South America (the Cocos, Nasca, Carnegie, and Sala y Gomez ridges).

In the Arctic Ocean, the Lomonosov and Mendeleev Ridges are of this type; the first stretches from the Siberian shelf to the Canadian shelf, is flat topped, and rises to a height of 2600 metres above the adjoining basins.

In addition to ridges the ocean floor is characterised by *rises* of various size and shape. In the Atlantic, for example, the Bermuda Plateau forms an oval elevation 1000 km long and 500 km wide, which rises 1000 m above the surrounding abyssal plain and lies at a depth of around 4800 m. The middle of the rise is a plateau with a very gentle relief.

West of Scotland there is the Rockall Plateau, oval in shape, 500 by 200 km, which rises from a mean ocean depth of 3000 m to a depth of 165 m above sea level. Around the Azores, at a depth of 2000 m is the Azores Plateau with a very broken surface.

The Rio Grande Plateau off the coast of Brazil is an oval block with an east-west elongation, 500 km × 300 km. Most of the rise is elevated 1500 m above the surrounding basins, while its centre rises by 2500 m.

In the Indian Ocean flat rises are the Crozet Plateau and the Kerguelen Plateau. The Crozet Plateau is 300 by 100 km and rises above the abyssal basins to a height of 4000 m. The Kerguelen Plateau trends eastward from Kerguelen Island for 1800 km.

In the Central Pacific, the Manihiki Plateau in Polynesia should be classed in this same type of rise. It is rectangular in shape, with sides 750 and 1000 km long, lies at depths between 2500 and 4000 m and is surrounded by deeps of 5500 m. Several volcanic islands rise from the plateau. In the same part of the Pacific gentle oval and linear rises of the floor are common and serve as the basement for chains of volcanic and coral islands. The latter, incidentally, are coral crowns on the tops of sunken volcanic cones; the abyssal nature of the islands is consequently the same in both cases. The Hawaiian Rise, which forms the basement of the Hawaiian Islands, is a very large one, about 3000 km long and 1000 km wide. It slopes very gently, the slope becoming gentler with depth, and passes imperceptibly into the bottom of the surrounding deep ocean basins. Some of these rises are surrounded by gentle peripheral depressions in the floor (trenches).

Local positive forms of relief are extremely widespread on the ocean floor, among which the following are usually distinguished *seamounts*, *volcanic islands*, *atolls*, and *guyots*. These forms are related because their bases are volcanic cones rising from the ocean floor. Seamounts are submarine volcanoes of conical shape with sharp peaks. Volcanic islands are the same volcanic cones, with their tops above sea level. Atolls are coral islands built upon submerged volcanic cones. Finally, guyots are also volcanic mountains, but with flat, truncated tops which, as we shall see later, were cut off by marine abrasion.

A feature of all these local elevations is their steep slopes, which become steeper toward the top and vary between 15° and 40°. Volcanic cones range in height from 1000 to 10 000 m that is to say, some of the highest mountain peaks on Earth, and higher than those on land, can be found among them. The diameter of the base of the cones varies from tens of kilometres to 200. The width of the flat tops of guyots varies from between several kilometres to 30 km. The depth of water above them varies between tens of metres and 2000 m. Submarine peaks are also found at various depths.

In the Atlantic Ocean, volcanic cones are dispersed on both sides of the mid-oceanic ridge, but there are few guyots. The base of the Bermuda Islands is a large volcanic mountain whose top is only 40 m below sea level. A group of seamounts rises from the Azores Plateau. In the zone of the mid-oceanic ridge in the South Atlantic, there are the volcanic islands of St. Paul, Ascension, St. Helena, Tristan da Cunha, Gough, and Bougvet, and a number of seamounts. In the Gulf of Guinea a chain of four volcanic islands stretches in a north-easterly direction.

In the Indian Ocean, far from all the local submarine elevations are yet known; in the north-eastern sector, however, more than a hundred seamounts have been found. A chain of them is distinguished between the Cocos (Keeling) Islands and Christmas Island. Another chain connects the Western Australian "aseismic" ridge with the continent of Australia. Belts of seamounts have been noted along both sides of the Eastern Indian Ridge. Only a few guyots have been found in the Indian Ocean.

The Pacific Ocean has more submarine and above-water volcanoes than other oceans. The total number of these local features is at least 10 000. A great number of guyots is also known. Around ten have been found in the central part of the Gulf of Alaska. The flat tops of most of these guyots lie at depths between 700 and 2500 m. The bases of the cones are generally between 30 and 40 km in diameter. The guyots found in the Aleutian Trench are at a greater depth (on the average at a depth 1800 m greater than in the Gulf of Alaska).

Another cluster of seamounts and guyots has been found in the area of the Pacific adjoining the coasts of California and Mexico. Here, the chains of seamounts stretch mainly along the above-mentioned escarpments on the ocean floor (particularly along the Murray Escarpment).

Seamounts and guyots are common in the central Pacific: in Polynesia, the Marshall and Hawaiian Islands, and the eastern part of the Carolines. The number of guyots in this part of the ocean, like that of conical seamounts, is hundreds. All these elevations are predominantly united in linear chains, the predominant trend of which is north-westerly. This trend is also evidenced in the ranges of volcanic and coral islands, but some chains trend east-west or north-south. The chains are as long as 2500 km and each of them consists of hundreds of submarine elevations and islands. The guyots in this part of the ocean rise from a depth of 5500 m and their tops are at depths from 300 to 2000 m (predominantly between 1000 and 1800 m).

In conclusion let us note certain topographical features of the *floors of marginal and inland seas*, lying beyond the continental shelf. In them we encounter the same topographical elements as characterise the ocean. Beyond the coastal shelf there is a continental slope and then abyssal plains. They are not, however, as deep as the oceanic ones. These seas also have submarine ridges and separate seamounts.

As an example, let us take the topography of the floor of the Mediterranean (see Fig. 108). This sea is divided into two halves, western and eastern. The western half includes two closed basins, the Algiers-Provence and the Tyrrhenian. In general their floor is an abyssal plain, but a weakly defined ridge runs along the axis of the Algiers-Provence Basin. These two basins are surrounded by continental slopes and lie at depths of 2600 and 3500 m. A volcanic cone 2840 m high has been found in the central part of the Tyrrhenian Basin.

In the eastern part of the Mediterranean, Central and Levant basins have been distinguished. The floor of the first lies at a depth around 3500 m and has a hilly relief. The second is outlined by a 2500-metre isobath. Between the basins there is a plateau at a depth around 1500 m. The Crete-Rhodes island arc, which bounds the Aegean Sea on the south, is fringed by the Hel-

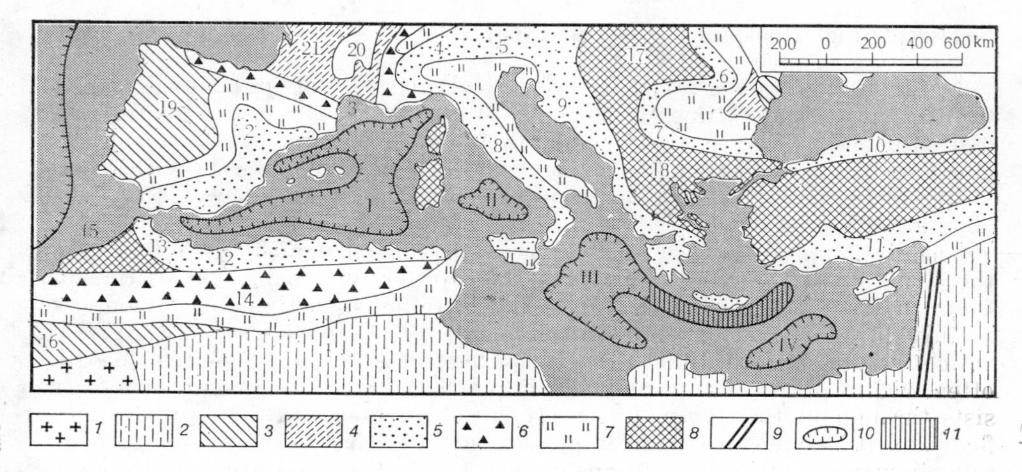

Fig. 108. Schematic diagram of the structure of the Mediterranean (after Beloussov):
1—outcrops of the Precambrian basement (shields); *2*—syneclises on Precambrian folded basement; *3*—Hercynian folded zones; *4*—syneclises on the Hercynian folded basement; *5*—Alpine geosynclines; *6*—Alpine parageosynclines; *7*—Alpine foredeeps; *8*—median masses in Alpine geosynclines; *9*—graben; *10*—marine and ocean basins (I—Algiers-Provence; II-Tyrrhenian; III—Central; IV—Levante); *11*—Hellenic Trench.
Figures on the map:
1—Betic Cordillera; *2*—Celtiberic geosyncline; *3*—Pyrenees; *4*—Western Alps; *5*—Eastern Alps; *6*—Carpathians; *7*—Balkans; *8*—Apennines; *9*—Dinaric Alps; *10*—Anatolides (Pontic Mts.); *11*—Tauric Mts.; *12*—Tell Atlas; *13*—Er Rif; *14*—the High Plateaux; *15*—Moroccan Meseta; *16*—High Atlas; *17*—Hungarian depression; *18*—Macedonian-Rhodope massif; *19*—Spanish Meseta; *20*—Central Plateau of France; *21*—Aquitaine Basin

lenic Trench with depths as much as 5000 m. This trench occupies a position in relation to the island arc similar to that of deep oceanic trenches.

A division of the floor into a shelf, continental slope, and abyssal plain at depths between 3000 and 4000 m has been observed in the Gulf of Mexico. The Caribbean Sea is divided into three abyssal basins by two submarine "aseismic" ridges.

In a number of cases, however, in marginal and inland seas we find very complicated floor relief which, as far as can be judged, is a submarine continuation of the relief of the adjoining sectors of the land. Examples are the Celebes and Sulu Seas and part of the Western Caroline Basin. In the east of the Celebes Sea, on the margins of the Sulu Sea, and to the north of New Guinea, a topography is observed composed of highs and lows, tens of kilometres wide, with an amplitude of 2000 m. This relief lies as deep as 2000 m. It is, undoubtedly, an extension of the ridges and depressions observed on the adjacent land areas of the Philippines and New Guinea and is in sharp contrast with the abyssal plains (lying alongside in the central part of the Sulu Sea, for example, at depths of more than 2500 m).

The sea between the Mariana arc and the Philippines, and also the seas of Melanesia (the Coral, Fiji, and Tasman Seas) also have complicated relief divided into numerous ridges and basins. A very complex relief is observed in the Sea of Marmara and the Aegean. There are many basins and steep ridges on their bottoms, a relief similar to that of the adjacent regions of the Balkan Peninsula and Asia Minor.

Sediments and Rocks of the Ocean Floor

In Chapter 17 we surveyed the geological structure of continental margins that are at the same time the margins of oceans. Two types of margin were distinguished, Atlantic and Pacific; their features were described and conclusions drawn as to their geological history. We shall not repeat this here, but shall immediately turn to the interior regions of oceans, i.e., to the bed of the world ocean.

Seismic techniques have made it possible to distinguish three "layers" in the structure of the crust underlying the interiors of oceans. Let us consider them in descending order. Layer I, the top one, is sedimentary, composed mainly of loose, unconsolidated sediments. The upper part of Layer II, which is accessible to direct observation, consists mainly of plateau basalts, often with a pillow structure; it is supposed that the same composition persists throughout this layer. Layer III is not exposed anywhere on the ocean floor and its composition can be judged only from velocity of propagation of seismic waves and isolated lumps of rock that have penetrated the overlying strata and are considered erratics from it. This evidence indicates that layer III is composed of a complex of various basic and ultrabasic igneous rocks—gabbro, serpentinized peridotites, and pyroxenites, and the metamorphic derivative of basic rocks, amphibolite. The Oceanic type Layers II and III, judging by the geophysical data, occur not only in the oceans, but also in marginal and interior seas, where their depth does not exceed 2000 m. Since this chapter is concerned only with the geological data on the structure of the ocean floor, Layer III, which is still inaccessible to direct geological observation, will not be considered.

The first (sedimentary) layer of the oceanic crust, a layer of loose sediments in the interiors of oceans, is very thin. In the basins of the Atlantic Ocean its thickness does not exceed 1000 m. In areas of the Pacific remote from the coast, it does not attain 500 m. Over considerable areas of the ocean floor there is almost no loose sediments, and patches of the top of the plateau-basalt Layer II are exposed on the surface of the floor. The sediments become thinner toward the crest of mid-ocean ridges, and are almost completely absent in the zone of the crest proper, being preserved only in certain depressions. In deep ocean basins sediments are developed everywhere where the floor is flat, but are absent in many sectors within areas with a hilly relief. Such partially bare areas are comparatively rare in the Atlantic and Indian Oceans, but occupy a considerable part of the surface of the floor in the Pacific. It has been established by means of seismic profiling that the hilly topography of the top of Layer II continues in the region of basins with a flat floor, but that there it is buried beneath sediments which, lying horizontally, fill up its irregularities.

The thickness of sediments increases to 3000 or 4000 m on the continental rise and continental slope. They become even thicker in closed and semi-closed marginal and inland seas, varying in thickness between 3000 and 8000 m; they become as thick as 14 km in the Black Sea, however, and 20 km in the southern part of the Caspian Sea.

The composition of the sediments depends on the distance from sources of terrigenous material and on the depth of deposition. On the continental slope and in basins adjacent to it terrigenous material plays the main role and is spread over great distances from the nearest continent (in places as far as 1000 km), and is deposited at depths of more than 5000 m. This long-distance transport is associated with the turbidity currents that run down the continental slope. A rhythmic bedding is observed in sediments of this type, i.e., a frequent alternation of coarse and fine material. The reason for this is the mechanism of turbidity currents. After a current has run down the slope and the water has become calm again, the suspended sand and clay material settles out in order of decreasing specific gravity of the particles: the heavier ones of the coarse fraction fall to the bottom first, and then finer and finer particles. This elutriation of sediments recurs with each new turbidity current.

Pelagic biogenic material plays an ever-increasing role in sediments with distance from a source of terrigenous particles, and ultimately becomes the sole component. Its composition depends on the depth of deposition, since there is a critical depth at which carbonates dissolve. In present-day oceans this critical level is at a depth of around 4000 m. At depths less than that, therefore, present-day pelagic sediments consist of calcareous mud and at greater depths, of siliceous mud and abyssal red clay. It can be thought that the critical depth of carbonate dissolution was approximately the same in the past, and on that basis we may appraise the depth of deposition of some ancient sediment or other.

Volcanic material plays a substantial role in the composition of oceanic sediments, especially near to active volcanoes.

Among both calcareous muds and terrigenous sediments, compact siliceous intercalations are encountered.

In the central areas of oceans, within deep oceanic basins, there are isolated limited sectors of shallow water, e.g., above seamounts or guyots. In these sectors shallow-water sediments accordingly accumulate: coral, algal, or detritus limestones, coarse sands, etc.

The organic residues (mainly foraminifera and diatoms) found in oceanic sediments help us to judge their age. The boreholes that have been drilled through the sedimentary oceanic layer in many places have made it possible to construct its stratigraphic section.

A very interesting discovery was the fact that rocks older than the Jurassic have not been found anywhere in the series of loose sediments on the floor of the present-day oceans. The oldest fossils studied belong to the Late Jurassic, but the thickness of the sediments occurring in some regions beneath the strata containing a Late Jurassic fauna permits us to think that at the base of the series there are somewhere also Middle Jurassic rocks, which would thus prove to be the oldest rocks at the base of the sedimentary series in the ocean. Their distribution, however, is very limited, and their presence is only so far probable in three regions: in the Atlantic, north of San Salvador Island (in the Bahamas); in the Pacific, in the area of the submarine Shatsky Ride (east of Japan and the Mariana Islands); and in the Indian Ocean, west of Australia. In other regions, younger rocks occur at the base of the sedimentary series, and are directly underlain by plateau basalts of Layer II.

A certain pattern is observed in the distribution of oceanic sediments of different age at the base of the sedimentary series. The oldest are most remote from the mid-oceanic ridges and give way to younger and younger sediments toward their crests. On the periphery of the Atlantic Ocean, for example, in addition to the above-mentioned Jurassic rocks, which occupy a small area, predominantly Cretaceous sediments lie at the base of the sedimentary series; at a distance of 1200 km from the crest of the Mid-Atlantic Ridge, however, they are followed by Palaeogene sediments, which, the closer we

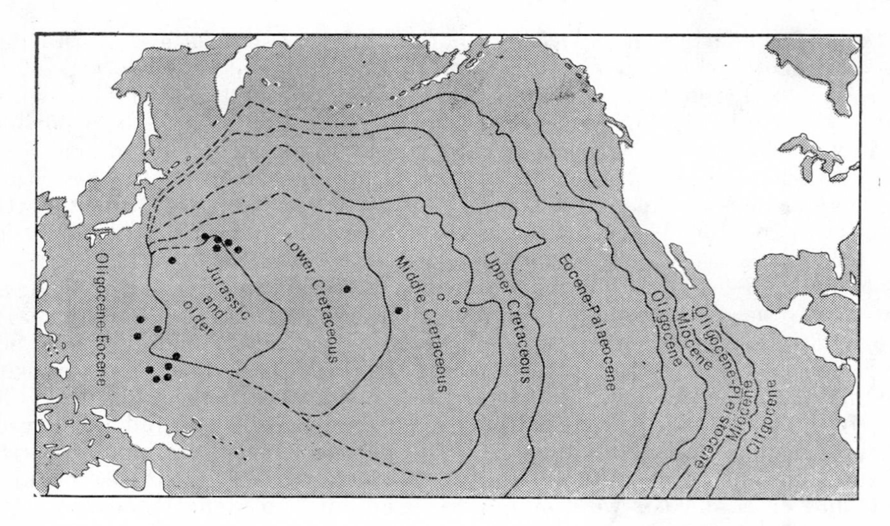

Fig. 109. Age of rocks at the base of the sedimentary series in the North Pacific (after Fischer, 1970)

come to the crest, give way to Neogene and Quaternary sediments. In the Pacific, the change from Cretaceous sediments at the base of the sedimentary series to Tertiary sediments occurs at a distance of 2000 km or more from the axis of the East Pacific Rise. The change in the age of the oldest sediments with distance from the crest of a mid-oceanic ridge is illustrated in Figure 109.

The intercalations of hard siliceous rocks mentioned above are mainly restricted to the Middle Eocene.

Despite their great distance from a mid-oceanic ridge, the marginal seas on the west of the Pacific Ocean are characterised by the young age of the oldest sediments. Only in the Tasman Sea have Upper Cretaceous sediments been found. In the Philippine Sea, the Sea of Okhotsk, and in the Bering Sea, the oldest sediments are of Palaeogene age, and in the Sea of Japan and in the Coral Sea the sedimentary sequence begins with the Miocene.

Observations that indicate a change in the depth of a deep oceanic basin during deposition of the sedimentary series arouse great interest. These changes, as far as can be judged from the data available, tend toward an increase in the depth of the ocean, or of some of its regions at least, during Mesozoic and Cenozoic times.

It has been discovered in a number of places, for example, that the red abyssal clay that makes up the top of the sedimentary series is underlain at deeper horizons by calcareous mud. This succession can, of course, be interpreted as the result of a change (in connection with a change in the water's regime of currents, temperature conditions, and salinity) in the critical level at which carbonate particles dissolve. It is more likely, however, that we are faced with evidence of an increase in the depth of the basin, in connection with which the surface of sedimentation sank from a zone above the "critical level" to one below. On the Shatsky Ridge, in the Pacific Ocean, such a change in the character of rocks occurred at the end of the Early Cretaceous. In the other areas of the same ocean, a similar change occurred in the Late Cretaceous, Palaeogene, or Neogene. That they happened at different times supports the view that they were caused by an increase in the ocean's depth, rather than disturbances in the circulation and temperature regime.

While this evidence can, nevertheless, be interpreted differently, there is another group of facts that quite unambiguously indicates that in many areas where considerable depths are now observed there had earlier been either shallows or even land.

Study of the section of the scarp that bounds the marginal Blake Plateau on the east (see Fig. 95) has shown that shallow-water limestones of the Aptian and Neocomian stage occur at present depths between 3000 and 4800 m. As these limestones must have been formed in shelf conditions at depths of only tens of metres, the unavoidable conclusion is that the edge of the Blake Plateau has subsided by nearly 5000 m since Cretaceous time with a corresponding increase in the depth of the ocean. Earlier, when considering the geological structure of the surface of the Blake Plateau, we established that the inner part sank by 1000 m or rather more at the end of the Cretaceous and beginning of the Palaeogene. The rate of plateau's subsidence was consequently uneven, the outer edge sinking faster than the interior.

The Jurassic sandstones of clearly continental origin, on the eastern scarp of the Demerara Plateau in the Atlantic, are now at a depth of 4400 m. North-east of Newfoundland, where depth is now as much as 1800 m, the sediments up to and including the Senomanian stage prove to be shallow-water ones, while those of the Maestrichtian stage and Tertiary already represent deep-water sediments. An approximately similar subsidence of the floor (by 1900 m) has been noted for the Rockall Plateau to the west of Scotland, where there are sands of continental origin at the base of the sedimentary section.

Signs of past shallow conditions are especially numerous in the Indian Ocean. Glauconite sands, phosphate deposits, sediments with a shallow-water fauna, and even lignites have been discovered in several boreholes. These depositions of a shallow sea and marshy lagoons now occur at various depths in the ocean, down to 6000 m. They belong to various stratigraphic levels ranging from the Albian stage to Miocene. West of Australia, for instance, where there are now abyssal plains, subsidence since the Early Cretaceous has reached 5000 to 6000 m. On the Naturaliste Plateau, the subsidence is more than 3000 m, and about 1200 m on the West Australian Ridge, where subsidence began in Eocene time. The amplitude of subsidence is around 1200 m. On the Eastern Indian Ridge, there are signs of subsidence along its whole length. In the north, subsidence reached 3600 m in the middle, where the sea is now 1600 to 2500 m deep, the section begins in descending order with Pleistocene, Neogene, and Palaeogene deep-water sediments down to the Lower Eocene; below them lie glauconite sediments of the Lower Eocene and Palaeocene, which are evidence of a much shallower depth of the basin; the section is underlain by lagoonal sediments of Palaeocene age and lignite.

On the Maldive Ridge in the same ocean, at the bottom of the sedimentary section there are shallow-water sediments of the Middle Palaeocene to Middle Eocene. The overlying sediments have a deep-water character, and the depth in the sea is now 2000 metres. On these two ridges, consequently, from the Eocene (in places Miocene) to the present, the ocean's depth has increased by 1600 to 2000 m.

The sediments found on the flat tops of guyots, and the sections on coral islands discovered by drilling and studied by geophysical techniques present special interest.

On many guyots in the central Pacific coarse clastic, clearly extremely shallow-water sediments have been found lying on the flat surface of their volcanic bases. There is no doubt that the flat tops of guyots have an abrasive origin and were originally at sea level. They are now at depths down to 2000 m. The age of the oldest sediments on guyots has proved to be Early Cretaceous. From that we may conclude that the central Pacific, where guyots are located, has become 2000 m deeper since the Early Cretaceous.

Drilling on Eniwetok Atoll in the Marshall Islands disclosed a basaltic basement (top of a volcano) at a depth around 1400 m. For 190 m from the surface there are coral limestones of Recent, Pleistocene, and Pliocene age. Below them lie Miocene coral limestones, below which, at a depth of 885 m are Eocene limestones, which in turn rest on the basaltic basement. No Oligocene fossils have been found there. Since the colonial corals that form islands cannot live at depths exceeding a few tens of metres, this section is evidence of a sinking of the island's basement (presumably together with the surrounding bottom) by 1400 m beginning in Eocene time.

Remnants of Eocene coral reefs have been discovered at a depth of 1000 metres on the top of a guyot in the submarine Tuamotu Ridge, near the archipelago of the same name. Colonial corals from Recent to the Neogene have been found on the Nasca Ridge off Peru at a depth of 400 m. According to the geophysical data, the coral reefs are 900-1000 m thick on Funafuti Island, 1300 m thick on Bikini Atoll, up to 2000 m thick on Kwajalein Island, and 770 m thick on Nunufetau Island.

There are numerous signs of sinking of the floor and an increase in depth of the basins of the Mediterranean type of sea. They include, for example, the presence of lagoonal salt-bearing sediments at the base of these seas, sedimentary section, which suggests extremely shallow, semicontinental conditions. The sedimentary section of the Gulf of Mexico begins with a Jurassic salt-bearing formation. At that time there were shallow lagoons on the site of the Gulf. But the Cretaceous sediments already have a deep-water look. The base of the series now lies at 12 000 m, the sediments are as thick as 8000 to 9000 m, and the present-day sea is more than 3000 m deep. According to the geophysical data, the sedimentary series of the Gulf of Mexico is underlain by a second layer with the same physical properties as are characteristic of Layer II beneath the ocean.

In the Mediterranean Sea the loose sediments lying on the basement to which a basaltic composition is also ascribed from the geophysical data begin with a salt-bearing suite that is considerably younger (Late Miocene and Early Pliocene). At present, the depth of this part of the sea is more than 2700 m. In both the Gulf of Mexico and the Mediterranean their depth has increased by approximately 3000 m in the latter the deepening occurred in a much shorter time than in the former.

On the Aves Ridge in the Caribbean Sea, shallow-water sediments have been found covering the interval from the Eocene to the Early Miocene. In the Middle Miocene they were succeeded by deep-water sediments, and since then there has been a subsidence of 1400 m. In the same Sea, Middle Eocene shallow-water limestones have been raised from the Beata Ridge from a depth of 2500 m. The upper Oligocene rocks in the same section have a deep-water character.

The second (plateau-basalt) layer of the oceanic crust. As we have indicated Layer II is composed, at the surface, of plateau basalt, and it is suggested that this composition is retained right through to the underside of the layer. In many places boreholes have reached the underside and specimens of its rocks have been raised in the cores. The basalt is mostly of tholeiitic type, i.e., pyroxene, and similar in general to continental plateau basalts, but differs markedly in a lower content of potassium and several trace elements (Table 4). It should also be noted that there are also alkaline (olivine) varieties among oceanic basalts, mainly restricted to submarine and surface elevations—seamounts, guyots, and volcanic islands. The alkaline oceanic basalts are similar to the terrestrial ones of rift zones.

Table 4. Composition of continental and oceanic basalts (compiled by B.G. Lutz)

Element	Oceanic tholeiite	Oceanic alkaline olivine basalt	Continental tholeiite
	Silicate analyses (% of weight)		
SiO_2	49.34	47.41	48.9
TiO_2	1.49	2.87	1.8
Al_2O_3	17.04	18.02	15.7
Fe_2O_3	1.99	4.17	—
FeO	6.82	5.80	10.7
MnO	0.17	0.16	0.1
MgO	7.19	4.79	8.7
CaO	11.72	8.65	10.8
Na_2O	2.73	3.99	2.3
K_2O	0.16	1.66	1.0
	Content of minor lithophile elements (ppm)		
Rb	8.0	33	20
Li	9	11	10
Sr	130	815	450
Ba	14	498	244
Th	0.18	4.61	0.4
U	0.4	1.14	0.2

The velocity of propagation of longitudinal seismic waves in Layer II is about 5.5 km/s. According to seismic data, its thickness varies over most of the ocean floor between 1000 and 2500 m, but rises to 10 or 15 km beneath aseismic ridges. The plateau-basalt layer is underlain by the third layer of the oceanic crust, in which seismic velocities rise to 6.5 or 7.0 km/s. The Mid-oceanic ridges have a specific structure: the thickness of Layer II increases to 5000-7000 m beneath their crests while Layer III wedges out, so that Layer II rests directly on the upper mantle, which has rather anomalous properties there. These features of deep structure will be discussed later.

The age of the basalts composing the surface of Layer II has been determined in a number of places by the radioactive K-Ar method. In some cases it has proved to be similar to that of the sediment directly overlying them. In addition to this lowering of the age of the rocks occurring at the base of the sedimentary series toward the crest of the ridge, the underlying basalts also become younger and younger: the Jurassic and the Cretaceous basalts that make up the basement of deep ocean basins remote from mid-oceanic ridges give way on the sides of the ridges to Palaeogene and Neogene basalts, while the youngest, Quaternary basalts predominate along the crest. Thus, there is a facial change of rocks toward the crest of a mid-oceanic ridge: sediments are followed by volcanic rocks, the oldest of which seemingly cover the whole floor of the ocean. As we approach present geological time, the zone of distribution of plateau basalts becomes narrower and narrower, concentrating closer and closer to the axis of the ridge.

From the point of view of ordinary "continental geology" such an explanation of the relations observed between sedimentary and basaltic rocks seems more natural: we shall see further that now a great number of scientists interpret the same facts quite differently, i.e., as a hypothesis of "spreading the ocean floor" (see Chap. 25).

In the Pacific Ocean the youngest volcanism is concentrated along the number of ranges of volcanic islands and submarine volcanoes (south and easterly Pacific rise) as well as along the mid-oceanic ridge. The age of the volcanic lavas on the islands, determined by radioactive methods, lies within several million years ago. The ranges are linear, which points to their association with crustal faults. The strike of the faults is predominantly north-western.

The rare boreholes that have penetrated Layer II to some depth have revealed an alternation of basalts and sedimentary rocks. This alternation has been found near California, in the Arabian Sea, on the Eastern Indian Ridge, and in a number of other places. This feature of the top of Layer II points to its possibly containing intercalations of sedimentary rocks at greater depths as well. The sedimentary rocks could have been metamorphosed and become indistinguishable from basalts as regards the velocity of propagation of seismic waves. They could even have melted into the basalt. Near the crest of the Mid-Atlantic Ridge, for example, Lower Miocene foraminifera have been found in basaltic glass, which suggests the fusion of sedimentary rocks in basalt. The sediments in Layer II, if they exist, must naturally be older than those lying above the basalts, and could prove to be Palaeozoic or older. Whether or not there are intercalations of sedimentary or parametamorphic rocks in Layer II is of significance in principle for understanding the history of the oceans, but such an answer can only be found by means of underwater drilling through the plateau basalt, which is so far technically beyond the bounds of feasibility.

CHAPTER 21

Tectonics of the Ocean Floor

Structures of the Ocean Floor

Although it is still difficult to study the mode of occurrence of rocks on the ocean floor, there is a circumstance, mentioned already, that considerably eases the task. The accumulation of sediments and processes of destruction in ocean depths are so slow, that the results of vertical tectonic movements are well expressed and long preserved in the topography of the floor. Regions of tectonic uplift are expressed in upwarped features of the floor, and regions of tectonic downwarping in its basins; the block tectonic structure is characterised by the intricately broken relief topography of the floor. In just the same way, volcanic forms are also well preserved in the floor's relief if they are hidden beneath the water.

We can consequently use the topography of the ocean floor fully to characterise tectonic and volcanic structures. The proviso must be made, however, that in the main the tectonic structures formed by vertical movements of the crust can be distinguished in this way. Structures related to horizontal movements can only be established when they have caused marked offsets of the separate parts of previously formed single features of the relief.

A survey of the relief of the ocean floor does not in itself indicate the age of any particular structure; to understand the history of oceanic structures, additional data has to be sought. Some of these data will be employed below.

The structures observable on the ocean floor can be divided into several categories by their size and order of importance. The structures of the first category, the biggest, include the *deep ocean basins* and *mid-oceanic ridges.* They are each other's opposites in almost the same way as platforms are the opposites of geosynclines on continents. The structure of the former indicates tectonic stability, while the latter, on the contrary, are evidence in their configuration of the great mobility of the crust.

Deep ocean basins occupy the greater part of the area of the bed of the ocean. In the Atlantic, they stretch as two strips on each side of the Mid-Atlantic Ridge. The biggest basins located west of the ridge are the North America, Brazil, and Argentina Basins (see the schematic tectonic map of the Earth attached to the book). East of the ridge the biggest basins are the Canary, Angola, and Cape Basins. In the Indian Ocean there are seven large basins: the Arabian, Somali, Madagascar, Mozambique, Central Indian, Western Australia, and Crozet Basins. The main area of the Pacific Ocean is occupied by the following basins: North-west Pacific, North-east Pacific, Central Pacific, South Pacific, and Peru Basins. A number of large basins are also known in the Arctic and Antarctic Oceans.

A feature of all the basins mentioned above is the almost completely horizontal position of their floor everywhere where there are sediments and where the basins are not complicated by the structures of the second and third categories, i.e., "aseismic" ridges or isolated smaller elevations. The seismic data indicate an almost horizontal bedding of strata within the sedimentary series beneath the flat floor.

In considering "aseismic" ridges and small elevations as secondary structures complicating basins, and taking into account that the quiescent areas occupy a much greater area than these secondary structures, which always have local character, we can classify these basins as a kind of "oceanic platform", that has undergone almost no dislocations since their formation.

On the whole the mid-oceanic ridges in the relief are gentle swells 1000 to 3000 km wide and 2000 to 3000 m higher than the neighbouring basins. Low scarps (under a few hundred metres) have been found on the slopes of mid-oceanic ridges, which can be interpreted as normal or slash faults. A very broken relief is typical of the axial zone of almost all mid-oceanic ridges, except the East Pacific Rise. Along the axis of a median ridge there is a chain of long, narrow depressions that attain a width of some tens of kilometres and have a bottom 1500 to 2000 m below the surrounding zone of the crest. The slopes of these depressions are very steep. The basins are adjoined on both sides by the highest zones of the ridge, which have a most complex relief, being divided into many short upwarpings and depressions with sharp falls in height measurable in hundreds of metres. All these feautres of the structure of the axial belt of mid-oceanic ridges should apparently be interpreted as a manifestation of intensive block tectonics, the axial basins representing grabens, and the median ridge on both sides of them being cut up by faults into elevated and depressed blocks.

The whole aggregate of structural features characterising mid-oceanic ridges allows us to classify them as analogues of continental rift belts. As we have seen (Chap. 17), the latter are also, on the whole, broad, gently sloping tectonic arches complicated by grabens along their axis. The width, and height of the arches and the widths and depths of the grabens of continental rift belts are close to the corresponding dimensions of mid-oceanic ridges.

The similarity of these oceanic and continental structures is also confirmed by the fact that they are directly connected with one another in the area of the Gulf of Aden, where the Mid-Indian Ridge joins the continental Arabian-African rift belt. Although the submarine ridge itself dies out as a bulge in the relief of the floor at the entrance to the Gulf of Aden, longitudinal depressions continue along the bottom of the Gulf, and their chain is linked with the longitudinal depression in the bottom of the Red Sea, about which there is no doubt that it is a graben. The continental uplifts that fringe the Gulf of Aden on the north and south and are spurs of the Nubian-Arabian arch along whose axis the Red Sea graben extends, can be seen as a continuation of the Mid-Indian Ridge.

The connection of the mid-oceanic ridge and the continental rift belt is not only expressed in tectonic structure, but is also manifested in volcanism. Mid-oceanic ridges consist of basalts, and in many places along their axial zones there are volcanoes that are still active, which erupt both tholeiitic and alkali basalts. Basaltic outpourings are also vigorous in continental rift belts but in them alkali basalts greatly predominate.

There are consequently adequate grounds for believing that mid-oceanic ridges are *oceanic rift belts.*

The arches, on which the grabens that compose the Arabian-African rift belt are located, were formed at the end of the Mesozoic and Palaeogene. The grabens began to subside at the end of the Palaeogene, but mainly in the Miocene, and their most intensive subsidence occurred at the end of the Pliocene and in the Pleistocene.

From the direct link of the system of mid-oceanic ridges and the continental rift belt, we may suppose that the uplift of a median ridge also occurred in the Late Mesozoic and Palaeogene, while the grabens along its axis were formed in the Neogene and Quaternary.

The fact that Iceland is a direct continuation of the Mid-Atlantic Ridge and may be considered a part of the ridge raised above sea level is exceptionally important for establishing the nature of mid-oceanic ridges and the chronology of their development. The main structure of Iceland, as we know, is a series of Miocene and Pliocene tholeiitic and alkali plateau basalts formed under continental conditions. In the Pleistocene the series was intersected by a graben trending, in general, north-south, to which subsequent volcanic activity has mainly been restricted. The grabens that complicate the submarine part of the Mid-Atlantic Ridge are probably of the same young age.

The longest continental rift belt, it must specially be noted, is a little more than 6000 km long, while the total length of the oceanic rift belt, united in a continuous system of mid-oceanic ridges is 60 000 km.

The category next in importance among the structures of the ocean floor includes, primarily, the "aseismic" ridges, viz. the Walvis Ridge in the Atlantic, the Mascarene, Maldive, Eastern Indian, and Western Australian Ridges in the Indian Ocean, a number of very long east-west ridge-faults in the eastern Pacific (the Mendocino, Murray, Clarion, and Clipperton Escarpments), and the Cocos, Sala-y-Gomez, Nasca, and West Chilean Ridges in the south-east of the same ocean. A number of the ridges of Melanesia should possibly be classed in the same category. In the Arctic Ocean, the Lomonosov and Mendeleev Ridges are "aseismic".

The rectilinear character of almost all the ridges listed above clearly suggests that they are connected with crustal faults. Other evidence for this connection is the fact that these ridges generally separate sectors of the bottom of different depth. The Maldive Ridge, for example, runs along the boundary between the deep Arabian Basin and a shallower sector of the Indian Ocean. The West Australia Basin is much deeper north of the West Australian Ridge than south of it. And we have already mentioned the different depths of the floor on either side of the ridge-faults in the eastern Pacific.

Further evidence for a link between these ridges and faults is the character of their structural continuations on continents. On the extension of the Walvis Ridge in South-West Africa and Angola, 11 ring-shaped volcanic structures composed of alkali lavas are known over a stretch of nearly 1500 km. The Lukapa graben in which intrusions of basic rocks and kimberlite pipes occur also lies in this extension. All the igneous rocks were emplaced after the Permian, but before the Late Cretaceous. They undoubtedly lie along the same fault system. Since their belt is a terrestrial continuation of the Walvis Ridge, one can suppose that the ridge is associated with a fault and was originally formed before the Late Cretaceous.

If we continue the trend of the Maldive Ridge northward, we come to the huge plateau basalt field of the Deccan in the region of Bombay. The plateau basalts must have poured out onto the surface along deep fissures; their age is Late Cretaceous to Eocene. Although a direct link between the submarine ridge and the Deccan has not been demonstrated, it is highly probable that

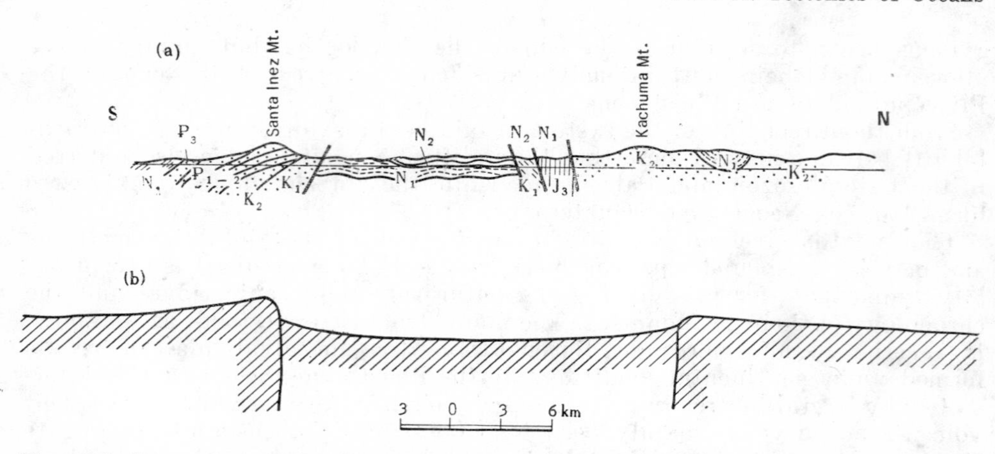

Fig. 110. Comparison of profile across the submarine Murray Escarpment with the profile through the Transverse Ranges in Southern California (after Beloussov):
a—geological cross-section of the Transverse Range; *b*—schematic profile across the Murrary Escarpment

both the submarine ridge and the plateau basalts are associated with a single fault system.

The arcuate shape of the Mascarene Ridge also does not contradict its connection with a fault. This ridge may fully correspond to a fault bounding the Mascarene Basin and separating it from surrounding sectors of the floor of the Indian Ocean.

All these "aseismic" ridges have flat tops and steep slopes. Allowing for that and their association with faults, they should be considered horsts confined to rupture sutures between large segments of deep oceanic basins.

By their topography the east-west "aseismic" ridge-faults in the East Pacific are one-sided horsts, because one side slopes gently while the other is steep. The Murray Escarpment consists of two single-sided horsts separated by a graben. At least two ridges extend into the North American continent. The Clarion ridge-fault continues in Mexico and a major fault separating Palaeozoic metamorphic rocks to the south from young volcanic rocks to the north lies on its continuation. There is along the fault a chain of volcanoes, both active and recently extinct.

The double Murray Ridge is an extension of the east-west Transverse Ranges of Southern California, which bound the Central Valley of California on the south. When we compare the topography of the submarine Murray Escarpment with the geological structure of the Transverse Ranges, we find a noticeable similarity (Fig. 110). The northern and southern flanks of the Transverse Ranges slope gently. They are composed of Cretaceous and Eocene sediments dipping toward neighbouring depressions. The axis of the ridge is occupied by a graben in which Miocene and Pliocene sediments have survived. Here we see, consequently, two single-sided uplifts that are separated by a graben and face each other on their steep sides. The topography of the Murray ridge-fault reflects the same structure. An exception is that on it the contrasts of relief are less than the amplitude of the tectonic relief on land. The Transverse Ranges began to rise at the end of the Late Cretaceous and retained the form of an arch until the Miocene. The longitudinal graben along its axis was formed at the end of the Miocene. From the direct link between these two structures, it is natural to suppose that the submarine Murray ridge-fault has the same history.

On the continuation of the Mendocino Escarpment, the whole system of the Cordilleras is crossed by a belt of Pliocene lavas. These are the lavas of the Maler Plateau and the basin of the Snake River. To the north and south the zone occupied by lavas narrows considerably and Miocene and Eocene lavas begin to predominate. It cannot be ruled out that on the continuation of the Mendocino Range the Cordilleras are cut by a fault buried beneath the lavas.

This same category of structures includes numerous submarine rises of round, oval, or irregular shape. These are the Rio Grande Rise in the Atlantic; the Crozet and Kerguelen Rises in the Indian Ocean; and, probably, a number of submarine rises in Melanesia in the South-west Pacific. They all have steep slopes and in general a smooth surface.

Other members of this category of structures are the linear ranges of submarine and surface volcanoes united by a common foot to form a gently sloping swell, developed in the Pacific Ocean. They include the range of the Hawaiian Islands, and the many submarine volcanoes belonging to it, the range of the Emperor seamounts, and the ranges of islands and seamounts of the Marshall, Line, Tuamotu, of the Cook, Tubuai and others. The outward expression of these structures has a volcanic, rather than a tectonic origin, it is true, but the fact that the volcanic ranges are linear suggests their association with crustal faults. We have already mentioned the predominantly north-western trend of these faults in the Central Pacific. The age of the eruptions within these ranges is Quaternary and Pliocene; the faults must therefore also be extremely young. The ranges of volcanoes in the Central Pacific, interestingly, run in a single belt along the axis of the ocean from the southern top of South America in the south-east to Kamchatka in the north-west. In the south-east, this belt cuts across the East Pacific Rise (near Easter Island) and hence is younger than the rise itself.

These volcanic ranges differ from "aseismic" ridges in having gentle slopes, which merge imperceptibly into the bottom of the surrounding basins, and a narrow crest crowned by a chain of volcanoes.

Finally, the last category of structures of the ocean floor should include the seamounts and guyots that stud all oceans in random fashion and are particularly numerous in the Pacific. They are completely volcanic in nature.

Although deep-sea trenches were discussed earlier, when we described the regimes of the continental margin, we must revert to them once more here. Normal faults have been found on the sides of them that suggest their involvement in the formation of these structures. The strata on the bottoms of trenches lie horizontally and abut against their steep sides.

The data on the time of formation of deep trenches are rather controversial. Their development continued into recent geological time, since they truncate Pleistocene continental structures in places, as is observable, for example, in the Philippines and Japan, but the beginning of their formation has to be assigned to an earlier age, probably the beginning of the Neogene.

It has been pointed out in the geological literature that the formation of trenches is related in time with the appearance of strong andesitic volcanism on adjoining island arcs and in the coastal ranges of the neighbouring continents. As we have seen, such volcanism has been developing since Miocene time.

The *relief of marginal and inland seas* is most complex. In seas like the Caribbean, Mediterranean, and Aegean, and in the seas of Indonesia and Melanesia, the floor is divided into elevated and depressed sectors of linear and irregular shape. This complex relief should be considered a reflection of block structure of the floor and vertical displacements of the blocks. This suggestion is supported by the fact that this relief proves in many cases to be a submarine continuation of analogous relief on the neighbouring land. One can find examples in all the seas listed above. Several separate relatively shallow areas of the floor in the open ocean also have the same finely dissected block structure; the whole of the North Atlantic, for example, is an area of widespread block tectonics.

Let us also refer once more to the considerable development of diapiric salt domes in seas and on the shelf regions of the ocean. These diapiric domes raise, deform, and pierce the overlying marine sediments.

General Palaeogeographic Data

As was pointed out in Chapter 17, all the pre-Mesozoic structures of the continents on margins of the Atlantic type are truncated by the edge of the shelf and do not continue further into the ocean. The Precambrian, Caledonian, and Hercynian folded zones of Western Europe are thus truncated by the edge of the Atlantic shelf of Eurasia. Folded zones of the same age are also truncated similarly along the opposite, North American coast of the Atlantic.

At the same time, on the continents fringed by an Atlantic-type margin, there is indirect geological evidence for the existence of continental conditions beyond their limits in the Palaeozoic and beginning of the Mesozoic, where ocean basins are now located. This evidence consists (1) in the transport of sialic clastic material landward from the present-day ocean to today's continents, and (2) in the faunal and floral exchange between the continents, for which land bridges were needed.

The first point indicates that, at least in part of the area now occupied by the ocean, there used to be sectors of land that underwent erosion; the sialic composition of the clastic material supplied by them suggests that the crust in them was of continental type. From the distribution of early Palaeozoic sedimentary rocks in the Scandinavian Mountains, for example, there was a considerable supply of clastic material from the north-west during their deposition, i.e., from the area now occupied by the Atlantic. The Ordovician sediments, consisting largely of limestone, become more terrigenous toward the coast in areas remote from the ocean.

On the other side of the Atlantic, in the Appalachians, there are signs of transport of sialic clastic material from the south-east in the Palaeozoic, i.e., once more, from an area now occupied by the ocean. A palaeotectonic profile of the Middle Ordovician of the Northern Appalachians is given as an example in Figure 111, from which it will be seen that the sediments become coarser to the south-east, i.e., toward the ocean.

In South Africa, during the Late Palaeozoic glaciation, some of the glaciers, judging from the direction of striae on the buried roches moutonnées, moved westward (from the side of the Indian Ocean) and, what is especially important, carried granite boulders with them. Transport of granite boulders from the side of the ocean is also noted in the Upper Palaeozoic in South Australia.

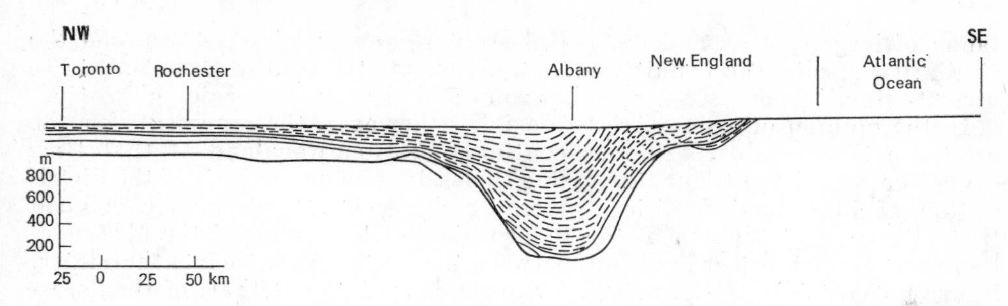

Fig. 111. Palaeotectonic profile of the Middle Ordovician sediments of the Appalachians (after Kay, 1942):
dots—sandstones; dashes—clay; and white areas—limestones

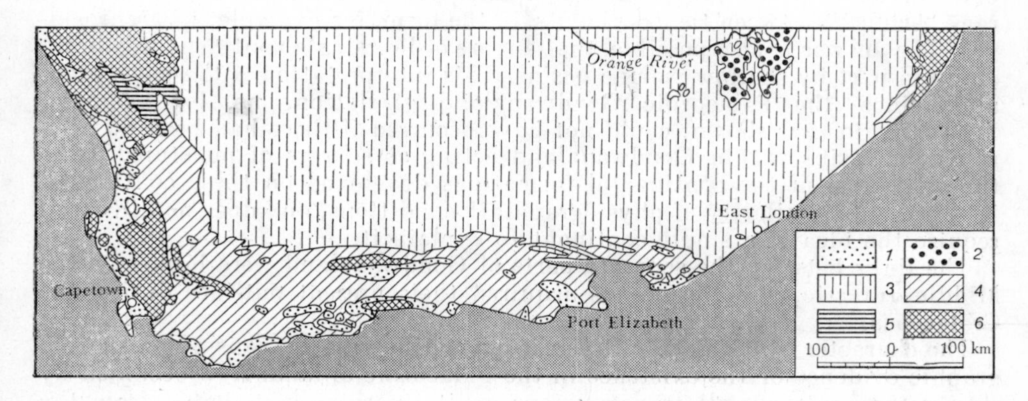

Fig. 112. Truncation by the Indian Ocean of the eastern extension of an Upper Palaeozoic-Mesozoic foredeep, filled with molasse of the Karroo formation (after du Toit, 1954):
1—Cretaceous and Cenozoic sediments; *2*—Rhaetian-Lias basalts and rhyolites; *3*—Rhaetian-Upper Carboniferous rocks (Karroo formation); *4*—Upper Silurian-Lower Carboniferous rocks (Cape formation); *5*—Cambrian (?) rocks; *6*—Precambrian rocks

The Cape Mountains, at the southern tip of Africa, consist predominantly of continental sediments of the Upper Palaeozoic and Lower Mesozoic that contain tillites (the Karroo formation). The sediments have the character of molasse, are very thick (up to 7000 m), and accumulated in a zone of downwarping, that is a typical foredeep in its shape and history. The uplift before which this foredeep developed, however, was located beyond the limits of the present-day continent, south of it as is indicated by the direction of the transport of clastic material. Farther to the east, along the strike, the belt of continental sediments of the Karroo formation is truncated by the edge of the Indian Ocean. Quite obviously, they must have continued into the area where the ocean now is (Fig. 112).

Evidence of the second kind long ago, as we know, led to the idea of existence of a single southern continent of Gondwanaland in the Upper Palaeozoic and Lower Mesozoic, which united the ancient platforms of South America, Africa, Arabia, Hindustan, Australia, and Antarctic. The idea of Gondwanaland arose in connection with the great similarity of the Upper Palaeozoic and Lower Mesozoic continental sediments that are common on these continents, and in connection with the great community of terrestrial fauna of large reptilia and flora of glossopteridales and equisetales, and the similarity of climatic conditions that led on all the southern continents to almost simultaneous Upper Palaeozoic and Lower Mesozoic glaciation.

In the opinion of a number of researches, however, the arguments for the existence of a single continuous continent of Gondwanaland at that time cannot be considered adequate. Basing themselves on an analysis of the distribution of individual families, genera, and species, they have concluded that although, at the end of the Palaeozoic and beginning of the Mesozoic, there was a much better land link between the southern continents than later, it was not absolute. H. Simpson, for example, points out that, of the Triassic reptilia known in South America, only 43% of the families and 8% of the genera have been found in Africa, and no identical species. Migration therefore took place, but there was some "filter" in the way that limited it. Instead of a whole continent, therefore, it is more probable that there were temporary "bridges" between the continents, in the form, for example, of archipelagos. The answer is also possible, that a more or less single continent was broken up in part by shallow marine basins. This interpretation is the more probable since the Palaeozoic Gondwanaland glaciation, as, for example, L.B. Rukhin has shown, could not have occurred if there had been no interior cold basins within Gondwanaland acting as sources of moisture. As to the similarity of the continental formations developed on these continents, it was due to similarity of physiographical conditions, and does not in itself require the idea of a single continent to explain it.

The links between the separate parts of Gondwanaland were broken in the Middle Jurassic, and the oceans acquired their present outlines during the Cretaceous.

The character of the Pacific Ocean is quite different. There is no palaeogeographic evidence of the existence in the Palaeozoic of lands now occupied by it. On the contrary, all transgressions on its margins distributed from the side of the ocean; so it was in the North American Cordilleras, the South American Andes, Japan, and Eastern Australia. One must therefore assume that, as early as the Palaeozoic, there was a large marine basin in the area of the Pacific Ocean.

For later periods, the Mesozoic and Cenozoic, there are signs in some of the regions surrounding the Pacific of the transport of clastic material from the direction of the present-day ocean. Such observations have been made in the Andes, the Cordilleras, and Japan, and on Kamchatka. Figure 113 gives a palaeogeographic scheme of Southern California for Tertiary time, which shows that the eroded Catalina uplift was located at that time beyond today's land. But, because deep-water sediments of the same Mesozoic-Cenozoic age have been found on the ocean floor not far from the coast, there is no ground to suppose that large land masses of some sort existed on the site of the Pacific in Mesozoic-Cenozoic times. The small marginal uplifts could have been eroded and then have subsided.

Large sectors of land were located until recent geological times on the sites of many marginal and interior seas, as has been established from analysis of the direction of transport and the composition of sediments, and on the basis of more general palaeogeographic reconstructions. The history of the Mediterranean throws especially interesting light on this.

Study of the structure of the young folded zones surrounding the Western Mediterranean (the Betic Cordillera, Maritime Alps, Apennines, and Atlas and Rif Mts.) makes it possible to get a quite full idea of the history of this

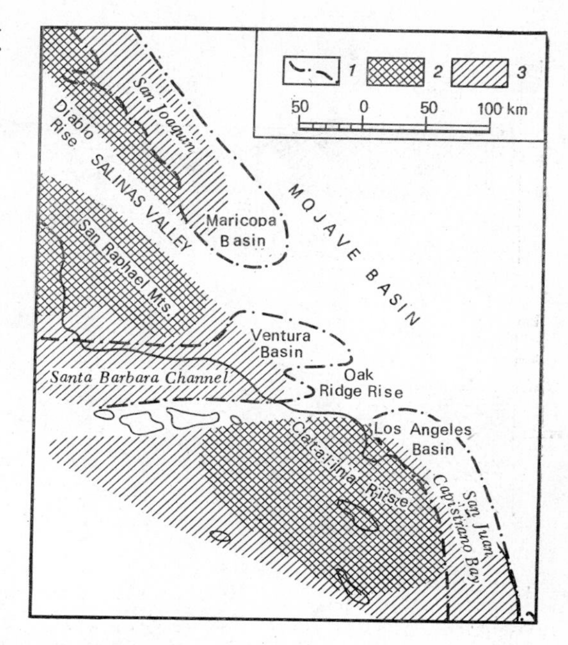

Fig. 113. Palaeogeography of Southern California in the Tertiary (after Reed and Holyster, 1936):

1—boundaries of areas of accumulation and erosion; 2—uplifts; 3—regions of subsidence

part of the sea. In the structure of these surrounding folded zones, predominantly in the Betic Cordillera in the south of Spain, there are tectonic nappes composed of Palaeozoic metamorphosed rocks enclosing granite intrusions. These tectonic nappes must have had their roots in the area of the present-day sea, from which they moved "centrifugally", in various directions. Hence, it can be concluded that in Palaeozoic time the western part of the Mediterranean had a crust of continental type composed at the top of metamorphic rocks and granites.

The development, around the same basin, of a Permian and Triassic "germanic" facies consisting of red lagoonal, gypsum and salt-bearing sediments, suggests that slow downwarping of the basin began then. It probably continued, and may have achieved considerable amplitude, during the Jurassic and Cretaceous. The Palaeogene in the surrounding folded zones, however, consists of a flysch formation containing huge volumes of sialic clastic material derived from an area now occupied by the sea, and this circumstance clearly points to the origin, on the site of the present-day sea, of a considerable uplift composed of rocks of the continental crust. The uplift increased and intensified; by the end of the Palaeogene it had taken in the troughs in which flysch had accumulated. These became involved in the uplift and the flysch was crumpled into folds. In the area of the uplift's development the above-mentioned tectonic nappes of Palaeozoic rocks developed. The Mediterranean uplift was apparently eroded down to the Palaeozoic rocks and sheets of these slid down the slopes in the form of Helvetian-type nappes.

The zones of downwarping then migrated outward, toward the platforms fringing the folded zones, and coarse molasse sediments began to accumulate in the troughs. River valleys have survived since then in the surrounding mountain ranges, the structure of which indicates that the rivers then ran not toward the Mediterranean Sea, as at present, but away from the sea, to the interior of the continent. This has been observed in the Betic Cordil-

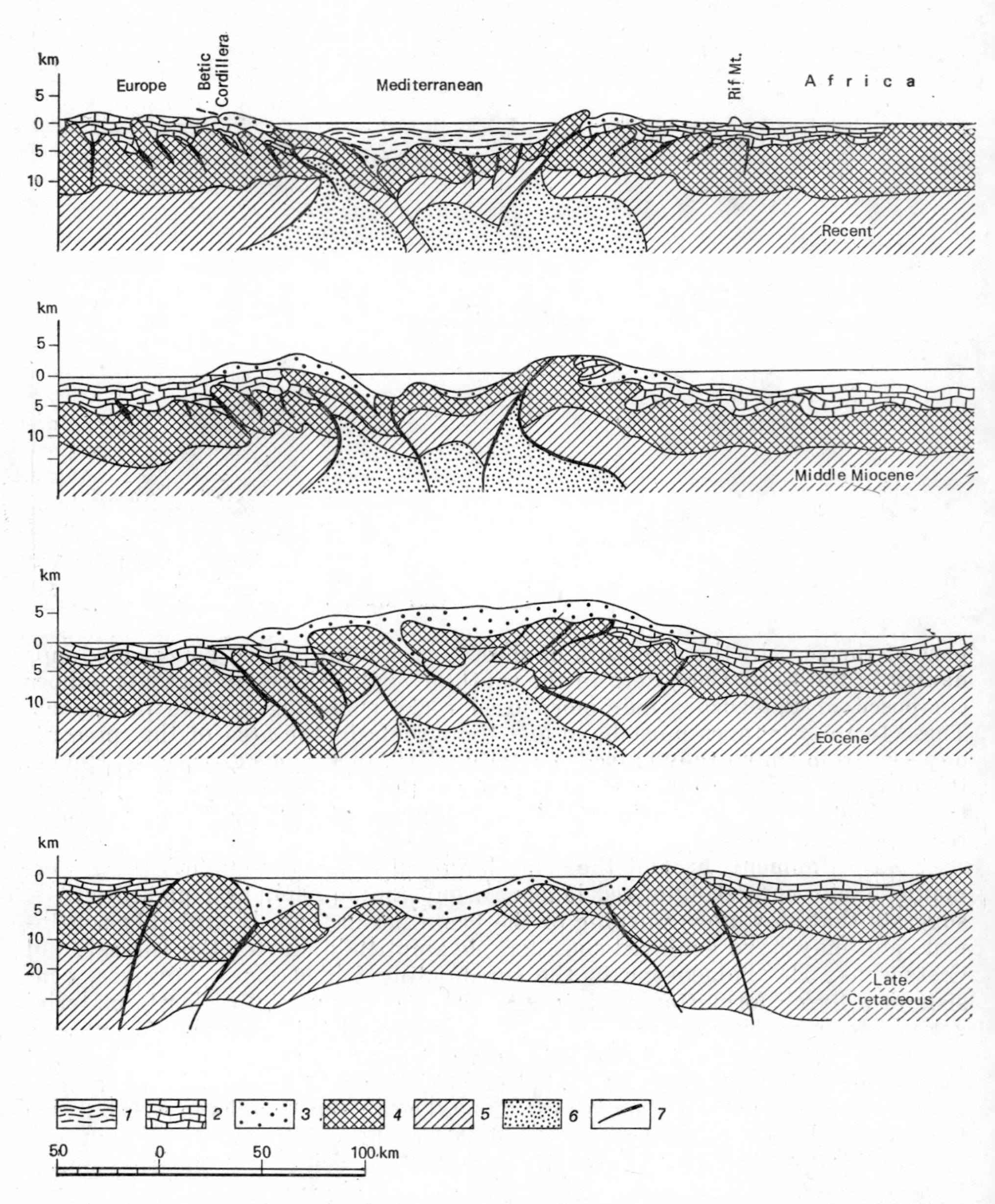

Fig. 114. History of the western part of the Mediterranean (the Alboran Sea) (after Glangeaud, 1962):

1—Miocene to Recent sediments; *2*—Mesozoic and Eocene sediments (in outer zones); *3*—Mesozoic and Eocene flysch; *4*—Upper Jurassic; *5*—Lower Jurassic; *6*—upper mantle; *7*—faults lera, Maritime Alps, and certain other regions. On the coast of the Tyrrhenian Sea, on the western side of the Apennines, Early and Middle Miocene conglomerates have been found containing pebbles of metamorphic and granitic rocks derived from where the present-day sea is. Only at the end of the Middle Miocene did sediments appear along the present coast that suggest the onset of downwarping in the region of today's sea. The rivers then began to run toward the sea, but initially only shallow lagoons were formed on the site. We have seen that boreholes have encountered Upper Miocene and Pliocene evaporites on the floor of the Mediterranean. The basin deepened in the Pleistocene, when the sea took on its modern look.

The whole tectonic history of the Western Mediterranean to the end of the Middle Miocene fully corresponds to the normal pattern of development of oscillatory movements in a geosyncline (Alpine in this case). The uplift formed at the beginning of the Palaeogene on the site of today's sea can be considered a central uplift; the flysch troughs on the site of present-day ridges were marginal troughs, while the zones of molasse accumulation were foredeeps. The shifting of troughs is a common wave-like process characteristic of geosynclines. The sinking of the present-day sea is a neotectonic phenomenon, discordantly imposed on the structural zones of the geosyncline: the edges of the basin (i.e., the present-day coasts) discordantly truncate the structures of the surrounding ridges, which we can regard as surviving peripheral parts of a central uplift. This superimposed, young subsidence was accompanied by changes in the structure of the crust: continental crust gave way to suboceanic, in which loose young sediments are directly underlain by the "basaltic" layer. The history of the western Mediterranean, as envisaged by the French geologist Glangeaud, is shown schematically in Figure 114.

Study of the geology of Japan, on the one hand, and the Maritime Territory of the USSR, on the other hand, lead to the conclusion that there was land on the site of the present-day Sea of Japan during Mesozoic and Palaeogene times that was only occasionally flooded to a slight degree by a shallow sea. More often, closed basins filled with continental coal-bearing sediments were formed. In Japan, marine Mesozoic and Palaeogene sediments occupy only the outer part of the country, facing the Pacific Ocean. As one moves inward and closer to the coast of the present Sea of Japan, these sediments become coarser and pass into continental formations or wedge out altogether. There are considerable differences in the forms of the marine fossil fauna between the Maritime Territory and Japan, which suggest a separation of the basins. For the Palaeogene, signs have been found of the transport of granitic clastic material toward the Sikhote-Alin Range from the area now occupied by the Sea of Japan. Only in the Miocene did marine sediments develop on the coasts of the Sea of Japan, suggesting the origin of a marine basin. The subsidence on the site of the Sea of Japan in the Miocene was accompanied with strong volcanism.

It has been established from a number of similar signs that other marginal seas along the western margin of the Pacific also sank during the Neogene and that there was earlier land on their sites with crust of continental type.

Some Conclusions about the Geological History of Oceans

Let us try and draw some conclusions from what has been said above about the geological history of the oceans, and for that let us use the methods of geological deduction that stem from the years of experience of continental geology.

This reservation is necessary, because much has been said recently about the need to approach the geology of the oceans with yardsticks quite different from those normal for "continental" geology.

The first conclusion is that up to and including Lower Mesozoic time (or rather up to and including the Early Jurassic) there were sectors of land composed of sialic rocks, including granites, on the site of those oceans that are fringed by Atlantic-type peripheries. Such sectors existed in areas now occupied by the Atlantic and Indian oceans. In the Pacific Ocean, with its own type of periphery, there is no evidence for the past existence of large sectors of land.

In this connection, there long ago arose an idea of two types of ocean: the Pacific Ocean is considered "primary" or, in any case, very ancient, older than the Phanerozoic, whereas the Atlantic and Indian oceans are classed as "secondary" ones formed early in the Mesozoic on the site of former land. The mechanism of their "secondary" formation, however, has been differently interpreted at various times. At first, subsidence of the land was assumed, with its transformation into an ocean. Recently many researchers have been of the opinion that the Atlantic and Indian Oceans did not exist before the Mesozoic because the continents surrounding them were joined together in a single continent, and that the formation of these oceans is associated with the breaking-up of this single continent and the horizontal moving apart of the detached continental blocks, which have become the present-day continents. According to this viewpoint, the single original continent was washed on all sides by the waters of the Pacific Ocean. After the break-up of the continent and moving apart of its segments the area of the Pacific was reduced.

We shall postpone critical consideration of this view.

The next conclusion is that the regions now occupied by the oceans became the arena, at the beginning of the Mesozoic, of volcanic activity of enormous intensity, but chemically uniform, which led to the whole floor of the ocean being covered by a series of plateau basalts. The composition of Layer III is quite unknown to us but it is highly probable that it constitutes a genetic whole with Layer II and is composed of hypabyssal intrusive basic and ultrabasic rocks forming the aggregate of the deep "roots" of the outpoured plateau basalts.

The ubiquitous outpouring of basalts must have ended by the Middle Jurassic, when sediments began to be laid down on the surface of the basalts in certain areas of the oceans. Outpouring of basalts, however, continued after that, only the area involved in volcanic processes gradually decreased. The volcanism was concentrated in an ever narrower zone along mid-ocean ridges. Parallel with the reduction in area of volcanism the area of sedimentation expanded, and younger and younger sediments rose higher up the slope of the median ridges. The youngest basalts—Pliocene and Quaternary—are developed along the crest of the ridges, and sediments are almost completely absent there.

The time of the formation of the median ridge itself is not exactly known. Basing ourselves on its link with continental rifts, we can assume that it began to rise in the Mesozoic. The formation of axial grabens on median ridges, as on continents, is a later phenomenon that took place in Neogene and Quaternary times. Then too the block structures that complicate the axial zone of median ridges were formed, and faulted escarpments arose on the slopes of the ridges.

The question of the time of formation of numerous faults that intersect the trend of median ridges and along which individual sectors of the ridges are offset in relation to each other is quite unclear. If we consider them strike-slip or transcurrent faults of some sort[5], they must be very young, because they offset the youngest structural elements on the crest of the ridges. Another point of view, however, can also be put forward, that the median ridge was formed at once in the form of separate segments, echelon-like, from the very beginning. This interpretation does not contradict observations on the continents, where such initial, imbricate arrangements of the separate structural elements of rift belts are a common phenomenon.

Beyond the limits of the median ridges, deep ocean basins have survived. Their further development has consisted in their being cut by faults along which the separate segments were displaced upward or downward by hundreds of metres. Some of the faults continued into the neighbouring continents, but have been weakly expressed there. In many cases block ridges rose along the faults. The age of some of them has been established as Late Cretaceous and Palaeogene. In other cases faults have served as a channel for new outpourings of basaltic lava onto the floor of the basins, and chains of volcanoes have arisen along the faults. The basement of the ranges of volcanoes is the predominantly young (Neogene and Quaternary) faults in the Central Pacific. Let us recall that since the end of the Jurassic, and especially in the Neogene, the faulting accompanied by volcanism also occurred on the periphery of the oceans, where it is of Pacific type. But andesite and andesite-basalt lavas were erupted on the ocean margins while only basalts were extruded in the central regions of the ocean.

During the Mesozoic and later, separate elevations and volcanic cones in the form of seamounts and guyots rose on the floor of deep ocean basins. All these structures have complicated the topography of the basins, but vast areas of them have nonetheless remained quite unaltered by these complications, and an extremely quiescent bedding of the sedimentary strata has been preserved on them.

All the phenomena listed above occurred on the background of a much more general process of deepening of the ocean. Signs of that are seen in the structure of guyots and coral islands and in the section of many sectors of the floor where younger deep-water sediments overlay older shallow-water ones. The reverse process, i.e., transformation of a deep basin into a shallow one, interestingly, has not been observed anywhere in the ocean.

The scale of deepening of the ocean varies from place to place, but it can be established that, for example, in the Central Pacific, the basin has become approximately 2000 m deeper since the Early Cretaceous. Where there are now depths around 5000 m, the bottom was not more than 3000 m deep in Early Cretaceous time; and in some areas of the Indian Ocean the subsidence of the crust and deepening of the ocean have been as much as 5000 to 6000 m since Cretaceous time.

The downwarping of the floor in the inner regions of the ocean during the Cretaceous period and later can be ranked with the subsiding of the oceanic periphery mentioned in Chapter 17 (when we were considering the endogenous regimes of continental margins). This subsidence occurred quickly on the whole Atlantic-type peripheries but much more vigorously on Pacific-type ones. For both types of margin, however, subsidence controlled the main ten-

[5] We shall see later that a number of workers assume the existence on the ocean floor of transcurrent, strike-slip dislocations of a special sort, called "transform faults", found nowhere on the continents in addition to ordinary ones.

dency of geological development. We see now that sinking of the crust is characteristic of the whole area of the oceans as well as of the peripheries. We can trace it by a number of indications for the whole time since the beginning of the Cretaceous.

To sum up, as briefly as possible, everything we have learned about the oceans we can regard them as extensive regions where the sialic rocks common on the continents are absent, and as regions of crustal subsidence, mass manifestation of basaltic volcanism, and intensive splitting up. The oceans manifested these features of their development during the Mesozoic and Cenozoic. The earlier history of the regions now occupied by the oceans is not known. On the basis of certain evidence, we can only say that prior to the early Mesozoic continental conditions existed, and rocks of continental type were developed on the site of the Atlantic and Indian oceans.

We have seen that both inland and marginal seas also exhibit signs of subsidence of the crust in the structure of their bottoms. Quite recently, in geological terms, there were sectors of land on the sites of many of them composed of continental rocks. The formation of the basaltic layer of the crust and the transformation of the land into deep marine basins occurred mainly in the Neogene, and only in some seas (the Caribbean and the Gulf of Mexico), it had already begun in the Mesozoic.

PART III

The Earth's Internal Structure, Composition, and Deep Processes

CHAPTER 22

General Data on the Earth's Shape, Rotation, and Structure

The interior of the Earth inaccessible to direct geological methods is studied by means of geophysical and geochemical techniques. The geophysical methods used include seismic, gravimetric, geothermic, geomagnetic, and geoelectric techniques. Geochemistry widely employs laboratory study of the properties of rocks at high temperatures and pressures. The behaviour of matter in conditions not amenable to experimental reproduction is studied on the basis of solid state theory.

Information on the Earth's shape, its rotation round its axis, and tidal variations is also important for explaining a number of points about its internal structure. The data come from geodesy and astronomy. Those branches of astronomy that are concerned with the origin of the solar system, planets, and meteorites (cosmogony) also help explain the Earth's composition.

This chapter is devoted to certain general facts about the Earth's structure and composition. The make-up and structure of the tectonosphere, i.e., core and mantle, will be considered in detail later.

The Size and Shape of the Earth

The Earth has a complex shape. As a first approximation it can be taken as a sphere with a radius of 6371 km. The next approximation allows us to represent it as a bi-axial ellipsoid of rotation (the terrestrial ellipsoid) with an equatorial radius of 6378.16 km and a polar radius of 6356.78 km, from which it is obvious that the polar compression (flattening) of our planet is 1 : 298.25.

A third approximation leads to a complex figure, the geoid, which can be taken as a quite real but smoothed out figure of the Earth. Over the oceans the geoid coincides theoretically with the surface of the water, but over the continents it is a surface that is perpendicular everywhere to the plumb line. The geoid deviates from the terrestrial ellipsoid by a maximum of minus 160 m and plus 120 m, altering in height extremely smoothly. The shape of the geoid reflects the distribution of masses of various density in the body of the Earth and therefore has not only a geodetic but also a geophysical significance. No unambiguous connection, however, has been established between the geoid and the relief of the surface; it may be that it reflects the distribution of mass at very great depth, in the lower mantle.

The Earth's Rotation

The Earth's rotation round its axis presents geotectonic interest because changes in the rate of rotation can give rise to stresses in the crust.

The tides caused in the Earth's solid, liquid, and gaseous envelopes by the attraction of the Moon and Sun are known to slow our planet's rotation down. This has not only been established theoretically; study of the conic skeleton (epitheca) of corals has revealed growth rings in their structure—coarse annual ones and fine daily ones (around 50 microns). Counting of these rings showed that there were 400 days in a year in Devonian times, rather than 365, i.e., the length of a day was 22 h. It can be calculated that the Earth's rotation has slowed by approximately 24 s per million years, which accords with the theory. At the beginning of the Palaeozoic—500 million years ago—days were 20.8 h long, and soon after the Earth's formation (4500 million years ago) it completed one rotation in 2h.

Comparison of the Earth's shape with that which a fluid body of its size would adopt if the force of gravity in its field were the same and it was rotating at the Earth's present rate, indicates that the terrestrial ellipsoid is close to the equilibrium figure of a rotating fluid body. This signifies (1) that the globe as a whole has a creep sufficient for it to be deformed by the effect of centrifugal force in the course of geological time, from which it follows (2) that the Earth's figure must change as its rate of rotation diminishes (its flattening must be lessened since the centrifugal force is diminishing). If the Earth had the properties of an ideal fluid, alteration of its figure would go hand in hand with the change in the value of the centrifugal force and stresses would not arise in its body, but since the material from which it is composed is viscous, the restructuring of its figure must to some extent lag behind the change in the rate of rotation, and that must lead to the rise of stresses. In fact, precise comparison of the theoretical figure of a rotating fluid body with the observed terrestrial ellipsoid leads to the conclusion that today's figure of the Earth corresponds to the Earth's rate of rotation 10 million years ago, i.e., at the beginning of the Pliocene, rather than to today's rate. There should therefore be a stress of between 1 and 1000 kg/cm^2, which is much lower than the strength of rocks, and is also lower than their creep.

Observations of the Earth's rotation indicate that rapid changes occur in it from time to time. This is reflected in the fact that 24 h get a thousandth of a second longer or shorter. These changes in the rate of rotation occur as "jumps" for one to three years. Significant "jumps" like this occurred, for example, in 1864, 1876, 1898, and 1920. In addition, seasonal variations in the Earth's rotation have been observed, in connection with which the day is approximately 0.001 s longer in January than in July. The stresses that arise in the globe because of these fluctuations in its rotation are infinitesimal. With a change of 0.01 s per annum, for example, which is greater than any fluctuation so far observed, the maximum acceleration that would occur at the equator would not exceed 0.01 milligal, which means that the weight of a body on the Earth's surface would alter by one hundred millionth (i.e., by one g in 100 t).

With upward and downward movement of blocks of the crust forces operating laterally should be generated, depending on whether the block moves nearer to the centre of rotation or further away from it, and whether its angular momentum alters. With an increase in angular momentum a block's pressure would have an easterly direction, and with a decrease a westerly direction. Calculations indicate that, if a block with an area of 1 km^2 moved vertically by 1 cm a year, that would give rise to a horizontal acceleration of two milligals, i.e., two millionths of the normal acceleration due to gravity at the Earth's surface (two gals per t).

If we allow for the fact that the continents float on a heavier substratum, often squeezing it aside, the flattening of the Earth gives rise to a force that pushes them toward the equator. The maximum pressure that would thus develop at the equator, if two continents running from here up to the poles were to "collide" there, would be 1 kg/cm^2.

The tidal forces that are gradually retarding the Earth's rotation must thus pull the continents across the surface of the globe in a westerly direction. The "forces of westerly drift", however, are extremely small. If they were sufficient to move America 50° to the west, as the theory of continental drift requires, they would bring the Earth's rotation to a halt within a year. In fact, as we have seen, the slowing of rotation proceeds one thousand millionth more slowly.

It is quite obvious that all the forces linked with the Earth's rotation enumerated above are too small to have any tectonic significance. For us a change in the position of the axis of rotation within the Earth is more important. If the change in inclination of the axis were measurable in degrees considerable forces would be generated, and in order to restore the Earth's equilibrium figure there would have to be changes that would shift any single point on the surface vertically by more than 1 km. The strain at any one point, however, would depend on how smoothly the deformation affected the surface. Such a tectonic effect could only be expected from this cause when an area is deformed unevenly because of non-uniformity of the crust and a local concentration of stresses is generated.

Tidal phenomena of interest are the tides caused in the solid body of the Earth by attraction of the Moon, and to a lesser extent of the Sun. Because of the Earth's rotation tidal waves with a rise and fall measurable in several tens of centimetres (to a maximum of 43 cm) move across its surface. As a result every point on the surface rises twice a day and falls twice a day. It is possible that this mechanism shakes rocks and causes cracks to form in them. Some workers see in it a possible cause of the general fissility of rocks. They also suggest that terrestrial tides may act as the "trigger" of earthquakes where the stresses generated by other, more powerful, deep-seated sources reach critical dimensions. But for all that it must be remembered that the smoothness of a deformation spread over a vast area leads in this case to a low mean degree of strain, which can only be heightened locally by the non-uniformity of the crust.

On the whole we may conclude that the tectonic role of phenomena connected with the Earth's rotation cannot be substantial.

The Earth's Inner Structure from the Data of General Seismology. The Earth's Seismicity

Basic data on the deep regions of the globe are obtained by using seismic methods that enable us to judge the velocity of the propagation of seismic waves at various depths.

The velocity of the propagation of seismic waves is determined by the medium's elastic properties (by the moduli of cubic compressibility and of rigidity) and its density, a relation expressed by the following formulas:

a) the rate of propagation of longitudinal (compressional or P) waves $v_p = \sqrt{\frac{\varkappa + 4/3\mu}{\rho}}$

b) the rate of propagation of transverse (shear or S) waves $v_s = \sqrt{\mu/\rho}$ where $\varkappa$ is the modulus of cubic compressibility (bulk modulus);

μ is the modulus of elasticity in shear (shear modulus);
ρ is the density.

Since density figures as the denominator in these formulas, seismic velocities should, it would seem, diminish as the density of the medium increases. In fact elastic moduli increase more rapidly still with increase in density, so that velocity increases rather than decreases with heightened density.

Information on the Earth's structure as a whole—from the surface to the centre—is obtained by the techniques of general seismology, while more precise data on the structure of the upper layers—to a depth of 100 to 150 km—are obtained mainly by the techniques of deep seismic sounding. The techniques of general seismology are based on recording natural earthquakes, while deep seismic sounding is linked with observation of the elastic waves propagated by artificial shocks (explosions).

As this section is concerned with the most general information on the Earth's structure, we are interested here in the data of general seismology that throw light on the distribution of seismic velocities of propagation along the Earth's radius.

Table 5. Distribution of seismic velocities with depth

Depth, km	v_p, km/s	v_s, km/s	Envelope	Layer
0-40 (continents) 5-12 (oceans)	5.5-7.4	3.2-4.3	crust	A
Mohorovičič discontinuity (Moho) (discontinuity of the first order)				
40/12-410	7.9-9.0	4.5-5.0	upper mantle	B
410-1000	9.0-11.4	5.0-6.4		C (Golitsyn)
Discontinuity of the second order				
1000-2700	11.4-13.6	5.4-7.3	lower mantle	D
2700-2900	13.6	7.3		
Discontinuity of the first order				
2900-4980	8.1-10.4	—	outer core	E
4980-5120	10.4-9.5	—		F
Discontinuity of the first order				
5120-6370	11.2-11.3	?	inner core	G

Table 5 is a summary of data on the change in velocity of propagation of longitudinal and transverse seismic waves with depth.

Let us proceed to a discussion of the table. At the top of the first column depths of 0-40 km are indicated for continents (40 km being the mean thickness of the continental crust), and depths of 5 to 12 km for oceanic crust (5 km being the mean depth of the water layer). In the "envelope" column the accepted names of the seismic layers are given, and in the "layer" column the alphabetic indicators of these layers suggested by the Australian seismologist Bullen. In the Soviet literature layer C is often called the "Golitsyn layer" after B.B. Golitsyn, the founder of Russian seismology.

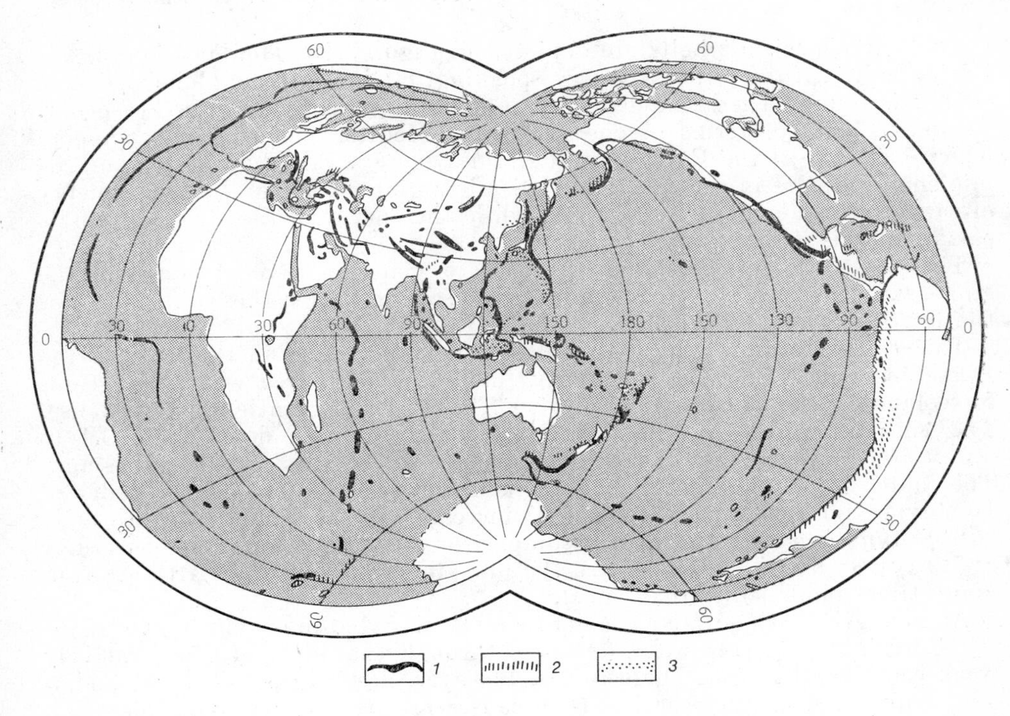

Fig. 115. Diagram of the Earth's seismicity (after Savarensky *et al.*, 1949):
Foci: *1*—shallow; *2*—intermediate; *3*—deep

Discontinuities of the first order occur where there is an abrupt change or jump in seismic velocities, i.e., between the crust and upper mantle, between the lower mantle and the core, and between the outer and inner core. A discontinuity of the second order is a level at which the derivative of seismic velocities changes, i.e., the growth rate of these velocities with depth. Such a discontinuity is found between the upper and lower mantle. In fact the velocity of longitudinal (P) waves increases in the upper mantle by 3.5 km/s/1000 km of depth, while it is 2.2 km/s in 1700 km in the lower mantle, i.e., an increase of only 1.3 km/s per 1000 km.

Transverse (S) waves are not propagated in the outer core, fading there, absorbed by the medium, which indicates that the medium of the outer core is close in properties to a liquid, while that of the mantle and crust is, on the whole, solid. As for the inner core, its state is not clear. It may be that it is solid and that shear waves are not transmitted through it because they are screened by the outer core.

As our table contains only the most general information on the distribution of elastic properties in the material of the globe, it ignores both the internal division of the crust and upper mantle, and the change of these properties horizontally. These more detailed data will be treated in the next chapter.

General seismology yields a picture of the Earth's present seismicity, which reflects an important aspect of the tectonic activity of the interior.

The Earth's seismic belts are depicted in Figure 115. The main one is the Pacific belt, within which, roughly speaking, 80% of all the globe's seismic energy is found. There the strongest and most frequent earthquakes occur. The next belt, in terms of energy, is the Mediterranean-Himalayan one, which has 15% of the Earth's total seismic energy. Finally, 5% of the seismic energy occurs in the remaining seismic regions. The total annual energy of earthquakes amounts to 10^{26} ergs, while the energy of the strongest single earthquake is 10^{24} ergs.

The most violent earthquakes generate general waves throughout the globe, with periods from a few minutes to several hours; these are known as the Earth's "free oscillations".

The whole picture of the distribution of elastic waves in the body of the Earth in connection with seismicity (including both free waves and tidal forces) indicates a considerable solidity. The crust and mantle are very dense. The shear modulus within the crust increases with depth from 0.1 to 4.0×10^{11} dynes/cm^2 at the base of the mantle it rises to 3×10^{12} dynes/cm^2. The outer core apparently has a much lower rigidity (of the order of 10^{10} dynes/cm^2), but rigidity again increases in the inner core to $2 \times \times 10^{12}$ dynes/cm^2. By the way of comparison steel has a modulus of rigidity of 7.9×10^{11} dynes/cm^2. The mantle and inner core of the Earth are thus many times more solid than steel.

At first glance these facts seem to contradict what was said above about the Earth's capacity to alter its shape through the effect of comparatively weak centrifugal forces. These forces, however, are long-period ones, and it is the duration of their action that causes creep in the Earth's material. Elastic deformation, however, is extremely short-period, and during it the globe acts like a completely solid body. This dependence of properties on the duration of the mechanical action is characteristic of any rock, and of any physical body in general.

Gravimetric Data

The average density of the Earth is 5.52 g/cm^3. At the surface the density is much lower; sedimentary rocks have a density of 2.4 to 2.5 g/cm^3, granites and most metamorphic rocks of 2.7 g/cm^3, and the main igneous rocks (basalts) of 2.9 g/cm^3. The average density of the crust can be taken as 2.8 g/cm^3.

The lower densities at the surface mean that the density of the material in the inner envelopes must be much higher than the mean density for the Earth as a whole. From a comparison of its rate of rotation and flattening with the data on the velocity of seismic waves at various depths, and on the discontinuities within the Earth, the following densities are now considered most probable:

At the top of the upper mantle	3.1 to 3.5 g/cm^3
At the bottom of the upper mantle (At a depth of 1000 km)	4.5 g/cm^3
At the boundary of the core (2900 km)	5.6 g/cm^3
At the top of the core	10.0 g/cm^3
At the centre of the Earth	12.5 g/cm^3

According to current cosmogonical conceptions, the Earth was formed through the accretion of particles of a primary cloud of gas and dust that consisted of particles of various composition and various densities, which came together in a haphazard way that led to a quasi-uniform proto-Earth, the average density of whose material was everywhere the same. If we now observe a non-uniformity of density along the Earth's radius, that is because of a subsequent process of differentiation of its material that produced a concentration of the denser matter in the interior and of less dense matter in the outer envelopes.

We shall return below to the question of when and how this differentiation took place. Here it is important to point out the immensity of the Earth's gross potential gravitational energy, which is estimated at 2.25×10^{39} ergs. Of this some 1.5×10^{38} ergs have been dissipated through the differentiation of matter during the Earth's existence. It is clear, from a comparison with other forms of energy, that gravitational force indisputably takes first place in the Earth. Only thermal energy can compare with it (see below).

Arguing from the distribution of density and acceleration due to gravity, we can calcutate the scale of lithostatic pressure at various levels within the Earth:

Depth	40	410	1000	2900	4980	6370
Pressure (kbars)	10	140	350	1360	3200	3610

Pressures corresponding to depths of 1000 km, let us note, can be obtained from protracted experiments in modern laboratory conditions. Higher pressures have so far only been obtained by means of explosions and can only last experimentally for tiny intervals measurable in millionths of a second. This circumstance complicates direct transfer of the experimental results to the natural situation in which matter is under pressure for hundreds of millions and milliards of years.

The Earth is extraordinarily close to isostasy, i.e., to equilibrium in the distribution of mass. With isostasy the gross mass of the matter concentrated along the vertical below a unit of surface area is everywhere the same.

With isostasy the relief of the Earth's surface and distribution of mass in depth should be strictly interdependent. Since the mass below any unit of surface area must be the same, the average density of matter below elevations should be less than under depressions. A deviation from this dependence is a breach of isostasy, or an isostatic anomaly.

Very few isostatic anomalies are observed on the Earth's surface, and these do not exceed tens of milligals, i.e. the deviation from equilibrium in the distribution of mass is expressed quantitatively in a few hundred thousandths of this mass. The biggest anomalies are observed when sectors no larger than a few hundred kilometres in diameter are studied. Broader sectors prove on average to be well balanced. We shall return to relatively big local isostatic anomalies when we deal with the structure of the crust and upper mantle.

The Earth's Heat Regime

We know that there is a continuous flow of heat from the Earth's interior to its surface. The scale of this flow can be calculated if we know the thermal conductivity of rocks and the heat gradient: $Q = \lambda \text{ grad } T$, where Q is the heat flow, λ the thermal conductivity, and grad T the heat gradient. The thermal conductivity of rocks is of order of $n \times 10^{-3}$ calories/s/cm/°C (on average about 5×10^{-3} calories/s/cm/°C), and varies within narrow limits for the commonest rocks, but heat gradients vary quite widely, from 6 °C to 150 °C per km.

Calculated heat flows indicate that the commonest rate is 1.1×10^{-6} cal/cm^2, which is taken today as the "normal" flow. The Earth's average heat flow, however, is 1.5×10^{-6} cal/cm^2. The average flow, it should be noted, is identical for both continents and the ocean floor.

The Earth's gross annual loss of heat is of the order of 10^{28} ergs. The total amount of heat given off during its existence varies, according to the model of its thermal history adopted, within the limits $(0.1-0.8)10^{38}$ ergs. The Earth's thermal energy, we see, is between a half and two-thirds of its potential gravitational energy and half to one order of magnitude smaller than the fraction of the gravitational energy dissipated during density differentiation of the Earth's matter.

The main source of the Earth's internal heat is radioactive decay. Radioactive elements with a long half-life (uranium, thorium, and potassium) play the main role. Their content is higher in acid rocks, lower in basic ones, and much lower in ultrabasic rocks. Their average content in the main groups of igneous rocks is shown in Table 6.

Table 6. Average uranium, thorium and potassium content of the main groups of igneous rocks

Rock	U, ppm	Th, ppm	K, percentage
Granite	4.0	20	4.26
Gabbro	0.9	2.7	0.46
Eclogite	0.043	0.1	0.1
Dunite	0.001	0.1	0.001

The various rocks generate heat in varying amounts in accordance with their radioactive content (see Table 7).

Table 7. Heat generation by rocks

Rock	Heat generation, 10^{-13} cal/cm^3/s
Granite	5.7
Gabbro	1.2
Eclogite	0.1
Dunite	0.02

There are two further important sources of the Earth's thermal energy, namely: gravitational differentiation and tidal friction. The first may yield as much heat as radioactive sources, but a considerable amount of it was apparently dissipated into space, especially during the planet's formation when it was still a small body. Tidal friction is able to generate a quantity of heat approximately equal to 20% of the radiogenic heat.

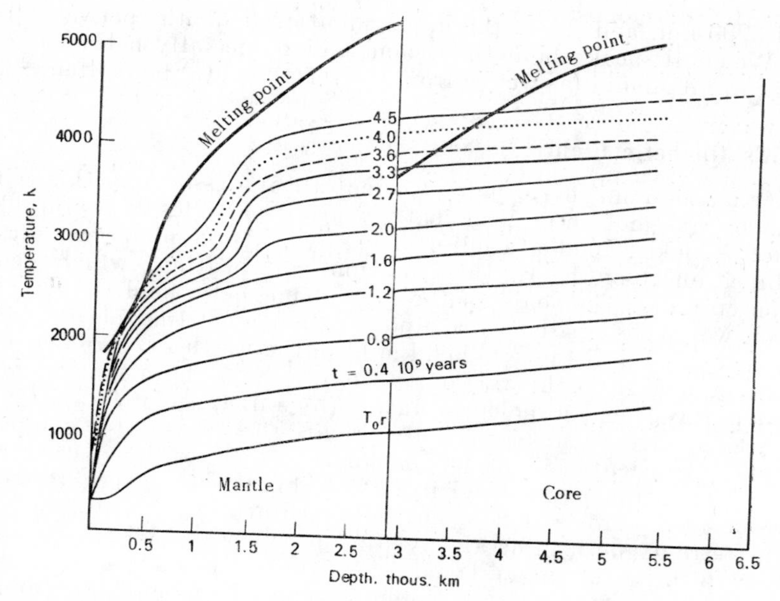

Fig. 116. Change in the Earth's internal temperature from its formation (T_0) to the present time (T = 4500 m.y.) (after Lyubimova, 1958)

Although much about the Earth's thermal regime is still unclear there are grounds for considering radioactive processes the main source of the heat flow now *observed*, and gravitational differentiation and tidal friction as supplementary factors. Most workers today, when estimating the Earth's total heat balance, are inclined to think that the globe's interior is very slowly getting hotter.

The Earth's present radial temperature distribution can be calculated from one or another set of assumptions as to its mode of formation, age, composition, content of radioactive elements, thermal conductivity, and certain other parameters. E.A. Lyubimova (1958) has made such calculations, based on a theory of the Earth's origin from a cold cloud of dust and gas and on the latest data for all the other factors. A diagram of the change in the Earth's internal temperature from its formation to the present time is given in Figure 116. Its temperature, we see, has risen at all levels deeper than 500 km below the surface throughout its existence. Its temperature also rose as well in the zone of the upper 500 km but only for 3800 million years from its formation, i.e. until 1200 million years ago. Since then cooling of the surface 500 km layer began simultaneously with continuing heating of the deeper interior. This polarization of the thermal evolution of the deep layers and those closer to the surface presents special interest. At present there is a high geothermal gradient down to around 1500 km, where the temperature reaches 4000 K. Deeper, the gradient diminishes considerably and the temperature does not exceed 4250 K at the boundary of the mantle and core (2900 km), and rises only to 4750 K at the centre of the Earth.

The diagram also shows the melting point curve for the Earth's matter at various depths. The curve of its present temperatures, as we see, intersects that of melting points in two regions: (a) in the outer core, at depth between

2900 and 3900 km, and (b) in the upper mantle, at depths between 150 and 450 km. We shall speak about the second region specially below; as regards the outer core, it should (as we see from the seismic data) be molten.

The Earth's Magnetic Field

The Earth has a magnetic field. Current theory associates this with the complex loop-like motion of electrically charged particles in its liquid outer core, which is looked upon as a self-exciting hydro-dynamo operating through the combined effect of the Earth's rotation and heating, which generates convection currents. The shape of the present magnetic field is very close to that which would be created by a bipolar bar magnet (magnetic dipole) located in the core at a distance of 436 km from the Earth's centre. This dipole is slightly inclined from the axis of rotation, so that the magnetic poles are located some way from the pole of rotation (by a distance of 11.5°).

The Earth's magnetic field not only orients the needle of a compass but also affects the orientation of ferromagnetic minerals (magnetite, titanomagnetite, haematite, pyrrhotine) in rocks. This effect can be exerted when solid ferromagnetic minerals float on the melt during the cooling of igneous rocks. The ferromagnetic minerals in sedimentary rocks may be oriented during the precipitation of mineral particles from a liquid. The rocks that react most strongly to the Earth's magnetic field are igneous basic and ultrabasic ones (basalts, gabbros, peridotites, and serpentinites) and sedimentary rocks (red continental sands).

When igneous rocks have cooled and sediments have lithified, the orientation of the ferromagnetic material is preserved. The techniques of palaeomagnetic research are based on this fact. Having determined the orientation of the ferromagnetic minerals we can establish the direction of the magnetic field that existed at the time a basaltic flow cooled or sandstone was deposited.

Only quite stably bedded igneous and sedimentary rocks, unaffected by tectonic deformation or metamorphic effects, can be used for investigations of this kind. Reliable inferences can therefore only be drawn from the rocks of ancient platforms. There is now quite extensive palaeomagnetic data on the rocks of the East European, Siberian, and North American platforms, and only relatively sporadic observations on the other platforms. Palaeomagnetic conclusions cannot, however, be drawn from single measurements, but must be based on statistical quantities. There are many factors that can distort the direction of residual magnetism. It is thought that the averaging of a great many carefully performed determinations, observing rigorous methodological requirements as regards the freshness of samples, establishment of the stability of the residual magnetism, etc., produces quite reliable results.

Palaeomagnetic determinations from rocks of various age have indicated that the Earth's field has not remained constant during geological history, and that its structure has altered.

Two phenomena are distinguished: (a) reversal or inversion of the magnetic field; and (b) migration or wandering of the magnetic poles.

The first of these phenomena consisting in this, that the magnetic field has reversed many times in the course of geological history, the north pole becoming the south and the south pole becoming the north. If the present-day magnetic field is taken as positive then geological history can be pic-

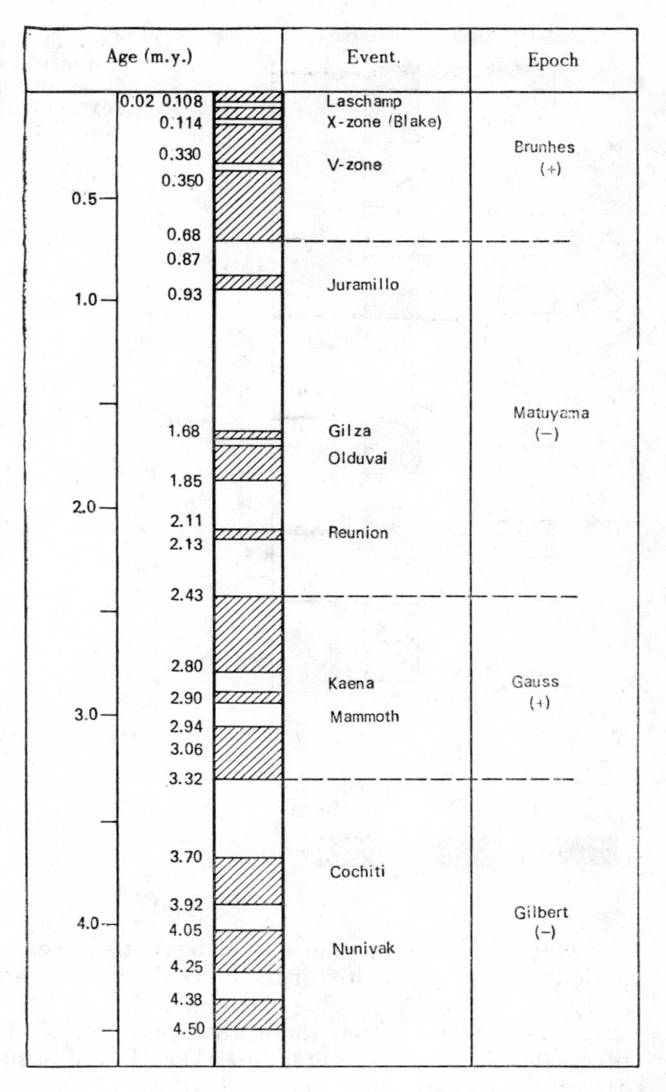

Fig. 117. Scale of magnetic reversals for the past 4500 m.y. (after Cox, 1973). Periods of a positive geomagnetic field are shaded

tured as an alternation of positive and negative magnetic epochs. The best studied reversals are the most recent ones that took place in the Quaternary Period and at the end of the Pliocene. They have been determined from basalts whose absolute age of extrusion has been established. It has been found that the current positive epoch, known as the Brunhes, began 700 000 years ago. Before it there was the negative Matuyama epoch of 1.7 million years duration. The Matuyama epoch was preceded by the positive Gauss epoch, which lasted 900 000 years; and before that there was the negative Gilbert epoch, which has been traced back in history for 1.2 million years. Within the epochs separate short-term "events" with a reversed field have been established. The whole scale consequently has a duration of approximately 4.5 million years (see Fig. 117).

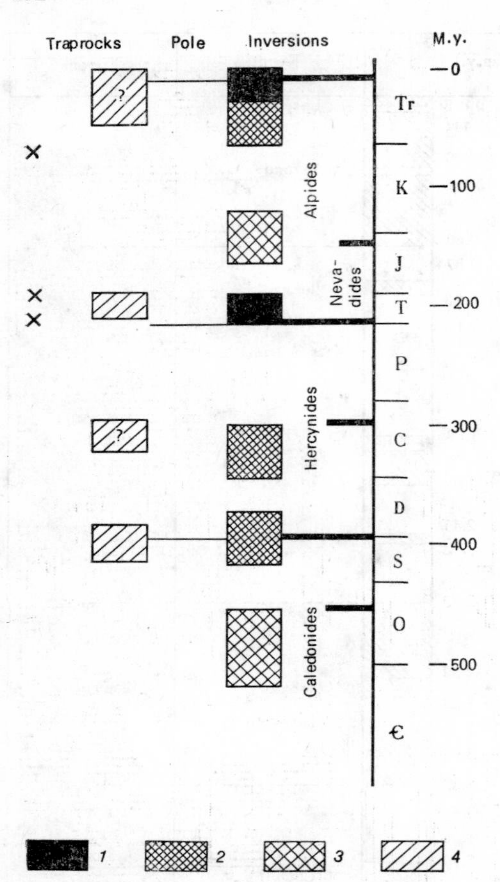

Fig. 118. Epochs of partial reversal and migration of the magnetic poles (after Schoenmann, 1973). The periodicity of reversals during their partial manifestations is:

1—less than a million years; *2*—two or three million years; *3*—four or five million years; *4*—epochs of major polar wandering

Reversals of the magnetic field have also been established for other, older periods of geological history, but their age has not been determined as exactly as for recent geological time.

It has been found that there were quiet periods in the Earth's history when reversals of the magnetic field either did not occur or seldom happened, and disturbed periods when the field altered its direction many times and frequently. Frequent reversals have been observed for the Late Cambrian and Early Ordovician, the Late Silurian and Early Devonian, the Carboniferous, the Middle and Late Triassic, the Early Cretaceous and in the Tertiary Period when the intermediate periods had a predominantly quiet magnetic field (see Fig. 118).

Migrations of the magnetic poles, i.e., changes of the Earth's magnetic axis, have been traced by palaeomagnetic data for the whole of the Phanerozoic, but the deeper we delve into the history of the crust, the less reliable the information becomes. The data on these migrations in recent times have been carefully analysed by Schoenmann Yu.M., on whose conclusions we base ourselves here.

The changes in residual magnetism in the rocks of the East European and Siberian platforms lead to the following conclusions.

Fig. 119. Migration of the magnetic pole (after Shoenmann, 1973):

I—region of polar wandering in the Early Palaeozoic; *II*—region of polar wandering in the Middle and Late Palaeozoic and Early Mesozoic

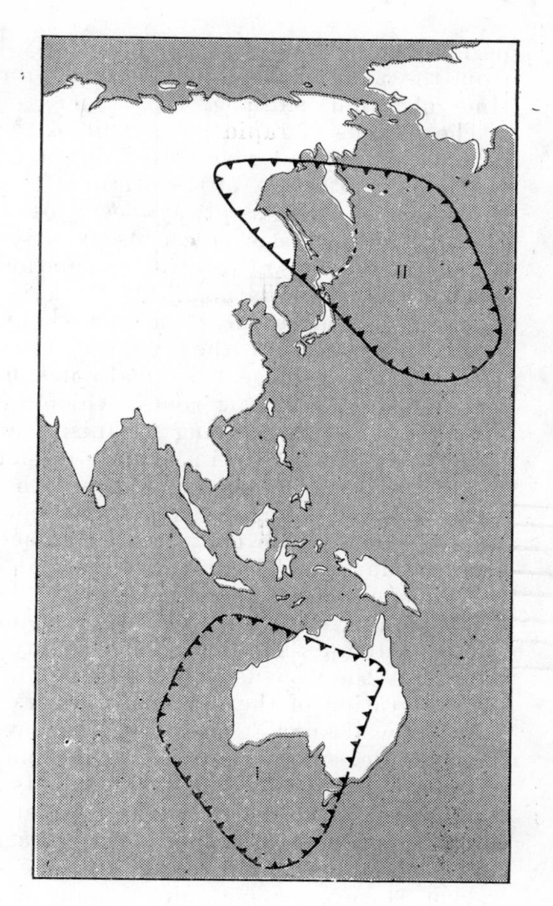

In the Early Palaeozoic (Cambrian, Ordovician, and Early Silurian), the north magnetic pole was located in the region of Australia, where it had apparently wandered in the area between Tasmania and New Guinea. In the Late Silurian and Early Devonian it shifted rapidly to the region of the Sea of Okhotsk, Japan, and the Maritime Territory of the USSR. It remained there, wandering within a limited area, until the beginning of the Triassic. In the Triassic there was a rapid shift to the present-day polar region, where it still remains (see Fig. 119).

The rocks of the North American platform yield roughly the same picture of shifting of the North Pole. The results obtained from the rocks of South America, too, do not differ from the foregoing, except that they indicate a rather earlier shifting of the pole to the present polar region (in the Late Carboniferous and Permian). Data from the rocks of India and Africa are so few that significance can hardly yet be attached to them.

In the results obtained from northern platforms the irregularity of polar movement catches the eye. Long periods of wandering within a limited area (in the Cambrian and Ordovician, Devonian and Carboniferous, and the Jurassic, Cretaceous, and Cenozoic) are followed by much shorter periods of rapid movement of the pole to a new region of "wandering". There were such periods in the Late Silurian and Early Devonian, and in the Triassic. Data from the South American platform point to the Late Carboniferous as a time of rapid shifting.

The periods of rapid movement of the poles, interestingly, coincide with periods of frequent reversal of the magnetic field (see Fig. 118).

The current theory of the origin of the Earth's magnetic field assumes that the magnetic axis has always been close to the planet's axis of rotation. Shifting of the magnetic pole consequently implies, in terms of this theory, a change in the axis of rotation in relation to the crust. It is improbable, from mechanical considerations, that the axis of rotation has moved so considerably in space. Rather the globe as a whole, or its outer envelopes, have shifted spatially relative to the axis of rotation.

Movement of the geographical poles, however, also implies a corresponding shifting of all climatic zones, which would affect the distribution of cold-loving and warmth-loving organisms and sites of the accumulation of sediments associated with a definite climate. Such sediments are, for example, the evaporites confined to regions with an arid climate, or coals, which require a moist climate, or glacial deposits linked with a cold climate. It would seem easy to compare the palaeomagnetic findings with the palaeoclimatic data, and by means of the latter to confirm or throw doubt on the phenomenon of polar migration.

The comparison, however, we must note, does not in fact prove unambiguous. Attempts have been made many times along this line for several decades, but the various workers have obtained opposite results, some that confirm migration of the poles, and others that refute it.

In his well-known book on general palaeogeography L.B. Rukhin (1959), for example, summarised the many findings on the distribution of plant and animal organisms by geological periods and concluded that the changes that took place in the distribution of organisms on the Earth's surface are inexplicable except by the idea of a migration of the geographical poles.

Other workers, however, have come to other conclusions. The American worker Stacey, for example, concluded from an analysis of the distribution of various organic forms of the Permian period that the position of climatic zones then did not differ from the present-day one, which implies that the poles were located at that time approximately where they are now.

Palaeoclimatic research has seemingly not yet reached the level needed to provide an unambiguous answer, but palaeomagnetic techniques, too, are still far from perfect. The whole problem of polar migration therefore remains one of those awaiting solution.

The answer, moreover, may be unexpected. It is not excluded, for example, that the shiftings of the palaeomagnetic poles have no relation to the position of the geographic poles. Such an independence of the two could occur if the magnetic field were not a dipole in earlier geological periods, as now, but more complex, i.e., that there were regional magnetic anomalies close in intensity to the main dipole field. In that case magnetic minerals would point to these "anomalies" and not to the poles of the Earth's axis of rotation.

The Composition of the Earth

The rocks developed on the surface or ejected onto it by volcanoes give an idea of the chemical and mineralogical composition of the crust and in part of the upper mantle. To judge the composition of the deeper envelopes

Table 8. Composition of the mantle and continental and oceanic crusts (compiled by B.G. Lutz)

Elements	Continents								Oceans		
	Chondrite	Garnet-peridotite	Mantle eclogite	Spinel-peridotite	Alpino-type ultrabasite	Granulite basaltic layer	Granitic-gneissic layer	Consolidated crust	Lherzolite (initial undifferentiated material)	Harzburgite (sediments of the melting of basalt)	Oceanic crust
					Silicate analysis (% wt.)						
SiO_2	47.04	44.04	45.26	44.88	40.49	58.2	65.6	61.9	45.7	45.0	48.65
TiO_2	0.14	0.16	0.45	0.11	0.16	0.8	0.5	0.6	0.2	0.1	1.40
Al_2O_3	3.09	2.12	14.78	1.90	1.49	16.0	14.9	15.5	3.7	1.7	16.52
Fe_2O_3	—	3.32	3.56	1.61	2.88	2.8	1.1	2.0	5.1	6.8	2.29
FeO	15.40	4.19	6.07	6.28	5.06	4.8	3.4	4.1	3.6	2.2	6.23
MnO	0.31	0.10	0.14	0.14	0.11	0.15	0.1	0.1	0.1	0.1	0.18
MgO	29.48	41.57	16.72	41.18	41.31	5.3	2.4	3.8	38.4	42.6	6.79
CaO	2.41	1.87	9.16	1.89	0.99	6.0	3.4	4.7	2.3	0.7	12.28
Na_2O	0.81	0.44	0.79	0.17	0.04	3.2	3.5	3.4	0.3	0.2	2.57
K_2O	0.11	0.10	0.19	0.03	0.01	2.0	3.7	2.9	0.1	0.1	0.37
					Content of rare lithophilic elements (ppm)						
Rb	3	2.9	5.5	0.75	0.24	35	160	100	0.52	0.54	—
Li	2	1.4	2.0	0.5	0.5	8	25	16	4.1	3.2	—
Sr	11	47	76	18	2.8	265	378	320	13.6	—	—
Ba	3.4	31	51	7.5	2.6	540	777	660	7.7	10.2	—
TR	5.2	16	20	7.9	4.04	86	200	143	10.6	11.5	—
Th	0.04	0.59	0.36	0.067	0.0153	2.7	14.6	8.65	0.698	0.641	—
U	0.014	0.778	0.13	0.023	0.0069	0.7	2.0	1.35	0.15	0.15	—
Zr	33	50	—	25	33	140	150	145	—	—	—

of the globe it is necessary to resort to other data, among which the chemical composition of meteorites has a major place. The assumption is that meteorites and the planets of the solar system were formed from one protoplanetic cloud of dust and gas, and that the average composition of meteorites should therefore be close to that of the Earth.

It is usually supposed that the group of meteorites known as chondrites is closest to the Earth's mean composition. One speaks of a "chondrite model of the Earth". The chemical composition of chondrites, which can be considered the Earth's initial material, is shown in Column 2 of Table 8.

The composition of the surface rocks of the crust, however, differs considerably from the Earth's average toward a higher enrichment by oxygen, silicon, sodium, potassium, aluminium, the rare earths, and radioactive elements, and a reduction in iron and magnesium content (Table 8, columns 8 and 9).

Originally, when the particles of the gas/dust cloud were coagulating into a single body, the latter would have been quasi-uniform. The difference between the composition of surface layers and the initial average composition observed now consequently serves as an indicator of the subsequent differentiation of the Earth's material. But, if the lighter components accumulated in the surface envelopes as a result of differentiation, where would the remaining products of this process, heavier than average, have accumulated?

The view now prevails that the initial mean composition of the Earth is preserved in the lower mantle. It has a composition closest to that of chondrites, and is the source from which both the lighter and heavier substances have differentiated. The lighter substances rise and form the upper mantle, and as a result of further differentiation of the mantle the crust is formed; the heavier substances sink, and their accumulation leads to the formation of the Earth's core.

The composition of the upper mantle and the continental and oceanic crusts is shown in Table 8. We shall discuss the difference between the two crusts later.

The composition of the core remains debatable. The traditional view assumes that it consists of iron and admixtures of nickel (around 7%). Another view attributes a silicate composition to it, enriched to a limited degree only by iron; it is further assumed that the silicates are in a "metallic" state because of the high pressures, i.e., their atoms are crushed and some of the electrons are in a free state as in metals. There is also an intermediate view that the outer core consists of silicates in a "metallic" state, while the inner core is iron.

In view of the increase in lithostatic pressure with depth the character of links must be altered and molecules must become increasingly simple. As far as can be judged from the tephra of volcanoes and kimberlite pipes, the upper mantle consists predominantly of peridotites and its main minerals are olivines (forsterite and fayalite). In the Golitsyn Layer a change in the ratio of oxygen and magnesium in olivine from ionic to covalent bonds can be assumed that is accompanied with a rise in density of approximately 18%. In addition, silica, represented higher up by quartz (density 2.53 g/cm^3) passes in this same zone into denser varieties, viz., coesite (2.93 g/cm^3) and stipoverite (stishovite) (4.35 g/cm^3). In the lower mantle the complex silicate molecules must be broken down into oxides. In the core separate atoms predominate.

Table 9. Absolute geological time scale

Era	Period	Subdivision	Absolute age of the beginning of the subdivision (in m.y.)
Cenozoic	Quaternary	Pleistocene	3
	Neogene	Pliocene	10
		Miocene	25
	Palaeogene	Oligocene	40
		Eocene	55
		Palaeocene	70
Mesozoic	Cretaceous	Late	100
		Early	136
	Jurassic	Late	155
		Middle	175
		Early	190
	Triassic	—	225
Palaeozoic	Permian	—	280
	Carboniferous	—	345
	Devonian	Late	360
		Middle	370
		Early	395
	Silurian	—	430
	Ordovician	Late	450
		Early	500
	Cambrian	—	570
Proterozoic	Late (Ryphean)	Late	1000
		Middle	1350
		Early	1600
	Middle	—	1900
	Early	—	2600
Archaean	—	—	3500

For differentiation to occur there must be the possibility for matter to move, i.e., light particles must rise to the surface and heavy ones sink. It is supposed that heating plays an essential role in heightening the mobility of the Earth's material. Although it was originally cold, the energy released by the collision of particles, and especially by the decay of radioactive elements in particles, led to heating. On the whole, however, the world was apparently never molten; if it had been liquid for any time density differentiation of its matter would have been completed in the liquid state, but the magmatic phenomena that have continued throughout geological history, and are still occurring, indicate that differentiation developed slowly throughout the world's life and has not been completed in our day. This protracted nature of differentiation in time has also made it possible for the gravitational energy released in the process to be the main energy source for all endogenous processes throughout the Earth's history.

The mechanism of differentiation will be considered below.

The Age of the Earth

In order to understand the nature of deep-seated processes it is important to know their rate of occurrence as well as their sequence, and that can only be ascertained from the facts about rocks' absolute age.

This brings us to the time scale of absolute age obtained by radioactive techniques (see Table 9). This scale is still extremely crude. For the Phanerozoic it indicates an absolute time scale partly for whole periods and partly for segments. For the Precambrian it is even less detailed, especially when we try to compile it for the whole world, as is done in the table. A rather more detailed scale is obtained for the Precambrian when it is constructed on regional material, for instance, for the Baltic, Canadian, or Ukrainian Shields, or for South Africa.

The oldest rocks so far observed in the crust have an age of 3500 ± 150 million years, but from the average composition of isotopes of lead in the crust its age (i.e., the age of the beginning of crystallization of its material after its separation in molten form from the mantle) has been fixed at 4500 million years. According to astronomical data, however, the Earth's age may exceed 500 million years.

CHAPTER 23

The Crust and Upper Mantle (Tectonosphere)

Everything known about tectonic, magmatic, and metamorphic phenomena indicates that they are directly linked with processes taking place within the Earth's crust and the upper mantle. It is in these envelopes that the direct causes of endogenous phenomena are concentrated. The significance of processes taking place at greater depths is not yet clear. It may be that it is very great; as we have seen, there are grounds for supposing that all the Earth's envelopes are involved in gravitational differentiation, which is the main source of energy for geological processes; these deeper processes, however, do not affect the surface directly but through the upper mantle and crust. They also give rise to movements in these envelopes and to transformations of matter that themselves lead directly to tectonic, magmatic, and metamorphic phenomena on the surface. Because of this role of the upper mantle and crust, it has been suggested that they should be joined together under the common title of the "Earth's tectonosphere".

This chapter is devoted to a more detailed review of the data on the structure and development of the tectonosphere and how they are pictured from the geological, geophysical, and geochemical facts.

A most important difference between the tectonosphere and the deeper envelopes is the horizontal disconformity of its properties, apparently associated with zones of various endogenous regimes. The structure of the tectonosphere is found to differ to some degree or another beneath various endogenous zones.

The Structure and Composition of the Crust

The geological, geophysical, and geochemical data point to the existence of two main types of crust on the Earth, viz. continental and oceanic.

Continental crust has an average thickness of around 40 km but its thickness varies from place to place between 20 and 70 km. It has been a common view for some time that its thickness is linked with the height of the surface relief, and that the crust is thicker below elevations and thinner below depressions. This view was originally based on gravimetric data, i.e., on the fact that the state of the Earth's crust is very close to one of the isostasy, which implies that there is an equal mass of matter below each equal area of the surface. Assuming that there are no horizontal disconformities in the lower-lying envelopes of the Earth and that the crust's average density is everywhere uniform, we have to conclude that isostasy can only operate if the crust is thicker under mountain ranges than under depressions.

The thickening under mountains must be very considerable. It is governed by the difference in density between the crust and the upper mantle. Since this difference is comparatively small (the average density of the crust is 2.8 g/cm^3 and that of the upper mantle 3.3 g/cm^3), the "roots" formed in the crust below mountains should wedge into the mantle to a depth several times the height of the range above sea level $\left(\text{if we take the average densities given above, the factor is } 5\frac{1}{2}\right)$. The compensation of surface relief by a corresponding bottom relief gives the isostatic model suggested in the last century by the English geodesist G.B. Airy.

Testing of the notion of "mountain roots" by the techniques of deep seismic sounding has shown that the Mohorovičič discontinuity is in fact at a greater depth beneath high mountains than below depressions. This natural link between the thickness of the crust and the height of the relief, however, is only maintained in comparatively few cases, viz. very high mountain ranges and deep depressions. Below the Pamirs, for example, the crust is 70 km thick, and beneath the Hungarian plain 20 km. But under relief of average height there are very great variations of crustal thickness not linked with it. On the Russian plain, for example, with an average crustal thickness of 35 to 40 km, there is an area (the Ukrainian crystalline shield) where the crust is 55 km thick. There is the same thickness of crust beneath the Great Caucasus, although they rise 3000 m on average above the Russian plain. Under the Ferghana depression the thickness of the crust is the same as under the neighbouring mountain ranges (50 km under both).

It can be concluded from this that isostasy is realised in a more complicated way than is suggested by Airy's model. It may often come about through horizontal changes in the density of the crust, i.e., in accordance with the isostatic model of another English geodesist, J. Pratt, who suggested that the crust had a dense basement and that relief was compensated by a reduction in average crustal density below elevations and an increase below depressions. Seismic sounding indicates that Pratt's model does not exist in its pure form, but it is possible to combine it with Airy's in one way or another. There are even better grounds, however, for supposing that isostasy does not operate simply within the crust alone but throughout the tectonosphere, i.e., in both the crust and the upper mantle taken together. In other words, the different ratio between relief and thickness of the crust points to the existence of horizontal density unconformities in the upper mantle as well as in the crust.

The state of the Earth's outer envelopes, however, it should be noted, can only be considered isostatic as a first approximation. The regions where present-day tectonic movements are distinguished by low intensity (ancient platforms and deep oceanic basins) are actually in a state very close to isostasy, but the position is different in regions of present-day tectonic activity; there are marked breaches of isostasy in them. Present-day vertical movements of the crust, interestingly, tend as a rule toward an intensification of anomalies rather than attenuation of them. Intensive upward and downward movements prove, in most cases, to be anti-isostatic; the present-day uplift of the Great Caucasus, for example, is anti-isostatic. If it were caused by forces tending to restore equilibrium there should be subsidence rather than rising, because it is overloaded. The foredeep to the north should also be rising instead of subsiding, since it is underloaded.

Regions recently freed of ice cover, like the Baltic and Canadian Shields, are moving in accordance with isostatic anomalies. On both shields the anomalies are negative and the shields are rising. Certain seas are also subsiding, tending to restore equilibrium; positive isostatic anomalies are observed, for example, in the Aegean and the Tyrrhenian Seas and the western part of the Mediterranean, where there has been a subsidence in recent geological time that is apparently continuing.

These relations between vertical movements of the crust and isostatic anomalies have to be considered in any attempt to clarify the causes of oscillatory movements. The anti-isostatic character of movements probably suggests horizontal overflow of material somewhere in the interior of the tectonosphere from troughs to uplifts. As for the rise of regions of recent glaciation in accordance with negative anomalies, it is explicable by their release from the weight of the ice, following which equilibrium has not yet been restored because of the high viscosity of the deep-lying matter. The sinking of seas, which is characterised by positive anomalies, suggests a consolidating of matter in the interior.

Three layers are usually distinguished in continental crust: sedimentary, "granitic", and "basaltic". Except for the sedimentary layer, these names are arbitrary and are based on comparison of the elastic properties of the material composing the respective layers with the properties of the commonest rocks. Nowadays these layers can be given names more closely corresponding to their actual composition.

The sedimentary layer, as we know, occurs intermittently, and varies in thickness from zero to 20 km, averaging 3000 m. We have discussed the composition of this layer and its dependence on tectonic conditions in earlier chapters.

According to the seismic data the "granitic" layer is characterised by P-wave velocities between 5.0 and 6.5 km/s. The composition of the upper part of this layer is known from its outcrops on the surface, mainly in ancient crystalline shields. Shields consist approximately of 50% granite, 40% gneisses and other metamorphic rocks of the amphibolite facies of metamorphism, and 10% rocks of the granulite and eclogite facies of metamorphism, plus weakly metamorphosed quartzites, phyllites, dolomites, and basic igneous rocks. It is supposed that the layer preserves the same composition right down to its base.

From these data it would evidently be correct to call this layer *granitic-gneissic* rather than granitic.

In most cases the granitic-gneissic layer varies in thickness from 8000 to 25 000 m, depending primarily on the thickness of the crust. On the non-shield parts of platforms it occupies roughly half the total thickness of the crust; on ancient crystalline shields its role is minor (20% to 30% of the total crust), while in the crust of young mountain ranges (the Pamirs and Caucasus) its role rises to 40%.

There are places on continents where it can be supposed that this layer is completely absent. There is none, apparently, on certain areas of the Baltic and Anabar Shields, where it has been eroded away and the "basaltic" layer is exposed on the surface. In addition the seismic data indicate the absence of a granitic-gneissic layer in certain deep tectonic depressions. There is none, for example, in the central part of the Caspian syneclise, where the sedimentary layers, which have a total thickness of around 18 km, were laid down directly on the "basaltic" layer.

Between the granitic-gneissic and "basaltic" layers there is a seismic discontinuity known as the Conrad discontinuity. Immediately below it the velocity of P-waves usually rises to 6.6 km/s or higher, and may be as high as 7.3 km/s at the base of the crust.

It has now been established that the "basaltic" layer (also known as the lower crust) consists predominantly of metamorphic rocks of the granulite facies, among which plagiogneisses play the main role with garnet and pyroxene in the nearly complete absence of micas. There, too, anorthosites and charnockite occur, and other intrusive basic rocks. This is the complex of rocks that is found in crystalline shields, where it is supposed that there is no granitic-gneissose layer. It would be more correct, therefore, to call this layer *granulitic-basic* rather than "basaltic".

It was pointed out, in Chapter 13, that the most favourable conditions for granitization are generally found at depths of 10 to 15 km which corresponds roughly to the boundary between the granitic-gneissic and granulitic-basic layers. Granite batholiths may be represented as layered, lens-shaped bodies occurring near the Conrad discontinuity and having both upper projections in the form of deep granite and granitic-gneissic diapirs and "roots" going deep down in narrow bands along fractures.

In the lowest part of the crust, judging from the xenoliths of kimberlite pipes, there is a not very thick layer of eclogites in a number of places, though not everywhere. Crustal eclogite, which consists of garnet and pyroxene, is the result of the cooling of basalt magma under pressure and has an extremely high density (3.6 g/cm^3). The origin of this layer has to be linked with the extrusion of basalt from the upper mantle, a process we shall discuss below. Continental crust consequently consists of four layers in places, rather than of three.

Oceanic crust differs from continental in being much thinner. The solid oceanic crust normally has a thickness around 6000 to 7000 m. Taking the average depth of the water cover as 5000 m, the base of the oceanic crust (the Moho) is 11 or 12 km deep.

Oceanic crust also differs from continental in composition. It has no granitic-gneissic layer. A thin sedimentary layer, no thicker than a few hundred metres, lies on the second "basaltic" layer (Layer II), which is usually 1000 to 15000 m thick. The velocity of P-waves in this layer is 5.0 to 5.5 km/s. Below it is a third layer, known also as the "oceanic" layer (or Layer III), the composition of which, as already mentioned, is not known, but which can be supposed, from various signs, to be quite complex and to have various basic and ultrabasic igneous rocks, viz. gabbro, peridotites, and pyroxenites, many of which would be in a serpentinized state. Some workers suggest a considerable role for amphibolites. Seismic velocities in Layer III are between 6.5 and 7.0 km/s.

Changes in the structure of the crust are observed in the zone of mid-ocean ridges, where there is a wedging out of Layer III and a considerable thickening of Layer II to 5000 m. In addition there is a significant increase in the total thickness of the oceanic crust (up to 15 000 and 20 000 m) below "aseismic" ridges. On the other hand the solid crust proves to be very thin (3000 to 4000 m) below deep oceanic trenches.

In addition to the continental and oceanic types of crust, there are other **intermediate types**: (a) suboceanic and (b) subcontinental.

Suboceanic crust is developed in internal and marginal seas whose depth does not exceed 2000 m. Its basement is similar to that of oceanic crust,

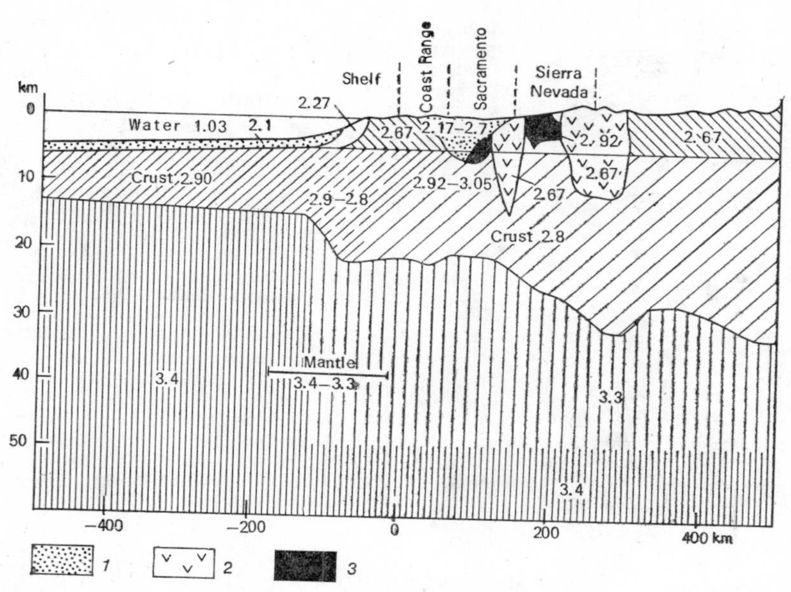

Fig. 120. Geophysical section of the transition zone to the ocean in California (after Worzel, 1965):

1—sedimentary rocks; *2*—granitic intrusions; *3*—basic intrusions. Matter of different density (g/cm3) according to the gravimetric data is hatched

differing only in the greater thickness of loose sediments, which carries between 3000 and 6000 m in many seas, but (as already said) attains 8000 m, and in some even 20 000 m, in some.

Subcontinental crust is characteristic of the margins of continents and island arcs. It has a general continental structure but differs in being thinner than typical continental crust. A general thinning is observed on the periphery of continents. In the central regions of North America, for example, the crust is around 40 km thick; near the Atlantic coast it thins to 30 km, and in the coastal zone of California to 18 km (see Fig. 120). In Eurasia there is a thinning of the average thickness of the crust from 50 km in the central regions to 35 km on the periphery. On island arcs the crust is 30 to 35 km thick. "Mountain roots" prove to be correspondingly less deep under continental margins and island arcs than under the centre of a continent.

Another difference of subcontinental crust is the absence of a distinctly marked Conrad discontinuity; the transition from the granitic-gneissic layer to the granulitic-basic one is gradual. This property is particularly marked on island arcs.

The normal division of the crust into two or three layers, mentioned above, is only a very rough scheme. Quite often the number of layers differing in velocity of seismic waves or in discontinuous surfaces from which seismic waves are reflected proves to be much greater. It is then difficult to determine which surface precisely should be considered the Conrad discontinuity.

Comparison of the data obtained by seismic techniques in various regions leads to the conclusion that the number of layers in the crust, their thickness, and the seismic velocities characteristic of them change at close distances. The Earth's crust proves to be divided into blocks of not very large dimen-

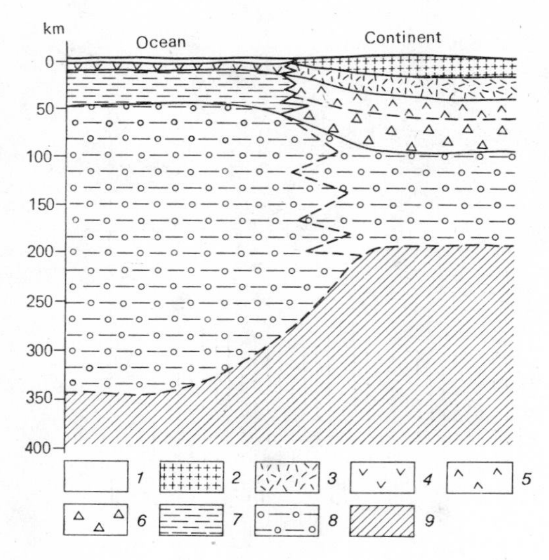

Fig. 121. Diagram of the transition from the continental tectonosphere to the oceanic (after Beloussov):

1—water; *2*—granitic-gneissic layer; *3*—granulite-basic layer; *4*—basaltic oceanic crust; *5*—predominantly Alpine-type ultrabasite; *6*—predominantly spinel ultrabasite; *7*—predominantly garnet ultrabasite; *8*—the asthenosphere (garnet ultrabasite whth inclusions of liquid basalt); *9*—the Golitsyn layer

sions (tens or a few hundreds of kilometres across) with a different internal structure. The same data indicate that the boundaries between these blocks are often expressed in the form of vertical fractures penetrating the whole crust.

The transition from continental to oceanic crust occurs within the limits of the continental slope where the depth of the sea of ocean is roughly 2000 m. At that depth the granitic-gneissic layer wedges out. The velocity of seismic waves in the granulitic-basic layer is the same as in Layer III of the oceanic crust. On old diagrams, therefore, these layers are merged into one, on the supposition that the "basaltic" layer of the continental crust continued directly into the basaltic crust of the oceans. We now know that the composition of oceanic crust and of the lower continental crust is different and consequently that no layer of the crust extends from a continent into the ocean; the continental crust breaks off completely at the continental slope, where it is replaced by oceanic crust of a different structure. The details of the transition from the one type of crust to the other, incidentally have not yet been adequately studied (see Fig. 121).

The Structure and Composition of the Upper Mantle

Immediately below the Mohorovičič discontinuity, both on the continents and in the oceans, seismic velocities rise to 8.0-8.2 km/s, velocities that are typical for the roof of the mantle. There are zones, however, where the roof of the mantle is structured differently. These zones are rifts, both oceanic (confined to mid-ocean ridges) and continental. In these spots there is a lens between the floor of the crust and the roof of the mantle up to 20 km thick, composed of material with seismic velocities intermediate between crustal and mantle (7.4-7.8 km/s), which is interpreted as a mixture of crustal and mantle material (see Fig. 122).

Seismic velocities increase with depth, and at some tens of kilometres below the Moho velocities up to 9.0 km/s may be met.

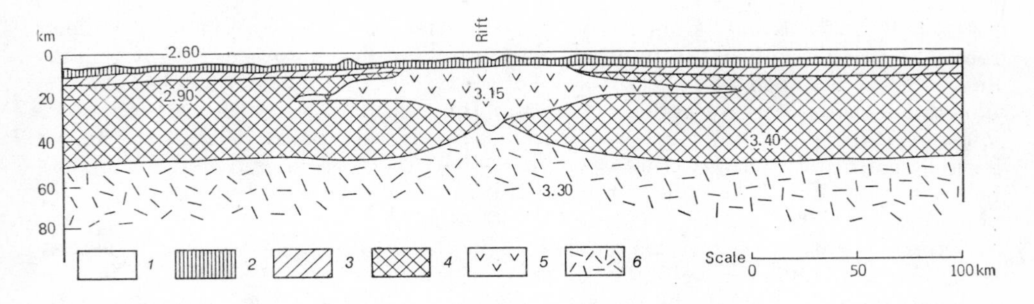

Fig. 122. Diagram of the structure of the tectonosphere under a mid-ocean ridge (after Talwani *et al.*, 1965):

1—water; *2*—Layer II of the oceanic crust; *3*—Layer III of the oceanic crust; *4*—substratum (P-wave velocity 8.1 km/s); *5*—lens of matter with intermediate seismic velocities (7.4 to 7.9 km/s); *6*—asthenosphere. Figures give the presumed density of the material (g/cm^3)

From observations of surface seismic waves it has been found that a layer of reduced seismic velocities ("Low Velocity Zone") begins at a depth close to 50 km beneath the oceans and varying between 80 and 120 km beneath the continents, in which the velocity of propagation of seismic waves is roughly 0.3 km/s slower than in the overlying layer of the mantle. The low velocity zone is also bounded below by a medium with higher seismic velocities. Elastic waves entering the low velocity zone are reflected, in accordance with the laws of wave propagation, from both the overlying and underlying layers, and are propagated mainly within this zone as in a channel. Such channels are known as *waveguides.* The low velocity zone is therefore called the *seismic waveguide*.

The waveguide, as we shall see, plays an extremely great role in the tectonosphere's development and in all endogenous geological processes.

Under the oceans the waveguide penetrates to a depth of 300 or 400 km; beneath the continents it is between 100 and 150 km thick.

Below it, in the Golitsyn layer, seismic velocities rise considerably.

Horizontal heterogeneties have been found in the waveguide clearly linked with the character of endogenous regimes. The waveguide is extremely weakly expressed under the most stable regions of the crust (ancient crystalline shields), and in places is apparently completely absent. Where it does exist below shields it begins at a depth of more than 100 km (on the Canadian Shield at 115 to 150 kilometres) and ends at a depth of 200 km; it affects only the rate of propagation of transverse (S) waves and does not affect longitudinal ones (P-waves). Under the non-shield part of platforms it shows up in its "normal" form, the roof being located at a depth around 100 km and its thickness reaching 150 km. Beneath regions of recent orogeny, however, like the Alps, the Caucasus, or the Tien Shan, the layer is expressed more strongly; its roof rises to a depth of 80 km, and it has a corresponding thickness, and its effect on the velocity of P-waves and S-waves is more appreciable. The low velocity layer is even more marked beneath rifts, where its roof is located at a depth of 50 to 60 km below the surface, and where it possibly links up with the lens of intermediate velocities lying between the crust and mantle. Finally, in present-day volcanic regions (e.g., the Kuril Islands) there are signs that the material with low seismic velocities projects up into the floor of the crust, and that its thickness exceeds

200 km. On the Kuril Islands P-wave velocities of 7.3 to 7.8 km/s have been recorded in the roof of the mantle, which are preserved to a depth of 80 km; and only at a depth of 125 km is a velocity of 8.1 km/s recorded. In neighbouring regions of the Pacific Ocean and the Sea of Okhotsk, however, velocities exceeding 8.0 km/s are registered immediately below the crust. As we have already pointed out, the waveguide beneath the oceans differs from that beneath the continents both in its shallow depth of occurrence and in its very great thickness, which exceeds 300 km.

Interesting data on the waveguide have been obtained for the Western Mediterranean. The low velocity zone beneath the sea lies at a depth of 50 km and sinks to 100 km at the surrounding coasts; beneath the Mediterranean, consequently, the upper mantle has a structure of oceanic type.

The structure of the upper mantle may be judged from a number of indications both direct and indirect.

Basalts occupy an enormous volume among the rocks on the Earth's surface, cover the whole floor of the ocean in a thick layer, and are common in the continental crust. There is no doubt that basaltic magma is formed deeper than the crust, in the mantle. It is impossible, however, to imagine that the upper mantle consists of basalts or of its intrusive analogue gabbro, since that would contradict the velocities of seismic waves (which are too high for gabbro in the upper mantle), density (which is also too high), and heat flow (which would have to be much higher for a gabbro composition than that observed).

Some of these contradictions can be overcome if it is assumed that the upper mantle consists of eclogite, which has the chemical composition of basalt but is much denser, and has a higher velocity of propagation of seismic waves. The difficulties in regard to heat flow, however, still remain. This supposition, moreover, is refuted by direct study of the composition of rocks of a mantle origin. These rocks form ultra-basic intrusions both on the continents and in the oceans, and are common in the form of xenoliths in kimberlite pipes and in basaltic extrusions. From this angle, kimberlite pipes are particularly interesting; judging from the diamonds and other minerals contained in them, which require high pressure for their formation, the xenoliths rise from depths between 70 and 280 km and can consequently bring up fragments from deep layers of the upper mantle.

Kimberlite pipes and basaltic extrusions on the continents contain inclusions of varying composition. This varied composition, however, proves to be identical for Siberia, Africa, Australia, and America, which suggests that the inclusions are not fortuitous, and that one may judge the composition of rocks deep in the interior from them. There are also rocks among the inclusions, of course, that have been torn from various layers of the crust, as well as rocks of a mantle origin. Xenoliths of sedimentary rocks, for instance, metamorphic rocks of the granulitic facies, and eclogite-like rocks coming from the lower parts of the crust, can be found among them. The mantle is represented by ultrabasic and basic rocks.

Study of the last-named rocks indicates that the mantle is extremely heterogeneous below the continents. The following rocks have been distinguished among mantle xenoliths: garnet lherzolite, garnet harzburgite, wehrlites, dunites, and pyroxenites and also eclogites (pyrope-diopsidic and pyrope-diopside-enstatic). As regards the main types only, all this variety can be reduced primarily to two rocks, viz. garnet peridotites and eclogites.

In addition to these typical deep rocks, less deep spinel peridotites are encountered, and even less deep, so-called Alpine-type, ultrabasic rocks of the same composition as rocks of the ophiolite formation in eugeosynclines.

Comparison of the chemical composition of these rocks with one another and with that of chondrites (which, as we have already said, may be considered the initial matter of the Earth) yields important pointers to the character of the processes taking place in the mantle (see Table 8). Chondrites are relatively rich in alkalis, alkali-earth, radioactive, and rare earth elements. The deepest garnet peridotites prove to be closest to them in composition. The not so deep spinel peridotites, and especially the Alpine-type hyperbasites, contain much less of these elements. But continental crust, on the contrary, is extremely rich not only in silica but also in alkali, alkali-earth, radioactive, and rare earth elements; by comparison with mantle rocks, however, it contains less magnesium, chromium, nickel, and several other elements. This allows us to view it as a product of the differentiation of mantle material, and to consider the hyperbasic mantle lying at relatively shallow depths (i.e., spinel and Alpine-type peridotites) as the residue of this differentiation.

It is interesting to determine which series of rocks of the upper mantle should be included in the differentiation for the crust to obtain the quantity of elements it has. The following depths are obtained for the lithophilic elements: silica 60 km, aluminium 140 km, calcium 50 km, sodium 180 km, potassium 1300 km. The last figure indicates particularly expressively the link between the composition of continental crust and the Earth's very deep interior.

Rocks representing the upper mantle beneath the oceans are found in the debris of the rift valleys of mid-ocean ridges. They are also peridotites, but of a much more complex composition than the ultrabasic rocks of the continental mantle. They are close in composition to the deepest garnet peridotites of the continental mantle, and like the latter are rich in alkalis, alkali-earth, radioactive, and rare earth elements (see Table 8). At the same time they form the highest layer of the mantle beneath the oceans, from which it follows that the oceanic mantle is much less differentiated than the continental, which is also confirmed by estimates of the depth of differentiation for individual elements (at a maximum this is 40 km for all lithospheric elements).

We must once again stress the existence of essential differences between oceanic and Alpine-type ultrabasites. The former, by their composition, are representatives of weakly differentiated, almost primordial chondrite material, while the latter should be regarded as the residuum of a long lasting differentiation of the same material, the supplementary components of which have been distinguished in the crust (see Table 8, columns 6 and 11). It is thereby impossible to equate these two groups of ultrabasites, which calls in question the validity of the now common views of the ophiolites of eugeosynclines as "ocean floor".

Peridotites, as the main rock of the upper mantle, satisfy all the geophysical parameters—density, seismic velocities, scale of heat flow—but we must satisfy ourselves that the peridotitic upper mantle can be the source of the basaltic lava that erupts onto the surface in such great quantities, and that the crust can have separated out from it through differentiation. The eclogites present in the upper mantle cannot be the immediate source of basalts because (1) eclogite undoubtedly plays a subsidiary role in the mantle, apparently forming only local accumulations among peridotites, and (2) mantle eclogites differ in composition from those encountered in the crust and are closer to ultrabasic rocks than to basalt.

It proves, however, that there is no difficulty in regarding garnet peridotites as the source of the formation of basalts. Numerous experiments have demonstrated that when peridotites are heated at the various confined pressures and temperatures usual for the upper mantle (from 1225 ° to 1500 °C) partial melting occurs, with the separation of liquid basalt. This, it must be stressed, is partial melting; full melting of peridotite can only occur (at the pressures existing in the upper mantle) at a temperature around 2000 °C, which is improbable in the upper mantle. With partial melting it is mainly the grains of garnet and pyroxene that melt, while olivine predominates in the residue.

Experiments have shown that with a different confined pressure basalts of various composition are melted; those melted at higher pressures contain more alkalis than those melted at lower ones. Attempts have been made to determine the absolute depth of formation of basaltic lavas of various composition. The Australian geochemists D. Green and A. Ringwood (1966), for instance, demonstrated by experiments that high-alumina olivine tholeiites would melt only at a depth around 30 km below the surface, alkali olivine basalts at depths between 35 and 60 km and picrites at a depth around 90 km. These depths, however, may differ substantially from those at which the corresponding varieties of basalt melt in nature. The depth of melting greatly depends on the water content of the peridotites and on how far melting has proceeded. The initial composition of the peridotite undergoing partial melting is also most important. Since the upper mantle beneath continents differs from that beneath oceans in degree of differentiation, the composition of oceanic basalts may differ from that of continental ones melted at the same depth.

From the observed surface geothermal gradient, and from the experimentally observed or calculated thermal conductivity of rocks of the crust and upper mantle, we can obtain an idea of the distribution of temperatures in the tectonosphere. Two curves of the distribution of temperature with depth from the surface to 115 km, as obtained by E.A. Lyubimova (1958) are depicted in Figure 123 (*1* and *2*). The difference between them is determined by the difference in the data on which they are based. The actual temperatures should lie between the two curves.

Contrasting of these curves with the temperatures needed for partial melting of peridotite at the pressures existing in the upper mantle indicates that such temperatures can be reached at a depth around 100 km below the surface. This is clear from the intersection of the respective curves in Figure 123. If the diagram were extended downward, we would find that there would be a reverse intersection at a depth around 400 km. In other words, partial melting of the peridotite composing the upper mantle, given the probable temperature distribution in the upper mantle, is possible at depths between 100 and 400 km.

These depths are close to the limits within which the low velocity zone is located, from which we may probably infer ideas about its nature. It may be thought that it is a layer of partial melting of peridotite and that it consists of a mixture of two phases, viz. a solid phase formed mainly of olivines, and a liquid one of basaltic composition. The existence of liquid in this layer would also lead to a lowering of velocities.

This view of the structure of the waveguide is confirmed by measurements

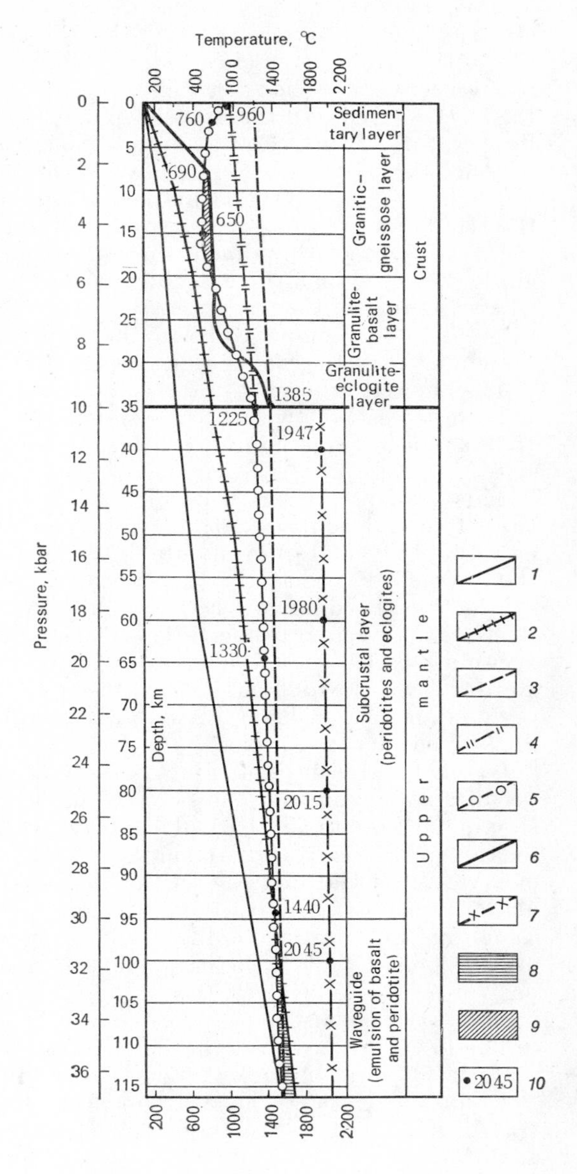

Fig. 123. Diagram of the thermal states of the tectonosphere (compiled by Beloussov from the data of Lyubimova, Tuttle, Bowen, Yoder, and Tilley):

1—minimum, "normal" (platform) temperature; *2*—maximum temperature; *3*—temperature of raised asthenoliths (adiabatic); *4*—temperature of dry melting in the crust; *5*—granitization temperature (in the crust) and melting temperature of basalt (in the upper mantle); *6*—temperature of the crust in the period of regional metamorphism and granitization ("geosynclinal" temperature); *7*—temperature of full melting of ultrabasic rocks of the upper mantle; *8*—region of the melting of basalt from peridotites (waveguide); *9*—region of granitization; *10*—temperature, °C

of the electrical conductivity of the material of the mantle at various depths. The electrical conductivity of a melt is considerably higher than that of solid rock. The difference may be as much as a factor of twenty or more. The earth currents induced by variations in the Earth's magnetic field, which penetrate hundreds of kilometres into the globe, are employed to measure the electrical conductivity of its interior.

At the point where a rise of the asthenosphere has been established by observations of surface seismic waves, there proves to be a layer of high electrical conductivity close to the surface. Temperature can also be judged from electrical conductivity. Beneath the Mid-Atlantic Ridge, for instance, where the asthenosphere is elevated, a temperature of 800 °C to 1000 °C has already been established at a depth of 15 to 20 km. Beneath the rift of the Great Basin (USA), where the asthenosphere is also close to the surface, there is a temperature of 1200 °C at a depth of 50 km, which is only encountered in areas beyond the rift at a depth of 150 km. A temperature of 1000° to 1100 °C is reached beneath the Baikal rift at a depth of 30 to 40 km, in the region of the Irkutsk amphitheatre the same temperature is found at a depth of 120 km.

In the light of these facts the various depths and thicknesses of the waveguide observed in different zones are determined by different heat regimes: where the roof of the waveguide is closer to the surface and where its thickness is greater, the temperature in the upper mantle is apparently higher than in zones where the waveguide lies deeper and is thinner.

The amount of fluid in the waveguide in relation to solid crystals may vary, judging from the lowering of seismic velocities, from 5 to 25%. We may take 15% as an average figure. The liquid seemingly forms a film enveloping the solid crystals.

This view of the structure of the low velocity zone leads to the conclusion that it should be differentiated by reduced density and viscosity. The presence of 15% of liquid should in fact lead to a lowering of the density of the waveguide's material by roughly 0.1 g/cm^3. The waveguide's density would thus become lower than that of the uppermost layers of the mantle itself that cover it, especially in its upper layers where the molten basalt accumulates. A situation of density inversion is created in the upper mantle similar to that which arises in the crust in connection with processes of regional metamorphism or with the presence of a thick layer of salt among denser sedimentary rocks.

Given the presence of liquid, viscosity should also be much lower, which permits us to consider the layer of partial melting as a zone in which there should be fundamental displacements of material with changes in the load on the Earth's surface. Attempts have been made to calculate the viscosity of layer of partial melting on that basis, studying the history of the reaction of the tectonosphere to lifting of the ice load after the melting of ice on the Canadian and Baltic Shields. E.V. Artyushkov (1968) obtained a viscosity of 10^{20} poises for the layer of partial melting; that of the overlying solid part of the tectonosphere (the uppermost layers of the mantle and the crust) has an average viscosity three orders of magnitude greater.

The lower viscosity of the layer of partial melting and the mobility of its material permit us to call this layer the *asthenosphere*, i.e., the geosphere "without strength". The asthenosphere is counterposed in its mobility to the solid part of the tectonosphere lying above it and comprising the crust and the layers of the upper mantle. This solid part of the tectonosphere is called the lithosphere. The term "substratum" has been suggested by E. Lyustikh (1965) to designate the solid uppermost layers of the mantle lying immediately below the Moho.

Let us now return to the question of the differentiation of the Earth's tectonosphere.

The basaltic crust of the oceans could have been formed directly by the melting of basalt in the asthenosphere and its rise to the surface as a relatively light material. The formation of continental crust cannot be explained so simply. There is a very great volume of acid rocks in it rich in silica, alkalis, and rare earths and radioactive elements. The origin of these volumes of granitic magma presents a difficult problem. Some of the granites were formed by the remelting of older granites already present in the crust or by the granitisation of sedimentary and metamorphic rocks by mobile chemical reagents forced up from lower layers of the crust by heating. This mechanism, however, is secondary, of course, and cannot explain the development of primary granites in the crust. It also does not explain the phenomenon of granitization observed in rocks of the granulite facies below which there are no granites capable of "mobilization". The obvious suggestion is that the granitic material that plays such a great role in the structure of continental crust has been differentiated from the peridotite mantle, but that could not be a direct result of either gravitational or liquational differentiation of basalts, melted from the mantle because (1) the volume of granites is too great for that, and (2) the ratio of the various elements indicates that a special selective mechanism operated that led to special enrichment of the crust by certain elements, including rare earths and radioactive elements, as well as by silica and alkalis. The processes of their special extraction from the mantle must have operated during the formation of the crust.

This point has been developed by B. Lutz. The mechanism proposed by him is based on the idea of acid magmatic leaching. The mantle contains a certain (albeit very small) quantity of water. At the same time, however, it also contains free hydrogen in connection with the strong reducing state there, the relative amount increasing with depth. The presence of hydrogen leads to an acid regime of aqueous solutions, with the result that they take up alkalis from the surrounding medium and are enriched by them. As the solutions rise, however, the hydrogen is oxidised and in part volatilised. As a result the acidity of the solutions falls, and in passing through the upper layers of the mantle they begin to dissolve acid components, primarily silica but also rare earths and radioactive elements. The deep solutions now acquire the composition needed for the processes of granitisation and regional metamorphism.

A result of oxidation is also heating of the solutions, which permits us to see in them not only the chemical reagent needed to form the granitic-gneissic layer of continental crust but also a source of energy for metamorphism and granitisation.

The same mechanism helps us to understand why basaltic volcanism occurs earlier in eugeosynclines than granitization and regional metamorphism, and not simultaneously with them. It is possible that basaltic magma absorbs the whole fluid phase during its melting so that the juvenile solutions that could produce regional metamorphism and granitisation are not separated out. If we suppose that the emulsion of basaltic films in the asthenosphere solidifies in the next stage (i.e., in the geosynclinal inversion stage), juvenile solutions could separate out and also rise into the crust. In this explanation there is an interesting link between the geochemical and tectonic processes; we have seen that the geosynclinal inversion stage, when granitisation and regional metamorphism occur, differs from the eugeosynclinal by smoothing out of the contrasts of block-undulatory oscillatory movements, and lowering of tectonic activity expressed in oscillatory movements. Could this decline in activity not happen in connection with the lowering of temperature in the asthenosphere that leads to the solidifying of drops of basalt?

The oceanic mantle does not extrude solutions rich in silica and alkalis. The reason may possibly have to be sought in the fact that its temperature is higher than that of the continental mantle, and that the solidifying of basaltic films, necessary for the separation out of deep solutions, does not occur in the oceanic asthenosphere.

The origin of the huge volumes of andesites erupted by the volcanoes of type-II island arcs is a special problem. It has been demonstrated experimentally that andesite can be melted from eclogite of a basaltic composition, and it turns out that the first substance to separate out during the melting of crustal eclogite with a rise of temperature has an andesite composition.

Another way can be suggested, however, for the formation of andesites. The average composition of continental crust is very close to that of andesites. The latter, consequently, may be formed by the melting of continental crust, in the course of which the material of its various layers becomes mixed.

All the points listed in this section are in essence only tentative, and further research is needed to resolve them.

The Heat Regime of the Tectonosphere

With the average values of heat flow of 1.5×10^{-6} cal/cm^2/s given above, and the most common value of 1.1×10^{-6} cal/cm^2/s, significant local variations of these values are observed, which correlate with present-day endogenous zones, and with how pronounced the asthenosphere is; in zones where it is more strongly expressed heat flow is more intensive, and where it is weakly expressed heat flow is weaker.

As for the continents, we must note first that there is an average heat flow of 0.9×10^{-6} cal/cm^2/s on ancient crystalline shields where the asthenosphere, as we have already mentioned, is either altogether absent or is very weakly expressed. On the non-shield sectors of ancient platforms it averages 1.1×10^{-6} cal/cm^2/s. In zones of weak orogeny on the sites of Palaeozoic geosynclines (such as the Urals and Appalachians) the intensity of the flow rises to 1.5×10^{-6} cal/cm^2/s but in the Tien Shan, where strong recent tectonic activization is observed and the asthenosphere is well expressed, it rises on average to 1.8×10^{-6} cal/cm^2/s. The average value is even higher in rift zones (around 2.0×10^{-6} cal/cm^2/s), beneath which the asthenosphere is much higher and where there is a lens of material with intermediate seismic velocities. Finally, the highest heat flow on continents is observed in regions of present-day volcanism where the asthenosphere is elevated in places right up to the floor of the crust. In these areas heat flow reaches $3.6 \times$ $\times 10^{-6}$ cal/cm^2/s.

In the oceans there is a very close relation between high heat flows and median ridges. The highest heat flows, reaching 8.0 units at some points, are observed on those ridges, although the average value of the flow for the rift valleys of mid-ocean ridges does not exceed 2.0. Heat flow slackens rapidly with distance from the crest of a ridge and falls to between 1.1×10^{-6} and 1.2×10^{-6} cal/cm^2/s.

A relatively high heat flow is characteristic of marginal seas (the Sea of Japan, Sea of Okhotsk, etc.), where its mean value reaches 2.1. For inland seas an average value of 1.3 has been determined.

B.G. Polyak and Yu.B. Smirnov (1968) have demonstrated a link between the rate of heat flow and age of folding, regional metamorphism, and granitisation in one zone or another. According to their data regions of Precambrian folding have an average heat flow of 0.93, Caledonian folded zones have a mean flow of 1.11, Hercynian of 1.24, Mesozoic of 1.42, and Cenozoic of 1.75.

These figures, however, can be interpreted rather differently. It can be estimated, from the values of the lithosphere's thermal conductivity, that from ten to twenty million years are sufficient for heat flow to traverse the path from the top of the asthenosphere to the surface of the Earth, which means that we are now observing heat at the surface generated in the upper mantle not earlier than the beginning of the Neogene, i.e., in the neotectonic period. This heat may, consequently, reflect, for example, the deep-seated processes that led to the development of the Cenozoic zones of folding, metamorphism, and granitization that were formed in the Neogene and later, but it may have no relation to processes that ended in the Mesozoic, Palaeozoic, or Precambrian. The heat associated with processes that took place in that remote time had long ago dissipated into outer space. It is therefore most probable that the correlation found by Polyak and Smirnov points to the present-day regime of the zones distinguished by them, or in any case to the recent regime. In fact, zones with a Precambrian age of intensive tectonic, metamorphic, and magmatic processes now form the stablest sectors of the Earth's crust (crystalline shields); zones of Palaeozoic endogenous activity, as a rule, have undergone tectonic activization, though relatively weakly, in the neotectonic epoch; strong recent orogenesis, however, combined with intensive volcanism, is confined to zones of the youngest endogenous activity, i.e., Mesozoic and Cenozoic.

It is interesting to calculate what fraction of heat flow should be attributed to the crust and what to a source in deeper shells of the globe. The simplest estimate for continental crust takes the following form. The granitic-gneissic layer, which is accorded a thickness of 8000 m, gives off heat of $0.25 \times \times 10^{-6}$ cal/cm^2/s, according to its average content of radioactive elements. The granulite basic layer, the thickness of which is taken as 32 km, raises heat flow by 0.2 of a unit. The crust, consequently, forms a fraction of heat flow equal to 0.45 of a unit. The upper layers of the mantle lying above the asthenosphere (the substratum), which consist of peridotite and are around 100 km thick, give off only 0.05 of a unit of heat flow; hence it follows that approximately half of the heat flow on platforms must come from below the lithosphere.

When we allow for the fact that the solid, consolidated oceanic crust consists of basalt and is around 5000 m thick, a discharge of heat at a rate of only 0.1 of a unit can be ascribed to it. The substratum does not, in any case, generate more heat than that; to ensure a flow of heat in the oceans at the same rate as on the continents, consequently, from regions lying deeper than the lithosphere, a heat of at least 1.0 unit (or more than double that on continents) must be ascribed to them.

The high heat flow in the oceans from below the lithosphere may be associated with the shallower occurrence of the top of the asthenosphere beneath oceans than beneath continents. If we assume that the temperature at the top of the asthenosphere is identical beneath oceans and continents (around 1500 °C), then its share in the formation of total heat flow will be greater in oceans than on continents, since the asthenosphere is closer to the surface beneath oceans. The higher occurrence and much greater thickness of the asthenosphere beneath oceans in turn is found to be linked with the lower degree of differentiation of the oceanic upper mantle compared with the continental. Because of the slighter differentiation radioactive generators of heat are not concentrated in the uppermost layers of the mantle and in the crust, but are dispersed throughout much greater thicknesses of the material of the mantle. In these circumstances the heating of the mantle should be stronger beneath the oceans than beneath continents. Estimates indicate that the temperature in the upper mantle up to 400 km beneath oceans should be 100 ° to 200 °C higher than at the same level beneath continents.

Researchers now have methods at their disposal for determining not only the rate of heat flow of today but also that which existed in any one region in past geological periods. Mineral palaeothermometers, whose temperatures of formation or structural and chemical variation are known are employed for this purpose. Under the pressure that exists in the granitic-gneissic layer of the crust, for example, α-quartz is converted into β-quartz at a temperature around 600 °C. It can consequently be established from a quartz's type of structure whether it was formed at a temperature above or below 600 °C. Feldspars are used as a perfect geological thermometer. They are known to consist of molecules of albite, orthoclase, and anorthite. The proportion of these molecules depends on temperature; by utilising that fact we can reliably distribute the whole series of rocks in order of rising or falling temperature.

By studying the minerals contained in metamorphic and igneous rocks from this angle we can establish at what temperature they crystallised, i.e., the temperature of the metamorphic or magmatic process. When these data are supplemented by a palaeogeographic reconstruction of the depth at which the rocks were located at the time they crystallised, we can measure the geothermal gradient that existed then and hence (knowing the rocks' mean thermal conductivity) the rate of heat flow as well.

Such determinations have been made for metamorphic series and for granites of various age. From the results obtained it follows that metamorphic processes and granitisation have always developed in the crust in conditions of heightened geothermal gradients exceeding the present-day normal gradient by at least a factor of three, and often even of five. Considering that the mean thermal conductivity of the rocks has not so far altered, we have to conclude that the heat flow was then that many times more intensive than today. It was consequently roughly of the rate characteristic of today's volcanic regions, or even higher.

Regional metamorphism and granitisation, however, do not occur always or everywhere. They are confined to geosynclines, and then only to a certain stage in their evolution, namely, the geosynclinal inversion region. That is a time of partial inversion and mountain formation and occupies only a certain interval in the history of a geosyncline, an interval also characterised by heightened heat flow and a heightened temperature in the Earth's crust. The structure of rocks formed during other stages of a geosyncline's development does not contain signs of the effect of such high temperatures. Since processes of regional metamorphism and granitisation are repeated in all endogenous cycles, it may be concluded that there has periodically been vigorous heating of the crust and that this increase in heat flow has each time been confined to the geosynclines active in the given cycle.

Certain Conclusions about the Tectonosphere's Structure and Evolution

The main conclusions from what has been said in this chapter that are essential for understanding the deep causes of endogenous processes may be formulated as follows. There are two tectonospheres differing in structure and composition, viz. continental and oceanic. The first is much more differentiated, and the end product of its differentiation is continental crust. During the formation of continental crust there was increased removal of silica, alkalis, rare earths, and radioactive elements from the mantle. The oceanic tectonosphere is less differentiated. The differences in structure of the two persist down to a depth of several hundred kilometres.

An important component of both tectonospheres is the layer of partial melting of the peridotites forming the upper mantle. This is the asthenosphere, which differs from overlying solid lithosphere in low density and viscosity.

A link has been found between endogenous regimes, rates of heat flow, and degree of expression of the asthenosphere. The most active endogenous processes go hand in hand with the highest rates of heat flow, the greatest thicknesses, and the shallowest occurrence of the asthenosphere, which affects the velocity of propagation of seismic waves to the greatest degree. Such an active asthenosphere is now found in volcanic and rift zones on the continents and in mid-ocean ridges in the oceans. In past periods it was also characteristic of geosynclines.

The quietest regimes, particularly those of the crystalline shields of ancient platforms, are accompanied with a much lower heat flow, and either a weakly expressed or completely absent asthenosphere. Hence there is a direct relation between the activity of endogenous processes and the heat regime of the interior. A specially high heat flow through the crust characterises stages of regional metamorphism and granitization in geosynclines. This last point allows us to think that the pulse-like character of endogenous regimes and the periodic succession of stages of excitation and weakening of endogenous activity is caused by corresponding periodic changes in the rate of heat flow. If we adopt Lutz's conception, then heightened transport of heat in the crust is caused by juvenile solutions rising in it from the asthenosphere. These solutions begin to flow every time when drops and films of basalt solidify in the asthenosphere, which means that periodic heating of the crust occurs with a phase delay in respect of the thermal processes in the asthenosphere; when the latter heats up, basalt is melted, which is extruded to the surface but apparently has little effect on the temperature of the crust; when, however, the asthenosphere cools down, solutions rise from it that, by being oxidised, become strongly heated and heat the crust sufficiently for regional metamorphism and granitization to occur. These same solutions also introduce the chemical reagents into the crust needed for these processes. The higher temperature of the oceanic tectonosphere is possibly the reason why granitizing solutions are not discharged from it.

The simultaneous manifestation of different endogenous regimes on the surface of continents points to heterogeneity in the Earth's heat field in one and the same period of time, and indicates that heat flows differ in intensity in various places. Rates of heat flow alter both in space and in time. Hence it is a major task of geotectonics to understand the spatial and temporal variations in the Earth's thermal field. The next job is to discover the mechanism that links heat flow with one endogenous regime or another.

CHAPTER 24

The Main Stages in the Development of Geotectonics

Geotectonics Before the Eighteenth Century

The development of geotectonics has always been closely linked with the progress of all the Earth sciences, initially with geology, and more recently with geology, geophysics, and geochemistry. The rise of scientific geology must be dated to the eighteenth century, when the mining industry began to develop and a practical need arose to understand the structure of the crust for prospecting and exploration.

Certain elementary observations of phenomena that we now class as geotectonic had, however, been made and described much earlier. Classical Greek thinkers had taught that the Earth's surface did not remain immutable.

The Middle Ages were a long period of stagnation in geology, as in the other natural sciences.

The freeing of geological science from the yoke of Church dogma began in the sixteenth century when the development of trade relations led to a need for metal to mint coinage and forge weapons, and for other purposes. Experience began to be gained in the study of structural forms, especially of veins, experience that was brought together at the beginning of the sixteenth century in the *De Re Metallica* of the German mineralogist and mining engineer Agricola (Georg Bauer) (1494-1555).

At roughly the same time the great scientist and artist Leonardo da Vinci (1452-1529) was in charge of building canals in Northern Italy; from observing the rock series opened up he expressed a number of ideas that were new for his time. He established, for example, that rocks occurred in the Earth in layers. Although most people then doubted that fossils were of organic origin and looked upon them as chance concentrations of mineral material, or a kind of "sport of Nature", Leonardo interpreted their significance correctly and contended from finds of remnants of ancient marine organisms in the ground that the seacoast had not remained fixed and that land and sea could have changed places.

In the seventeenth century the greatest figure in geology was Nicolaus Steno (Niels Stensen) (1638-1686), a Dane by birth who lived mainly in Italy. In 1669 he published a small work *Prodromus de solido intra solidum naturaliter contento*, the geological ideas, contained in this work, were based on his observations in the neighbourhood of Florence.

Steno established that layers of rocks could be traced across valleys in mountains, from which he formulated the principle of the extension of strata: a stratum did not end where it disappeared under other strata; its continuation could be discovered on the other side of a valley or hill. A second principle affirmed that strata were always formed in a horizontal position, and that if they were now inclined it was the result of subsequent disturbances. The cause of the disturbance Steno saw in the formation of voids in the Earth through the action of subterranean fire or water, after which the strata covering them were disrupted and their fragments given an inclined position. On the basis of that idea he described the geological history of Tuscany, breaking it down into six stages, each of which began with horizontal deposition of strata and ended with their partial collapse into recurrent underground voids.

At the turn of the seventeenth and eighteenth centuries G. Leibnitz (1646-1716) suggested for the first time that the Earth had originally been molten and had then gradually cooled. Cooling had been uneven, with the formation of irregularities and voids. So hills and valleys had arisen on the surface of the planet.

Although the work on the Earth of the English scientist Robert Hooke (1645-1703) appeared at the beginning of the eighteenth century, it really belonged to the preceding one and as it were completed it. Hooke's work was formally devoted to earthquakes but its content was much broader. For the first time the idea was expressed in scientific literature that the Earth's surface had experienced many changes. Although the Bible said that there had been only one major event in the Earth's history, the Great Flood, that was quite insufficient to explain all the traces of great changes imprinted in rock strata. Many reciprocal movements of land and sea had occurred as a result of some kind of eruption of undergound fire or earthquakes, in which connection large masses were raised above their normal level. It seemed improbable, but the peaks of the highest and most impressive mountains could have been under water and that they were the result of very strong earthquakes. Earthquakes could cause uplifting and sinking of the Earth's surface. Sinking occurred through breaking of the roof of a subterranean void when underground fire erupted to the outside. Earthquakes could shake the whole surface, overthrow mountains and turn them upside down, shift land from one place to another, bury surfaces and raise underground parts to the surface.

By the beginning of the eighteenth century, we see our science has accumulated enough data to affirm that the history of the Earth's surface was complex and that it had experienced many outward and inward changes.

Geotectonics in the Eighteenth Century

The eighteenth century was the time of the first broad regional geological investigations. Already at the beginning of the century, the English geologist J. Strachey had described the geological structure of certain coalfields in England in detail. The first drawings of the complicated structure of the strata in the Alps made from nature also appeared then. Regional investigations expanded, particularly in the middle and latter half of the century. The Petersburg Academy of Sciences organised several major expeditions, a result of which was the first general, natural historical descriptions (including geological ones) of both the European and the Asiatic parts of what is now the Soviet Union. Among those who took part in these expeditions were V.F. Zuev, N.P. Rychkov, I.I. Lepekhin, E. Laxmann, and especially P.S. Pallas (1741-1811), who is credited with several generalisations about the laws of the structure of mountain ranges.

About this same time H.B. de Saussure was studying the structure of the Alps, J.G. Lehmann and G.H. Füchsel described the geological structure of

Saxony and Thuringia, and Jean-Étienne Guettard and N. Desmarest were the first regional geologists in France.

Regional geological data were the basis for the initial scientific generalisations in that department of geology that we now refer to as geotectonics.

Here we must mention, above all, the work of the great Russian scientist Michael Lomonosov (1711-1765), who should rightly be credited with the first scientific geotectonic conceptions and considered the father of geotectonics in general.

Lomonovos's theory was set out in this work *On the Earth's Stata.* He had come to the conclusion that the forces altering the face of the Earth were both external and internal. The internal "action" was earthquakes, which led to elevation and sinking of the Earth's surface, changes in coastlines, and the development and disappearance of mountains.

As for the causes of earthquakes, "the force that raised such a load could not be ascribed to anything... other than the heat prevailing in the bowels of the Earth".

Lomonosov drew the following picture of change in the bedding of strata through the effect of internal fire:

> "The exceedingly hot material, which produces a strong, vast blaze, racks the surface lying above that very spot and seeks a path to the air through fissures. Then, having got free, it drops the severed anchorings by their own weight, which cannot restore the crushed parts to their proper positions and order. They fall in various sections like collapsed brick arches, higgledy-piggledy, one on top the other, sideways, crossways, edgewise, poking up; and such caving in takes up much room, leaving big gaps in places—from which mountains raise themselves above surface of the earth."

In addition to these rapid, catastrophic movements, however, Lomonosov also spoke of slow, imperceptible movements of the surface, calling them "an insensible and long-period rise and fall of the terrestrial surface".

His remark about "insensible" movements means that Lomonosov was the first to put forward an evolutionary notion of the Earth's development, though not in very clear form. He thus began to break with the ideas of the supporters of "universal catastrophes" widely held at the time. These ideas were an attempt to reconcile the existence of traces of numerous changes in the Earth's surface, on the one hand, with the brevity of life on Earth as asserted by Biblical teaching, on the other hand. The Church legend allotted a span of 6000 years for the whole of the Earth's existence. The famous French naturalist G. Buffon (1708-1788) provoked fierce attacks from Churchmen in the middle of the century for "daring" to extend the length of the Earth's existence to 75 000 years. No one was prepared to go further at that time. The whole succession of geological events imprinted in the structure of rocks had to be crammed into such a short interval of time. It was thus impossible to manage without enormous catastrophes caused by the action of unusual, supernatural forces that manifested themselves from time to time on the planet's surface. Lomonosov was the first to speak of the slowness of even a few geological processes.

The further evolution of general ideas in geotectonics is associated with the name of the Scotsman James Hutton (1726-1797). In his *Theory of the Earth* Hutton had already pointed out very clearly that geological phenomena should only be explained by the effect of those natural agents whose action was also observable now. The Earth developed according to the laws of physics and mechanics. All geological processes were interconnected and occurred in a definite sequence. The main role in movements of the crust was played by vertical movements, more particularly by the lifting force of the planet's internal heat. An elevation of a certain part of the surface occurred and formed a continent. It was broken up by external forces and the products of its destruction accumulated at the bottom of surrounding seas. After destruction of the first continent a new one was raised from the sea, composed of fragments and debris of the previous one. During elevation of the crust the original bedding of the strata was disturbed.

In coming forward as an active advocate of evolutionary ideas Hutton fought to extend the limits of geological time, which he considered infinite, saying that he could find neither traces of a beginning nor signs of an end in the economy of Nature.

Evolutionary ideas finally triumphed considerably after Lomonosov and Hutton, as we know, in the nineteenth century. Their confirmation is associated with the work of Sir Charles Lyell and Charles Darwin.

In 1777 P.S. Pallas addressed the Petersburg Academy of Sciences with a speech in which he set out his inferences on the structure of mountain ranges, mainly on the example of the Urals. In this view the spine of the range consisted of granites, which were covered by schists steeply inclined to the horizon. These were the oldest rocks, which belonged to the "primary formation". They did not contain fossilized organic remains. Then followed symmetrical bands of limestones of the "second formation", which did not occur so deeply and contained organic remains. Further away from the axis of the range there were "tertiary" loose rocks (sands, clays, and marls) with fossil remains. The structure of the mountains consequently resembled a fan. The cause of their uplift he saw in volcanic phenomena.

At the turn of the century the English geologist William Smith brought out the stratigraphic role of fossils. This discovery, which predetermined the subsequent progress of geology, made it possible to pass from "geognostic" investigations (in which only the composition of rocks was ascertained) to geological study and the division of rocks according to their age, to represent the sequence of geological processes correctly, and to decipher geological structures, i.e., gave an impetus to the development of historical geology as a science. It was this discover, too, that led to geologists' beginning, already in the early nineteenth century, to understand the geometry of folded structures. Instead of "disorder in the bedding of layers", ideas developed on the regular flexures of strata, i.e., on folds.

Geotectonics in the Nineteenth Century

The views of Lomonosov and Hutton on the character of crustal processes constituted the basis of the geotectonic conception that can be called the "uplift hypothesis". This hypothesis was developed in the nineteenth century and given more concrete shape by the work, in the main, of Leopold von Buch (1774-1852).

Von Buch had noticed, while studying the structure of volcanic cones, that they were formed of layers that sloped from the crater to the periphery. Not realising that this slope was primary, he decided that the dome-like structure of the cones was caused by the buoyant force of magma. He then transferred the structure of volcanoes so understood to any mountain chain in which there was, as a rule, a falling way of strata from the centre of the massif to the periphery. What was mistaken as regards volcanic cones proved to be close to the truth for mountain ranges.

Von Buch saw the direct cause of the uplift of mountains in the injection of various crystalline rocks of deep-seated origin, mainly granites. A number of Swiss geologists, developing his ideas, tried to explain not only the uplifting of mountains altogether but also the folding of strata. They suggested that magma, when injected into overlying strata and elevating them, at the same time pushed them aside, which led to the formation of folds.

The uplift hypothesis held a leading position in geotectonics until the middle of the nineteenth century; the facts accumulated on the structure of mountains ultimately, however, led to its being abandoned in the form in which it then existed. Study of mountain ranges indicated that they were folded zones that formed long belts of linear folds parallel to one another. The "uplift craters" hypothesis could not explain these extended folded zones. In ascribing an active role in the formation of uplifts and folding to the central granite massifs of the Alps, the hypothesis assumed for a start that the granites were younger than the country rocks. This assumption was refuted by observations. The granites proved to be older than the folded country rocks and, moreover, were themselves deformed together with the country rocks.

In the middle of the century a geotectonic hypothesis arose in place of the uplift one that prevailed in geology almost continuously up to and including the 1920's, namely, the hypothesis of the contraction, i.e., compression, of the Earth.

The substance of this hypothesis was predetermined by at least two circumstances: (a) the advances made in study of the geometry of folded structures, and (b) the universal recognition accorded to Laplace's hypothesis of the origin of the solar system. Study of folded regions led to recognition of the geometrical regularity of folds. It was natural to explain this regularity by the longitudinal compression of strata that had been previously bedded horizontally. Since folding was then pictured by geologists as the main form of disturbance of the bedding of strata, the new hypothesis had first of all to explain precisely this process of folding. Laplace's hypothesis postulated the origin of the solar system, including our planet, from a hot nebula, and the whole subsequent evolution of the planets was represented as the process of their cooling, solidifying, and contraction in volume.

The contraction hypothesis was first proposed in 1830 by the French geologist Élie de Beaumont (1798-1874) and published by him in full form in 1852. According to it the Earth was in a state of cooling and contraction. Its upper envelopes had already cooled as far as was possible, and their volume was no longer shrinking, but the interior, which had not yet quite solidified, continued to cool and contract. As a result the crust became too big for the shrinking core and was forced to "settle" and contract its surface, and buckle into folds.

While the preceding uplift hypothesis had considered vertical movements in the crust as primary, the new hypothesis consequently gave priority to horizontal forces of compression. Whereas uplifts and subsidences had played the main role in the first hypothesis, folding was the main process for the contraction theory.

According to the contraction hypothesis the whole crust was uniformly affected by compression forces, but observations indicated that some regions were crumpled and others remained undeformed. The reasons for this unevenness of the manifestations of folded deformations on the surface were explained by the famous Austrian geologist Eduard Suess (1831-1914) by the crust's being divided into rigid and plastic segments. The rigid segments were not deformed, or only weakly so, but as the globe contracted they were shifted and acted like a vice, squeezing the plastic zones between them so that folds were formed in them. On the basis of this conception Suess published a work on the tectonics of the Earth in several volumes under the title *The Face of the Earth.*

The authority of the contraction hypothesis increased greatly at the end of the century when big tectonic overlappings or mass overthrusts were discovered in folded zones. At the time these were interpreted as the result of the development of huge recumbent folds and therefore considered a sign of a great reduction in the original width of the folded zone (by hundreds of kilometres).

The time of the contraction theory's success was marked by great advances in various fields of geotectonics.

In the 1870s the Swiss geologist A. Heim published work on the mechanism of mountain-building in which the mechanical principles of folding were treated from the hypothesis of horizontal compression of the crust. A little later the French geologist Bertrand established the periodic character of the major epochs of folding from analysis of angular unconformities, and distinguished folded zones of various age, introducing a division of them into Precambrian (Huronian), Caledonian, Hercynian, and Alpine.

A theory of geosynclines developed on the basis of that still has very great value. The American geologist James Hall first drew attention, in the middle of the nineteenth century, to the fact that the thickness of sediments is greater in folded regions than where strata lie undisturbed, i.e., on platforms. A little later, in the 1870s, another American geologist, J.D. Dana, formulated the principles of the theory of geosynclines, which he considered to be large troughs formed in the crust by the action of horizontal compression forces. The troughs became filled with sediments and proved to be the most plastic zones of the crust. With further compression the sediments accumulated in them were buckled into folds. Dana suggested that geosynclines were always formed on the margins of continents as a consequence of the ocean floor's pressure on neighbouring continents when sinking during the general contraction of the earth.

The theory of geosynclines was developed further by E. Haug, E. Argand, and others.

The last geotectonic synthesis based on the contraction conception is to be found in the work of Hans Stille, which already belongs to the 1920s. In his *Principles of Comparative Tectonics* (1924) he treated vertical oscillatory movements of the crust as its rough warping or buckling under the action of horizontal compression forces manifested in its rigid segments. Fault dislocations and the rise of magma were also controlled by horizontal compression forces.

Against the background of the victorious march of the contraction hypothesis, views on the importance of slow vertical movements in the history of the crust developed more modestly. At the end of the 1860's the Russian hydrogeologist N. A. Golovkinsky published work in which he treated the forming of layered series of sedimentary rocks against the background of subsidence and uplift of the surface and shifts of the coastlines of seas.

The work of A.P. Karpinsky, devoted to the structure and history of the development of the European part of Russia, i.e., the Russian or East European platform, published at the very end of the nineteenth century, presents exceptional interest. Having studied facies of deposition of various age and compiled palaeogeographic maps of the platform, he reviewed the history of its oscillatory movements and noted that "the direction of the oscillations is almost always parallel to the ridges of the Caucasus and Urals".

The significance of Karpinsky's palaeogeographic generalizations was the greater in that they were published at a time of universal enthusiasm for folded structure and ignoring of any tectonic phenomena except folding, when non-geosynclinal (platform) regions seemed dead blocks lacking any development of their own and capable only of squeezing geosynclines and coarsely buckling under the pressure of the crust's general contraction. In addition to the platform's vertical movements, complying with proper regularities, Karpinsky established the existence of local dislocations in its body as well. Views on the tectonics of the East European platform were developed subsequently by A.P. Pavlov's work devoted mainly to the structure of the Zhiguli Hills.

The nineteenth century ended on the whole with the triumph of the contraction hypothesis. It greatly stimulated careful study of all the features of folded structures and taught geologists to think about the patterns of the crust's development as a whole, since it provided a basis for that by linking tectogenesis with the general cooling of the planet.

Geotectonics in the First Half of the Twentieth Century

In this section we shall be concerned with the main stages in the development of geotectonics from the beginning of the century to the 1960's.

The first three decades of our century were marked by a gradual decline in the leading position of the contraction hypothesis. By the end of the 1920's it had almost completely lost its scientific value.

The first blow to the hypothesis was the discovery of radioactivity at the turn of the century, which proved to be a powerful source of heat within the Earth. This discovery first threw doubt on the feasibility of the cooling required by the contraction hypothesis, and then completely refuted it.

Faith in Laplace's cosmogenic hypothesis was also undermined. Views began to develop by which the Earth was built up from a host of tiny cold particles and had never been molten, and by which its development had proceeded along the line of a heating up rather than of cooling.

Geologists also became disillusioned with the hypothesis from purely geological considerations as knowledge of the crust's structure became more detailed and broader. The structures observed proved much more diverse than it had previously been supposed, and it became clear that it was impossible to explain all this variety just by general contraction of the crust. Forces of horizontal tension must also have been involved in addition to compression in the formation of the crust, and also forces operating vertically. The history and distribution of all these structures, so varied in the conditions of their formation, proved so complicated that the crust had to be pictured as a mosaic of small sectors each subjected to the effect of its own special forces and developed in its own way. A blow also came from the problem that the hypothesis of contraction had been created to explain, namely that of the origin of folding, which turned out to be a genetically heterogeneous phenomenon obviously unconnected with horizontal compression in many of its forms. The pattern of development of the crust's oscillatory movements would not fit into the narrow scheme of general horizontal contraction in any way.

Because of the inability of the contraction hypothesis to explain all tectonic phenomena simply by general compression of the crust, a search began for new hypotheses that assumed, in particular, an alternation of compression and tension in the crust. At the beginning of the 1930s the American geologist Walter Bucher proposed a hypothesis of pulsation of the Earth, by which our planet's history was represented as a succession of phases of compression and tension. In the tension phase the crust was stretched, the more plastic sectors being stretched more strongly, with "necks" formed on their sites reflected on the surface in troughs. These were geosynclines. They became filled with sediments, and in the compression phase these same plastic zones were buckled into folds.

The Soviet scientists V.A. Obruchev and M.A. Usov suggested their own version of the pulsation hypothesis, in essence very similar to Bucher's, in 1940.

It proved, however, that an alternation of contraction and stretching instead of just contraction was too crude a scheme, and that it was impossible to fit the whole variety of tectonic processes into it that both followed each other and developed simultaneously in different sectors of the world.

Another trend that arose on the "debris" of the contraction hypothesis was based on the ideas of the old uplift hypothesis, but given a new content in accordance with the rising level of knowledge. This trend arose from the conviction that the observed spatial heterogeneity of tectonic processes could not be explained by horizontal forces, either of compression or tension, that affected the crust as a whole. The individual nature of the separate segments of the crust could only be explained if it were assumed that the tectonic forces causing this development were applied directly to the separate sectors and by-passed other, even neighbouring, sectors. But that could only happen if the forces were applied to each segment vertically.

As for the manifestations in the crust of signs of horizontal compression, expressed in folding, or horizontal tension, which were imprinted in normal faults, the horizontal forces causing them must be considered local, mainly secondary processes controlled by the crust's vertical movements.

The German geologist Erich Haarman published his *Theory of Oscillations* in 1930. He called vertical (oscillatory) movements of the crust primary tectonogenesis and folding caused by horizontal compression of strata secondary tectonogenesis. Oscillatory movements arose through the effect of processes of some sort in the subcrustal substratum, the nature of which was still unclear. As a result the crust was raised in places and sank in others. The slope of the flanks of uplifts and troughs was sufficient for soft rocks saturated with water to begin to slide from the uplift ("geotumor") into troughs ("geodepressions"). With the slumping of strata slopes were eroded at the top and on accumulating below were buckled into folds.

This is how the idea of "gravitational folding", first put forward by A. Reier in 1888, was maintained.

This idea of the primacy of vertical forces and the secondary nature of phenomena of horizontal compression and tension was very substantially developed by the work of the Dutch tectonist Van Bemmelen in the 1930's. He put forward a hypothesis of "undat'ons" (waves) in the crust that arose

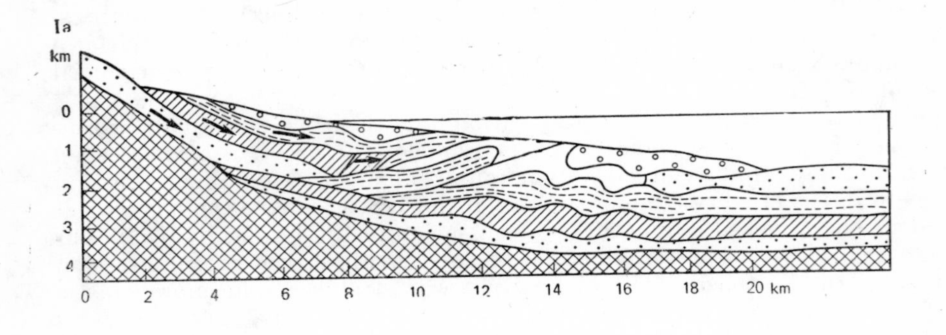
Ia
km
0
1
2
3
4
0
2
4
6
8
10
12
14
16
18
20 km

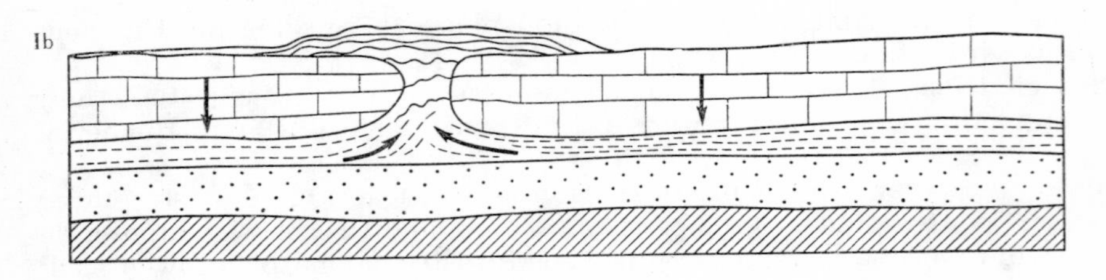
Ib

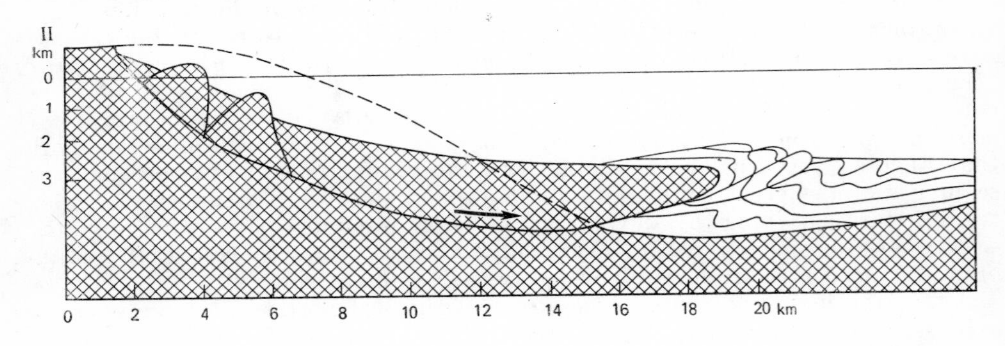
II
km
0
1
2
3
0
2
4
6
8
10
12
14
16
18
20 km

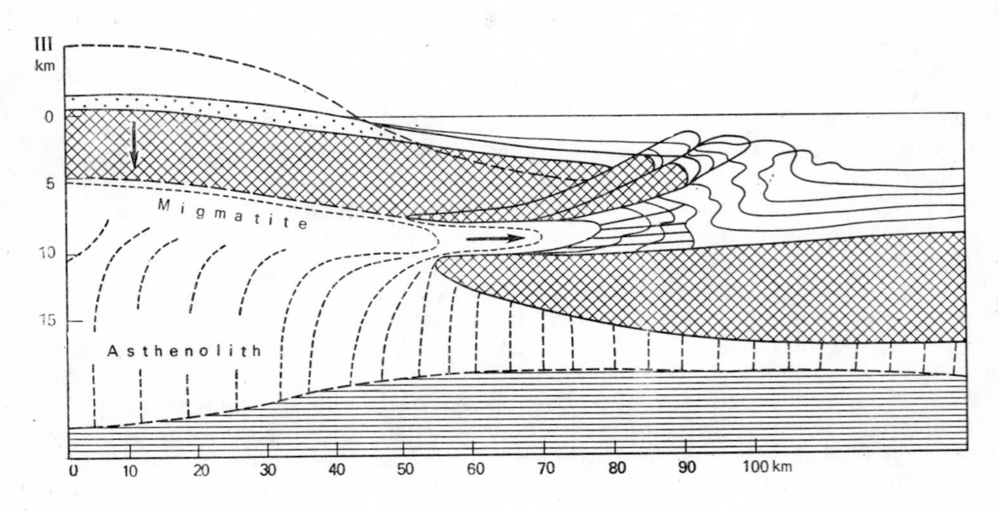
III
km
0
5
10
15
Migmatite
Asthenolith
0
10
20
30
40
50
60
70
80
90
100 km

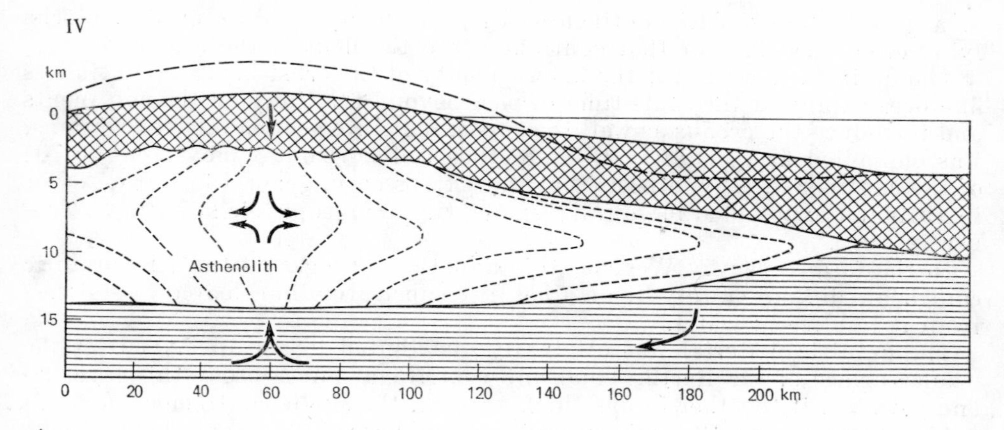

Fig. 124. Secondary (gravitational) tectonogenesis (after R. Van Bemmelen, 1956): *I*—epidermal (*a*—free sliding; *b*—squeezing); *II*—**der**mal; *III*—bathydermal; *IV*—subcrustal

as a result of uneven gravitational differentiation in the subcrustal matter that led to horizontal non-homogeneities or discontinuities; in some places the lighter matter floated up to the surface and formed a bulge, and in other places the heavier matter sank and troughs came into being in the places on the surface above them. Troughs became geosynclines. In the axial part of a geosyncline the basaltic layer of the crust was differentiated into acidic (granitic) material, which was raised to the surface, and ultrabasic (peridotite) material, which sank. A bulge was formed above the site of a granitic uplift, i.e., a median geanticline, while troughs were preserved on both sides of it. The subsequent evolution of a geosyncline consisted in broadening of the median geanticline and migration outward of the troughs fringing it. This was primary tectonogenesis. It disturbed the gravitational balance of the crust, which was restored by the movement of material from uplifts to troughs, a movement that could happen at various levels, namely: in sedimentary series, which became folded through sliding down the uplift's flanks; in the solid crystalline basement, when it was expressed in its splitting into flakes that were thrust one over the other; and in the plastic zone of granitization and migmatisation, where material was extruded to the surface on the edge of the uplift; and below the crust, where deep-lying matter flowed away from beneath the uplift (see Fig. 124).

Van Bemmelen explained the tectonic structure and development of Indonesia, the Alps, and a number of other folded regions in the light of his "undation" hypothesis.

All the geotectonic hypotheses so far mentioned were based completely on geological observations on the continents and in essence quite ignored the existence of oceans, although they were sometimes mentioned. This attitude to the oceans, which occupy more of the world's surface than the continents, was due to the almost complete absence then of definite facts of any kind on the structure of the ocean floor.

The position changed very slowly during the first three decades of our century because techniques for studying the ocean floor had still not been developed, but first information was obtained all the same which indicated that the crust of the oceans differed from that of the continents in the absence of a granitic layer. The continents began to be pictured as granite blocks lying on a basaltic floor that continued into the floor of the oceans.

The desire to account for the heterogeneity of the crust beneath continents and oceans, and at the same time to pass beyond the limits of the continents and introduce the oceans as well to some extent into geotectonic conceptions, was embodied first of all in the hypothesis of horizontal movement (drift) of the continents, a hypothesis that was expressed in general form by Taylor in 1910 and expounded more fully by the German geophysicist Alfred Wegener in 1915.

In the form given by Wegener this hypothesis suggested that the granitic continents did not simply lie on a basaltic substratum but floated isostatically in it like ice in water.

The granitic material was originally distributed in a thin layer over the whole surface of the Earth. Tidal and centrifugal forces broke this layer up and gathered it together in one thicker block, the single continent of Pangea (which happened in the Palaeozoic). In the Mesozoic the same forces broke the single continent into pieces. America was broken away from Eurasia and Africa and, outstripping the rest of the continental mass in its movement westward through the effect of tidal forces, moved ahead, which led to formation of the Atlantic Ocean. Meeting resistance to its movement from the basalt substratum, the American continent buckled into folds on the forward edge, with the result that the Cordilleras and Andes were formed.

Africa half separated from Asia and rotated clockwise somewhat, and between it and Hindustan an ocean was formed. Antarctica and Australia also separated from Pangea and moved away from each other. The island arcs that girdle Eurasia on the east represented fragments of the continent that were separated from it and left behind by its westward movement. The Mediterranean-Himalayan mountain zone was formed by the pressure arising between the continents through the effect of the centrifugal force of the Earth's rotation.

The hypothesis of continental drift was based on various arguments.

(1) It relied on geophysical data about the heterogeneity of continental and oceanic crust. (2) An important argument was the parallelism of the outlines of many continents, which encouraged the idea of fitting them together so as to form a single block with almost no gaps. The parallelism of the opposite coasts of the Atlantic Ocean was especially marked, which made it possible to fit the continents fringing it fully at will. The other continents, however, can also be fitted together almost without gaps if it is supposed that they not only drifted but also rotated. (3) The similarity in features of the geological structure of continents now divided by oceans was pointed out, a similarity that is discovered between South America and Africa, although the Atlantic Ocean lies between them. (4) Palaeoclimatic reconstructions were enlisted in support of the hypothesis.

Publication of this hypothesis caused a great sensation, but in the 1920's and 1930's the numerous discussions around the possibility of continental drift favoured its opponents. Many considered the hypothesis completely refuted, but that was too extreme a view. In fact there have always been supporters of it among eminent workers, who tried to improve it and bring it into line with new facts, and link it more soundly with deep-lying processes.

The improvements mainly concerned the causes of horizontal movement of the continents. The tidal and centrifugal forces adduced by Wegener were clearly insufficient to cause movement of the granitic continents over the basaltic substratum. The idea arose that they might be caused by convection currents in the subcrustal material due to radioactive heating in the interior and cooling at the surface; and various versions of the geotectonic "convection" hypothesis arose.

Vening-Meinez's hypothesis, for example, put forward in the 1950s, assumed that convection was expressed originally, when the Earth still had no core or crust, in the movement of matter in one direction across the centre of the globe, with a return current along its surface. During differentiation, acid (granitic) material was thereby carried away by an ascending current to a region where it ultimately formed a single, granitic continent. After the core developed, convection broke up into separate cells which led to a breaking up of the continent and pulling of its fragments apart to where the continents now are. Wherever the descending plumes of two neighbouring cells met, the crust was sucked down and a geosyncline formed, which was compressed by these descending flows so that the strata accumulated in it were buckled into folds. After attenuation of the convection current this thickened, downwarped, crushed sector of the crust rose isostatically, which led to the formation of a mountain range.

Similar ideas of the downwarping of the crust by descending convection currents were also expressed by Ernst Kraus.

David Griggs pointed out that if one assumed a threshold of plasticity in the subcrustal material, convection could develop unevenly, with an alternation of epochs of movement and quiescence, and he thought it possible to explain the periodic character of tectonogenesis in this way.

Holmes' hypothesis assumed that convection currents stretched the continents, causing geosynclinal troughs to form in them, and then when a continent reached the site of a descending plume during its drift the geosyncline was crushed and folded.

The conception of continental drift has not only been revived in an astonishing way but also became vastly popular in the 1960's, when it was incorporated in revised form in the latest ideas known as "sea-floor" spreading, "plate tectonics", or "global tectonics". Because of this turn in its fate, it will be well to postpone consideration of this hypothesis until we have familiarised ourselves with the latest views as a whole.

The parallelism in the outlines of the continents also gave rise, in addition to propositions about drift, to a hypothesis of a vast increase in the volume of the Earth. This is the hypothesis suggested by the Hungarian geophysicist Egyed in the 1950's, that the globe was originally so small that its surface was equal in area to the total area of the continents, which constituted a single whole covering all the Earth. The radius of the Earth then did not exceed 4000 km. The matter of the core was extremely dense, as high as 17 g/cm^3. This dense phase was formed in special cosmic conditions of some sort and proved to be unstable in the Earth, so that it gradually passed into a less dense phase with an increase in volume. The single continental cover cracked and separated into the individual continents, which moved further and further apart, while the mantle came to the surface between the continents, forming the floor of the oceans.

This hypothesis evoked many objections. The proposition underlying it, of an originally extremely dense core, is physically unacceptable. It is incomprehensible how matter, compacted somewhere in another place in conditions of much greater lithostatic pressure, could then, without changing its state, occur in the core of the Earth. In addition, an even increase in the surface of the globe cannot explain the location of the continents now observed. A uniform surface increase should divide all the continents into a number of small islands that would be distributed uniformly on the surface, while uneven expansion cannot be assumed because that would contradict the Earth's sphericity. Expansion, consequently, if it occurred, must have been accompanied by a horizontal shifting of the continents, i.e., the expansion hypothesis must be joined with that of continental drift.

To conclude this chapter let us briefly describe the main achievements of the Soviet geotectonic school in the first half of this century.

Geotectonics began to develop rapidly in the Soviet Union after the Great October Socialist Revolution. Soviet workers closely studied foreign experience, a study expressed in a number of critical surveys published by Borisyak, Milanovsky, Obruchev, and Strakhov in the 1920's and early 1930's.

In the years of the first five-year plans, when geological work greatly expanded in the USSR, Soviet tectonists directed their main efforts to studying the structure and history of the tectonic development of separate regions of their vast country. As a result of this regional research facts were accumulated that helped explain the pattern of tectonic processes, initially locally, and then more and more broadly.

A first-class example of regional tectonic research and summaries was Archangelsky's work on the structure of the USSR published in the twenties, thirties, and forties. In his last summaries he was already surveying the tectonics of the whole world and had come to formulate general patterns of the Earth's tectonic development. Broad light was also shed on the tectonics of the East European platform by N.S. Shatsky's work. A large number of tectonists studied the structure and patterns of development of the folded zones of the Soviet Union.

All this gave them a vast amount of data and enabled them to pose and answer the most varied questions on their basis: viz. concerning types of tectonic movements and their interrelations and patterns of development, the stages of tectonic history, tectonic regimes, the link between tectonics and magmatism, the mechanism of rock deformation, and so on. The experience of regional tectonic research accumulated gave Soviet workers (N.S. Shatsky and A.A. Bogdanov) the chance to take the initiative in the 1950's in compiling international tectonic maps, drawing tectonists from many countries into the work. A tectonic map of Europe was published first, and then maps of other continents were compiled or are being worked on.

The publication of M.M. Tetyaev's *Fundamentals of Geotectonics* in 1934 (second edition in 1941) deserves special mention as an important event. In this book Tetyaev tried to develop a general logical scheme of geotectonics as a science. His central idea was that the various types of tectonic movement were different forms of one process of geotectogenesis, in which in turn more and more general processes of the Earth's development as a cosmic body were embodied. This approach enabled him to formulate the tasks of geotectonics as a science in a new way. In addition to studying the separate forms of tectonic movement with which it had been mainly concerned up to then, attention was paid to clarifying the patterns that united these movements and to establishing the links and dependencies between various tectonic phenomena. Tetyaev distinguished oscillatory and folding forms of geotectonogenesis, magmatic movements, and a macrowave form, understanding by that those movements we now include in the concept of orogenesis, particularly major vertical movements of blocks of crust along faults. The task was to find the connections and dependencies between these forms of geotectogenesis. The posing of the problem of the unity of tectonic processes helped formulate the subject matter and tasks of geotectonics more fully and exactly, and also its position among the other branches of geology.

There was a contradiction between vertical oscillatory movements caused by vertical forces and folding movements that required horizontal forces for their formation. Tetyaev thought it possible to accept the proposition that folding occurred as the result of layer by layer horizontal flow of material through the effect of compression in a vertical direction. Strata were folded not because the area occupied by them was contracting, but because they themselves were becoming longer through being vertically crushed. The physical mechanism of this vertical spreading, however, was not sufficiently considered. Only the struggle of two tendencies in the Earth's matter was discussed, viz. extension and compression, or attraction and repulsion, which also led to the spreading of strata when they met. In Tetyaev's view the struggle of two opposing tendencies in general constituted the *leitmotif* of the Earth's evolution, the struggle being concretised in the behaviour on the one hand of the Earth's internal material, which tended to expand, and of the crust on the other hand, which counteracted this expansion. He suggested that this contradiction would become sharper in the course of time, since the crust was becoming increasingly thicker and stronger from one tectonic stage to another. While it was sufficient, at the present time, in order to crack the crust and eject some of the matter of the interior, for a small part of this matter to be mobilised and melted, later a much longer accumulation of energy of expansion and mobilization of matter from a greater depth would be needed.

The value of Tetyaev's work, however, was not in its exposition of concrete mechanisms of tectonic processes of one sort or another but in its development of the philosophical system of geotectonics and confirmation of geotectonics as an independent branch of knowledge. He also introduced geotectonics into the geology syllabus for the first time as a separate subject.

One can say, on the whole, that, on the general issues of geotectonics at the end of the forties, and in the fifties, most Soviet geotectonists supported the ideas of the primacy of vertical forces and movements and the secondary character of deformations requiring horizontally directed compressive and tensile forces. Other general trends were the following: rigour in the factual substantiation of theoretical generalizations; distrust of purely speculative constructs; a historical approach to the study of tectonic phenomena; and an endeavour to elucidate the patterns linking various tectonic processes together.

CHAPTER 25

Recent Views on Geotectonics

The following circumstances determined the special character of the development of geotectonics in the fifties and early sixties:
1. the progress made by geophysics and geochemistry in studying the structure and composition of the deep, previously inaccessible interior of the Earth;
2. progress in study of the sea floor;
3. the involvement in geotectonic problems of a quite new group of specialists with a training in physics and mathematics rather than geology, who used new methods and had a scientific slant new to geology.

The geophysical and geochemical penetration of the Earth's interior that had been slowly advancing over the preceding several decades culminated in the fifties and sixties in what might be called an explosion in the development of knowledge of the structure of subcrustal depths and of the processes taking place there. The barriers to understanding of the interior were suddenly broken down and instead of the enigmatic, deep-seated substance comprising remote, inaccessible "sima", we became quite familiar with a specific mantle composed of quite concrete rocks, and governed by quite normal physicochemical laws. The improved techniques made it possible to discover horizontal heterogeneities in the composition and properties of the matter down at least to the uppermost layers of the upper mantle. It thus proved possible for the first time to try and realise the long-standing desire to tie up the tectonic processes observed at the surface with the properties of the upper mantle, and so to understand geotecto-genesis as an indispensable part of the evolution, if not of the world as a whole, but at least of its outer shells.

As a result geotectonics ceased to be simply a geological discipline. It became a sector of a broader science of the deep interior of the Earth and of endogenous processes, a science that combined the techniques and data of geology, geophysics, and geochemistry. This united science of the Earth has been named "geonomy". It is in the state of emerging, but there is no doubt that it will develop and that its synthesising significance will grow.

A brief review of recent geophysical and geochemical data on the structure and matter of the Earth's interior was given in earlier chapters. This is the information that has to be used to disclose the deep causes of tectonic processes.

Until the early 1950's almost nothing was said about seas and oceans in the literature on terrestrial tectonics, and that in spite of the fact that the oceans cover two-thirds of the Earth's surface.

The absence of information on the structure of the ocean floor led to geology, and with it geotectonics, being a purely "continental" science. In such circumstances it could hardly be expected to create a generalizing geotectonic theory of any kind that could establish the patterns of development of the Earth as a whole.

In the mid-50's powerful new geophysical methods of studying the structure of the sea floor were developed. In the 1960's it proved possible to organise drilling in the deep ocean, at least through the series of loose sediments. The number of oceanographic research ships increased considerably, and the oceans began to be studied systematically and intensively, not haphazardly as before. As a result we have obtained reliable data over the past 15 or 20 years on the most important features of the ocean floor's structure.

The fact that people with a different training have come into geotectonics may not, at first glance, seem to be very important, but it actually was extremely so.

Physicists and mathematicians, who had previously done separate, mainly methodic jobs for geophysics, became involved in studying concrete problems of the Earth's structure and development. Being free of any ballast of continental geology's previous views, this group of workers approached these problems, including problems, in essence geological, that had not previously been tackled in this field. Problems that geologists had long considered their own were posed in quite a new way; at the same time they were studied in the way more customary for the new workers, namely by the methods of generalized physical models in which reality is schematized to the degree needed for mathematical expression. Where geologists relied on nuances, semitones, and not very clear differences in structure and processes, expressing it all in developed verbal form, the new workers looked for opportunities to make nature speak the language habitual for computers, i.e., a language in which any verbal half-tones were replaced by contrasted oppositions of affirmation and negation.

Has this been good or bad for geotectonics? Any fresh approach to a scientific problem, and any looking at it from a new angle and free of the prejudices and biases accumulated over a long period must, of course, be welcomed. The drive to introduce the quantitative element into views on endogenous processes, including tectonic ones, also has to be considered a quite positive tendency. The point, however, is how far the schematizing of natural processes needed for a quantitative approach is justified in any concrete case, how far it selects the features of those processes that are really most important to them, and how far it looks for these features. We shall try to answer those questions at the end of this chapter.

"The New Global Tectonics"

The system of views on the development of the tectonosphere that has become very popular in recent years is grouped under the common soubriquet of "the new global tectonics". Its components are the "theory of sea-floor spreading" and "plate tectonics". This system of views arose as a result of a striving to understand and correlate the new data obtained from study of the ocean floor.

The following initial data underlie this system.

1. The oldest sediments on the ocean floor are not older than the Jurassic, a circumstance that may be interpreted as an indication of the geological youth of the whole ocean floor.

2. The thickness of the sedimentary cover in the oceans thins toward the crest of mid-ocean ridges. The thinning is linked with the successive pinching

out of sediments from the bottom upward as we move toward the crest of a ridge, so that only the youngest sediments lie near its axis. As a rule this axial strip is 30 to 50 km wide and completely bare of sediments. Some determinations of the absolute age of the basalts underlying the sediments suggest that their age coincides in most cases with that of the sediments lying directly above them. The basalts consequently become younger and younger toward the axis of a mid-ocean ridge.

3. Study of the magnetic field of the oceans has shown that the flanks of median ridges are characterised by a linearity of anomalies in bands running parallel to the ridge. On some sectors of the bands a symmetry in the distribution of the anomalies is observed, with an axis of symmetry that coincides with the axis of the ridge. The bands of anomalies correspond to changes in the polarity of the magnetised rocks; rocks with a positive magnetisation, i.e., one corresponding to the present-day position of the magnetic poles, alternate with rocks that are negatively magnetised, i.e., with a reversal of polarity.

4. Heightened heat flow is associated with the crest of mid-ocean ridges, reaching record levels in places.

5. There is a lens of material with intermediate seismic velocities at shallow depths beneath ridges.

6. The axial zone of oceanic ridges is divided by transverse fractures into segments that are offset dextrally or sinistrally to each other. The foci of earthquakes are confined to transverse faults. Study of the foci's mechanism has indicated that their shifting tends to be in the opposite direction to the visible offset of the segments of the ridge (with a dextral offset of the segment, movement in the focus proves to be sinistral, and vice versa).

7. The seismic data have indicated that the "Benioff zones" confined to oceanic margins of Pacific type are accompanied to a depth of more than 700 km with a layer parallel to each zone about 100 km thick that differs from the surrounding medium in a significantly lower value of the absorption of transverse seismic waves.

8. A number of palaeomagnetic and palaeogeographic findings, and the parallelism of the outlines of the continents (primarily those bordering the Atlantic Ocean) can be interpreted as signs of the former existence of a single continent, and of its subsequent breaking up and the drifting of its parts.

The concrete content of the views combined by the "new global tectonics" and based on these initial data is as follows.

There is a process of constant renewal of the oceanic lithosphere, which is why we see it so young although basins filled with water are very old. The new material of the lithosphere rises in molten form from the asthenosphere along the axis of a mid-ocean ridge along a fracture to which the rift valley at the surface corresponds. This rise of a mantle melt is the cause of the high heat flow at the surface and of the development of a zone of intermediate seismic velocities between the crust and the mantle, where the melt becomes mixed with the solid shells of the lithosphere.

The mantle material, having filled the fracture, cools, forming a new block of the lithosphere. While cooling it becomes magnetised (positively or negatively) in accordance with the geomagnetic field existing at the time. This block is later caught up in the movement involving all the lithosphere of the oceans as a whole, namely its horizontal migration away from the mid-ocean ridge. The block splits into two halves along the ridge, and these halves, which are as thick as the lithosphere, i.e., as much as 100 km, move

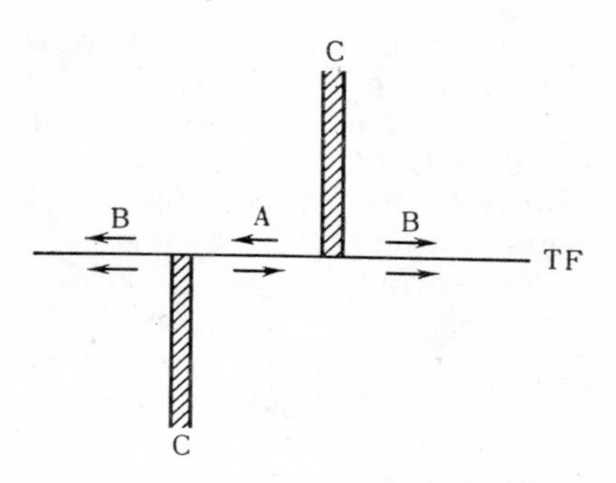

Fig. 125. Diagram of a transform fault (after Wilson):

C—the axis of 'spreading' of the lithosphere, severed by a transform fault. The arrows indicate the trend of 'spreading': on sector *A* the relative shift of the two lithospheric plates is sinistral, although the apparent shift of the axis of spreading is dextral; on sector *B* the plates, separated by the fault, are moving in parallel; earthquakes therefore arise only on sector *A*

apart. A new fracture opens up between them that is filled in turn with molten material from the mantle. If, during this time, the direction of the geomagnetic field alters, this new block of the lithosphere will accordingly be magnetised in another polarity after cooling.

The lithosphere of the ocean is thus being continuously formed, gathering into vertical blocks that subsequently become united with each other as the ocean floor "spreads". A geomagnetic time scale is imprinted as it were in fossilised form in the picture of the linear magnetic anomalies; the oldest sectors of the floor, with magnetic anomalies corresponding to this old epoch, prove to be farthest removed from the crest of the ridge, while younger blocks of the oceanic lithosphere, with anomalies belonging to younger geomagnetic epochs, are closer to it.

The lithosphere is displaced in bands divided by fractures. Since each band moves away from the rest, and the position of the next one in the transition along its trend proves to be shifted to the right or left across the fault, this mechanism may provide an explanation of the non-conformity of the apparent shift of the crest with the direction of the shift in the foci of earthquakes located along this fault. Faults with these properties have come to be called transform faults (see Fig. 125). Each band of the lithosphere moves as a rigid undeformable plate. The rate of movement of plates can be measured by the distance of the magnetic anomalies (whose absolute age is known) from the axis of the ridge. The absolute geomagnetic scale substantiated by facts does not (as we said above) go further back than four million years or so. Allowing for the fact that the anomalies closest to the axis of a ridge belong to the Brunhes, Matuyama, Gauss, and Gilbert epochs, and relying on their absolute duration, the rate of movement of the plates has been calculated. It turns out to be different in different regions: namely, 1 cm a year for the North Atlantic, 1.5 cm a year for the South Atlantic and the Carlsberg Ridge, and 4.5 cm a year for the Pacific Ocean. These figures indicate the rate of movement of each plate separately; the rate of renewal of the oceanic lithosphere is consequently twice as fast. The figures, incidentally, also vary greatly from place to place along a ridge. Attempts to measure the rate of movement of plates for earlier epochs are based on extrapolation of the geomagnetic time scale back in time, which the supporters of the "new global tectonics" think permissible on the grounds of the age of the sediments lying on the basalts, and from measurements of the thickness of the sedimentary series with distance from the axis of the ridge.

When there is no deep ocean trench between an ocean and the continent, i.e., where the margin of the ocean is of the Atlantic type, the plate of oceanic lithosphere and the continent and its lithosphere represent a single whole.

In this case the continent moves together with the oceanic plate as a single "raft". North and South America, for example, on the one hand, and Eurasia and Africa, on the other hand, move together with the lithospheric plates of the Atlantic Ocean in opposite directions on both sides away from the Mid-Atlantic Ridge, along which new blocks of lithosphere are all the time being formed.

The surface of the globe, however, is not increasing, so that the development of new lithosphere in one place must consequently be compensated by its annihilation in others. This annihilation is postulated on oceanic margins of Pacific type, where the ocean is fringed with deep trenches. It is said that the oceanic lithosphere sinks down into the mantle there, following a downward slope along "Benioff zones" (see Fig. 126). The low absorption of transverse waves in these zones is also explained by this same similarity: the lithosphere is more solid than the material of the mantle and should therefore absorb elastic waves less. The sinking of the lithosphere (which is termed subduction in the literature abroad) is explained by its cooling with distance from the median ridge and the increase in its density that occurs in that connection, which forces it to subside into the mantle. The subsidence of cold lithosphere into the mantle is accompanied with a transformation of basalt into eclogite, which further increases its density. But in gradually penetrating a zone of high temperatures the lithospheric plate is heated and melts. The maximum age of the ocean floor (150 million years) roughly indicates the length of the full cycle of its renewal.

Starting from the disposition of mid-ocean ridges as centres of lithosphere spreading and of oceanic margins of Pacific type as zones of its subduction, the authors of this conception considered that the whole surface of the Earth could be divided into eight very large plates of the first order, including both ocean floor and the corresponding continents moving together with them. The idea of the division of the lithosphere into a limited number of horizontally moving plates also constitutes what is called "plate tectonics". Present-day tectonic activity, by which is understood mainly seismic activity, occurs where neighbouring plates touch, either rubbing against each other or colliding. The inner sectors of plates, however, are tectonically quiescent. More recently, in face of the growing complexity of tectonic processes, the supporters of the "new global tectonics" have begun to divide

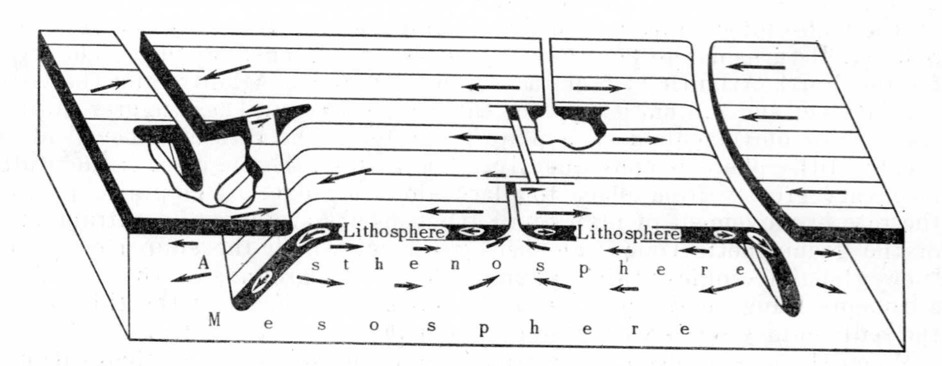

Fig. 126. General schematic diagram of the movement of the lithosphere according to the "new global tectonics" (after Isacks *et al.*, 1968)

the main plates up into an ever larger number of plates of the second and third order.

This whole conception was originally intended to explain what happens on the floor of the oceans, but it subsequently spread to the tectonics of continents and of inland and marginal seas. In relation to continents, the "new global tectonics" states the following.

Eugeosynclines originate in the oceanic lithosphere, melted directly from the mantle. Ophiolites, i.e., the oceanic crust and the siliceous rocks that usually accompany it, are deep oceanic sediments. The geosyncline develops in the course of subduction of the oceanic lithosphere, which under-rides the neighbouring continent. In one place a trough (deep trench) is formed, in another layers raked up the surface of the subducted plate are buckled into folds. The plate that is sinking into the mantle, however, melts (as already said). Because eclogites now play a significant part in its composition, their partial melting leads to the separation of andesite magma, which goes into the formation of an acid (granitic-gneissic) upper layer of the crust on the site of the geosyncline. As this light acid layer becomes thicker it rises to the surface and forms an elevation on the site of the geosyncline, which now has a continental crust.

Such is the development, primarily, of those geosynclines that are associated with oceanic margins of Pacific type, e.g., the Cordilleras and Andes, or the island arcs of the West Pacific. But according to these ideas, intercontinental geosynclines like the Mediterranean geosyncline of Southern Europe and North Africa also develop in an analogous way. The first stage of development of the eugeosyncline there is also connected with stretching, when the continental crust separates and an oceanic lithosphere is formed in the gap; ophiolites build up an oceanic crust on the surface of this oceanic lithosphere. In the second stage the ocean lithosphere is subducted with all the events that go with it. In these conditions, however, not only may the oceanic lithosphere move downward beneath the continental one, but the two blocks of continental lithosphere may collide. The relationship of India and Central Asia is considered a classic example of the collision of two continental plates. It is suggested that the Indian peninsula was earlier an independent continent and situated much further south of its present position. In moving northward it finally reached South Asia, and its pressure on the latter caused the uplifting of the Himalayas and other Central Asian mountains.

While there are median ridges in the oceans that are considered as the axes from which oceanic crust spreads, there are no such ridges in inland and marginal seas. The consolidated parts of the crust in these seas, however, do not differ in structure from oceanic. Since the "oceanic mechanism" is inapplicable in these cases, it is suggested that there is another mode as well with the oceanic lithosphere formed, namely through its areal diffused melting out from the mantle, which is accompanied with a moving apart of the blocks of continental lithosphere bordering the sea. It is suggested, for example, that the island arcs of the West Pacific earlier bordered on the neighbouring continent; then a fracture was formed between them and the continent, which widened as more and more new basic and ultrabasic intrusions were injected, ultimately forming an oceanic lithosphere there with a basaltic oceanic crust, while the island arcs moved away from the continent to their present positions. From this point of view the seas can be considered a geosyncline in the ophiolite stage of its development.

20—0601

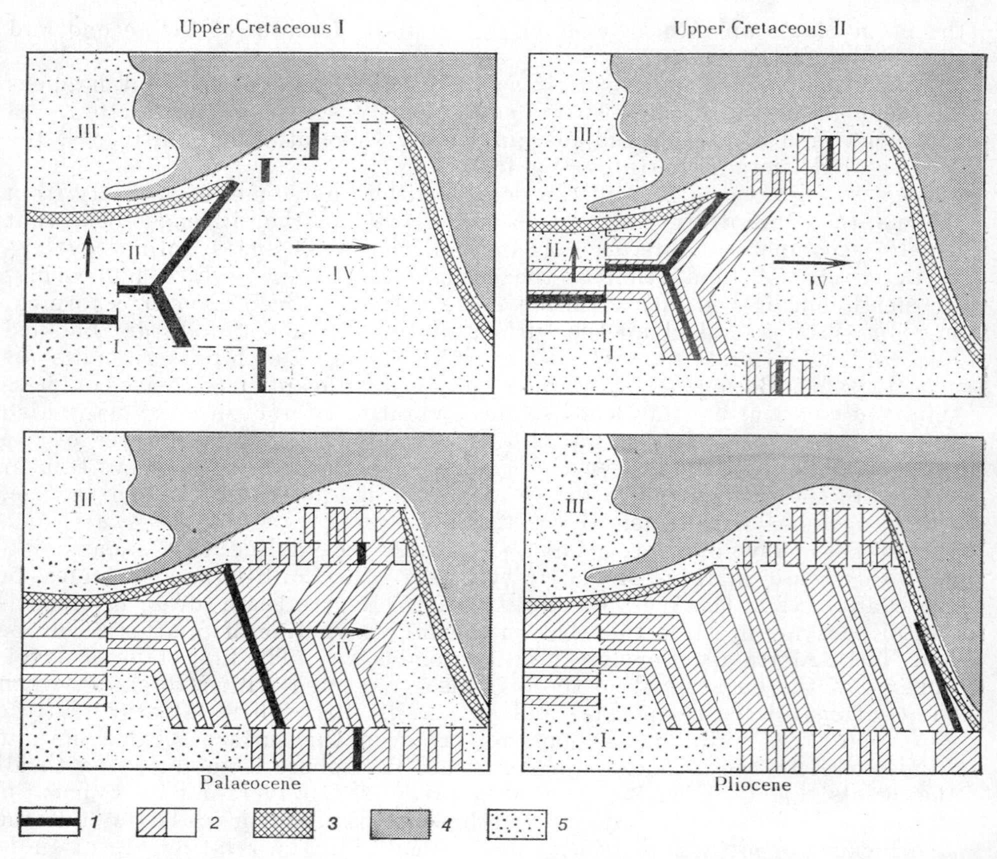

Fig. 127. History of "sea-floor spreading" in the North-east Pacific (after Pitman and Hayes, 1968):

I-IV—blocks of the lithosphere moving independently in the direction of the arrows; *1*—the axes of "spreading"; *2*—magnetic anomalies; *3*—deep trenches; *4*—the North American continent; *5*—lithospheric plates. The northward shift of the east-west axis had led by the Palaeocene to its being buried together with neighbouring anomalies beneath the Bering Sea; by the Pliocene the north-south axis, which had shifted, proved to be beneath the continent

It is suggested that the axes of sea floor spreading may alter their position from time to time. In the North Atlantic, for example, there proves to be an ancient axis of spreading coinciding with the Norwegian Basin. This axis died out at a certain stage, after which a new one developed on the site of the present-day median ridge. Complex migrations and jumps of the axis of spreading are also suggested for the North-east Pacific and several other regions (see Fig. 127).

As to the causes of the movements of lithospheric plates, the "new global tectonics" provides no answers. Convection currents in the mantle are usually adduced as an explanation. The existence of layering in the mantle, however, connected with a change in the composition of its matter from layer to layer, makes such currents improbable, and even simply impossible. It has been suggested that lithospheric plates creep along the asthenosphere through the effect of their own weight, since the crest of a median ridge rises above the surrounding basins and the plates could slide from it.

The unresolved character of the causes of the movements of course cannot, however, be considered a circumstance denying the fact of the movements itself, provided they exist.

Critique of the "New Global Tectonics"

In our critique of the ideas of the "new global tectonics" we shall touch first of all on general considerations and then turn to more particular ones.

We would first ask whether it is legitimate in general to draw conclusions about the patterns of development of the Earth's tectonosphere as a whole from the data on the structure of the ocean floor.

At first glance one should be able to answer that question positively since the oceans occupy more of the Earth's surface area than the continents. On the other hand, however, the history of the oceans is only accessible to study to any extent for the past 100 or 150 million years. No natural documents shedding light on earlier events have been found in their structure, while the history of the continents has been traced back for 3500 million years into the depths of geological time! The continents, consequently, are the only "windows" we have onto the ancient history of the tectonosphere, and if we want to explain its long-term, stable patterns of development we have primarily to study the continents.

What is observed in the oceans, possibly, is a feature of development proper only to the very latest, Mesocenozoic stage in the tectonosphere's history occupying not more than a twentieth of the whole length of geological history. It would be more logical, therefore, to consider that the data obtained on the oceans should be interpreted in the light of the facts obtained from the experience of continental geology rather than contrariwise. No one, however, will deny that the final synthesis of the tectonics of the whole globe must take into account the facts that have become known in recent years about the oceans. We would like to emphasise here simply that it is necessary in explaining the universal involvement of the ocean floor to retain proper (and therefore more!) respect for the conclusions of continental geology.

Various facts connected with the existence of large horizontal unhomogeneities not only in the crust but also in the upper mantle were cited above. The unhomogeneities are expressed in changes in both the composition and the structure of the upper mantle, and are also manifested in the varied course of certain geophysical, mainly geothermal, processes.

A discontinuity of the first order is expressed in the difference in composition and structure of the tectonosphere beneath the continents and oceans. The difference is so great that we have every right to speak of two types of tectonosphere, viz. continental and oceanic. The continental tectonosphere is more differentiated in composition than the oceanic. For the crust to have been supplied with certain elements observed in it, differentiation must have involved the mantle to a depth of hundreds of kilometres. The uppermost layers of the mantle (the substratum) have a residual composition following the separation of the crust from them. Owing to deep-running differentiation radioactive elements are concentrated in the uppermost layers of the tectonosphere, and in that connection the crust has a total heat flow of at least half of the heat generated. The asthenosphere is usually found at a depth around 100 km and is between 100 and 150 km thick. It is expressed to a different degree beneath different endogenous zones, which also have a heat flow of different intensity.

Since endogenous regimes give way to one another in time, it must be supposed that the magnitude of heat flow alters with them. Confirmation of that is to be found in palaeothermometric signs of higher ancient heat flows compared with present-day ones where geosynclines were earlier located and where quieter regimes have now developed. The fraction of the heat flow generated in the crust, however, has not apparently altered much. Its magnitude can only have altered in connection with a change in the rate of effusion of heat from regions lying deeper, at least, than the asthenosphere, since the degree to which latter is expressed (which varies with change in the endogenous regime itself) depends on the heat flow coming from deeper down.

There is thus a direct connection between the endogenous regimes of continents and very great depths measurable in hundreds of kilometres.

The tectonosphere beneath the oceans is very much less differentiated. It is sufficient, for oceanic crust to be formed, for a series of mantle rocks only a score or so of kilometres thick to be involved. Radioactive elements are therefore disseminated much deeper in the oceanic mantle than beneath continents, and the oceanic crust contributes only about a tenth in all of the total heat flow; the other nine-tenths come from the mantle. In this connection the temperature in the mantle beneath the oceans is higher than that in the continental mantle. The greater thickness of the asthenosphere, the top of which lies at a depth around 50 km, and the bottom around 350 to 400 km, is linked with the higher temperatures.

Can these differences in the structure and composition of the continental and oceanic tectonospheres, and the "roots" of continental endogenous regimes that go down hundreds of kilometres deep, be correlated with horizontal shifts of lithospheric plates and continental drift? The answer to this must be negative.

It was pointed out above that mean heat flow is the same on continents and in the oceans. If, however, the continental lithosphere were shifted, in the course of drift, to where an ocean used to be, heat flows on that continent should increase greatly because a much higher proportion of heat flow in the oceans is of deep, sublithospheric origin. Conversely, if the oceanic lithosphere were shifted to where a continent previously was, its heat flow should be much reduced for the same reason. Equality of mean heat flow on continents and oceans, consequently, contradicts both the shifting of lithospheric plates and continental drift in any form at all.

Since continental regimes are firmly linked with processes taking place at very great depths, greater than the depth of the asthenosphere, and since the way the latter is expressed also depends on these deep processes, a reverse shifting of endogenous regimes should be observed on the surface if the continental lithosphere is shifted in relation to the asthenosphere, because the continent would thereby have come above different zones of the mantle with different temperature regimes and deep-seated processes of a different character. No such shifting of endogenous zones has been observed. On the contrary, they display protracted spatial stability over long periods of geological time. The geosynclinal belt, for example, that encircles the Pacific Ocean has been in one and the same place (in terms of the co-ordinates of the continents surrounding the ocean) since the beginning of the Proterozoic. The Mediterranean geosynclinal belt was established later but on the whole there has been no directional shift of it since at least the Upper Proterozoic. The regions of uplift and subsidence on ancient platforms are also as stable.

The Moscow syneclise, for example, has not altered its position on the whole from the Palaeozoic to the present.

Much attention is paid in the hypothesis of continental drift to the parallelism of the coastlines of the Atlantic Ocean. This similarity enables the continents on both sides of the ocean to be shifted in imagination so that they fit quite exactly (though not fully). Given the fundamental objections to drift listed above, this argument cannot be accorded independent significance. The parallelism may be due to the boundaries between the continents and oceans having been formed by deep faults, which (as we know) are grouped in parallel systems.

As regards the possibility of joining up the other continents, the curvature of their contours (as Lyustikh has shown) is generally (and, apparently, regularly) so similar that there is always the possibility of joining any of them together, depending on how they are rotated.

The "new global tectonics" is unable in principle to explain very important features of the evolution of the tectonosphere, for example, such a regularity as the periodic character of endogenous processes. This conception, having at its disposal only certain data on the ocean's history for the past 100 to 150 million years, can naturally say nothing about endogenous cycles that have been manifesting themselves, as the data of continental geology indicate, for thousands of millions of years.

The "new global tectonics" completely ignores tectonic events taking place outside geosynclines, on ancient and young platforms. For it continents, as at the dawn of the contraction hypothesis, have again become dead blocks that can only be moved apart and shifted in position but that have no endogenous life of their own. The adherents of this hypothesis are therefore taken by surprise, for example, by the oscillatory movements taking place on platforms, and maintain complete silence about the signs of folding on them.

It also cannot be considered acceptable, however, in regard to geosynclines. First of all, it completely ignores the essential chemical difference between the rocks of the ocean floor and geosynclinal ophiolites. It is hard to understand how the oceanic lithosphere, which is very poor in potassium, could differentiate on melting into the material of the granitic-gneissic layer, so rich in potassium.

Let us note, too, the extraordinarily schematic character of the new ideas on tectonic processes in geosynclines, which are considered in the same way as in Dana's day, as troughs on the margin of a continent filled with sediments and crushed by the pressure coming from the ocean. It is as if no progress had been made in our understanding of geosynclines in the past hundred years, and as if we knew nothing about the complex inner differentiation of geosynclines, the sequence of the development of oscillatory movements within them, and types of folding.

Let us now pass to several more specialised objections.

As there are median ridges in all the oceans, appearing everywhere as axes of spreading of the lithosphere, we need to examine how spreading processes taking place simultaneously in the different oceans can be correlated. It will be readily understood that this can be done only if it is assumed that the mid-ocean ridges themselves are displaced in the spreading. In fact, if it is assumed, for example, that Africa and Eurasia have stood still and that North and South America, on the one hand, and Australia, on the other hand, are moving away from them respectively westward and eastward, it is necessary, in order to preserve symmetry of spreading, for both the Mid-Atlantic and Mid-Indian Ridges to be shifted simultaneously, always remaining half-way between the continents. For that to happen their rate of shift would have to be exactly half that at which America and Australia are moving. If the axis of the Mid-Atlantic Ridge is considered immobile, then the Mid-Indian Ridge must "run" ahead of an eastward-moving Africa at such a rate as to be always in the middle between it and Australia, "running" even faster eastward. In that case the rate of shifting of the Mid-Indian Ridge would be half the combined rate of movement of Africa and Australia. In moving, too, the median ridges must preserve their shape.

Simultaneously with these movements the African-Antarctic and Australian-Antarctic Ridges must be shifted to the south after Antarctica, which is moving southward even faster. The position in regard to Antarctica, however, is very puzzling, because it is almost completely surrounded by median ridges and seemingly cannot drift anywhere.

To this already quite complicated shifting there has still to be added the movement of the Indian sub-continent from south to north, from the region of the present-day central part of the Indian Ocean to the southern edge of the Asiatic continent. Africa must also experience additional movement, away from Europe, opening up room for the Mediterranean geosyncline, and rather later toward it again so as to cause mountain-forming in this same geosyncline, and rotate clockwise in order to form the graben of the Red Sea.

All this is very strange. The basic premise of the conception is that the position and movement of lithospheric plates is dictated by the position of the axes of spreading. The axes are the cause, the behaviour of the plates, the consequence. The axes are located where molten material rises from the mantle. It can be assumed, of course, that the spot where mantle material rises alters; the above-mentioned agreement between the rate of shifting and the rate of movement of the lithospheric plates, however, inevitably leads to cause and effect changing places: the position of the axes of spreading depends on the plates' movement. Having got into a vicious circle (the movement of the lithospheric plates depends on the position of the axis of spreading, and the position of the latter depends on the movement of the plates) we witness a breakdown of the conception's inner logic.

It is also, of course, hard to understand how India's meridional movement from south to north is combined with flow of a lithospheric plate eastward from the Mid-Indian Ridge, carrying Australia with it.

Summing up these remarks one can say that even from the geometrical angle the shifts proposed by the "new global tectonics", simple so long as they are considered for one region of spreading, become not only muddled and confused but also contradictory and simply improbable when one tries to picture their simultaneous development in two or more areas of spreading.

In Figure 109 we showed the distribution of sediments of various age in the basement of the sedimentary series for the northern half of the Pacific Ocean. The oldest layers are concentrated in a comparatively small area near the Marianna and Marshall Islands, and this area bounded on the north and east by ever younger rocks that consistently succeed one another. If that distribution of sediments corresponds, as is now asserted, to a shifting of bands of the ocean floor of different age, and if this shifting is the result of ocean floor spreading, then the material on a perimeter of vast extent must have been gathered into a small area in some way or other without deformation, which is obviously mechanically impossible.

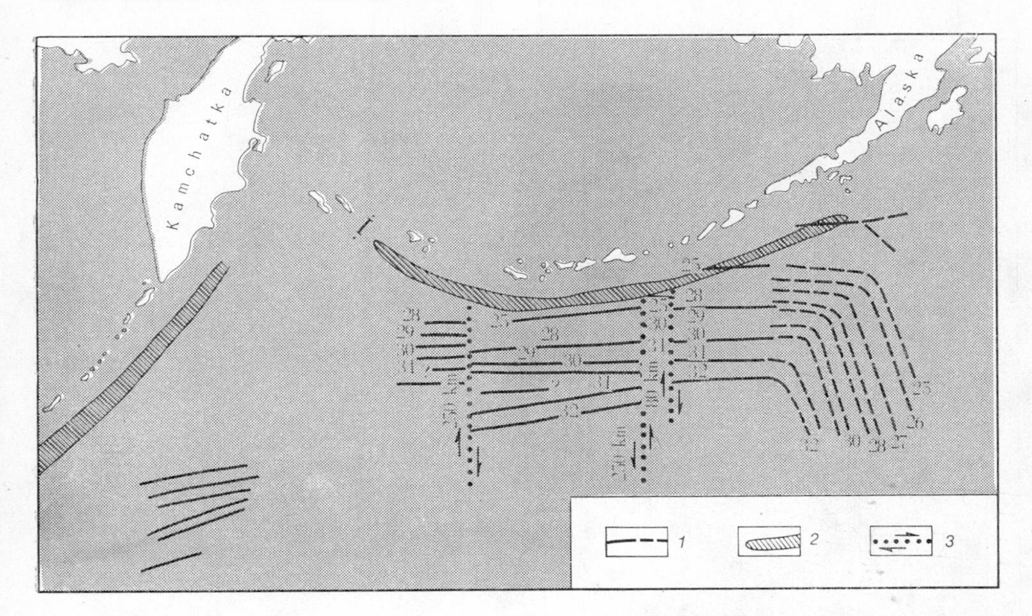

Fig. 128. Trend of magnetic anomalies in the north-eastern part of the Pacific Ocean (after Hayes and Heirtzler, 1968):

1—axes of positive anomalies; *2*—deep trench; *3*—transverse shifting of anomalies. The figures are the number of an anomaly on the scale adopted for such constructions

Although, according to the "new global tectonics", all deep trenches must subduct oceanic lithosphere, the Aleutian Trench clearly does not perform this function. The location of magnetic anomalies around it is the opposite of what this conception requires; the anomalies do not become older with closeness to the trench, as the authors of the conception themselves assert, but younger (see Fig. 128). To explain this, it is supposed that the median ridge, in gradually moving north, under-rode the Aleutian Trench, which thus continued to exist. Consequently, either the age identification of the magnetic anomalies is wrong, or the nature of deep ocean trenches is not that ascribed to them.

There is a host of more partial objections to the constructions of the "new global tectonics". One of them, for example, indicates that the distance between bands of magnetic anomalies is not proportional in most cases to the length of the geomagnetic epochs. If we take the length of the Brunhes epoch as unity, then the lengths of the three epochs (Brunhes, Matuyama, and Gauss) whose absolute values are known have the following ratio: 1.0 : 2.4 : 1.6. On the Reykjanes Ridge (the northern part of the Mid-Atlantic Ridge), for example, the distances between the anomalies closest to its axis, however, have the ratio 1.0 : 0.5 : 0.4. A breach of proportionality no less great is observed on the East Pacific Rise, which can apparently be explained from the angle of the "new global tectonics" by considerable changes in the rate of spreading over the past three million years. But if the ratio is broken in most observed cases, that would seem to remove the factual basis for considering linear anomalies a "fossilised time scale" and also for extrapolating this scale into the depths of geological history.

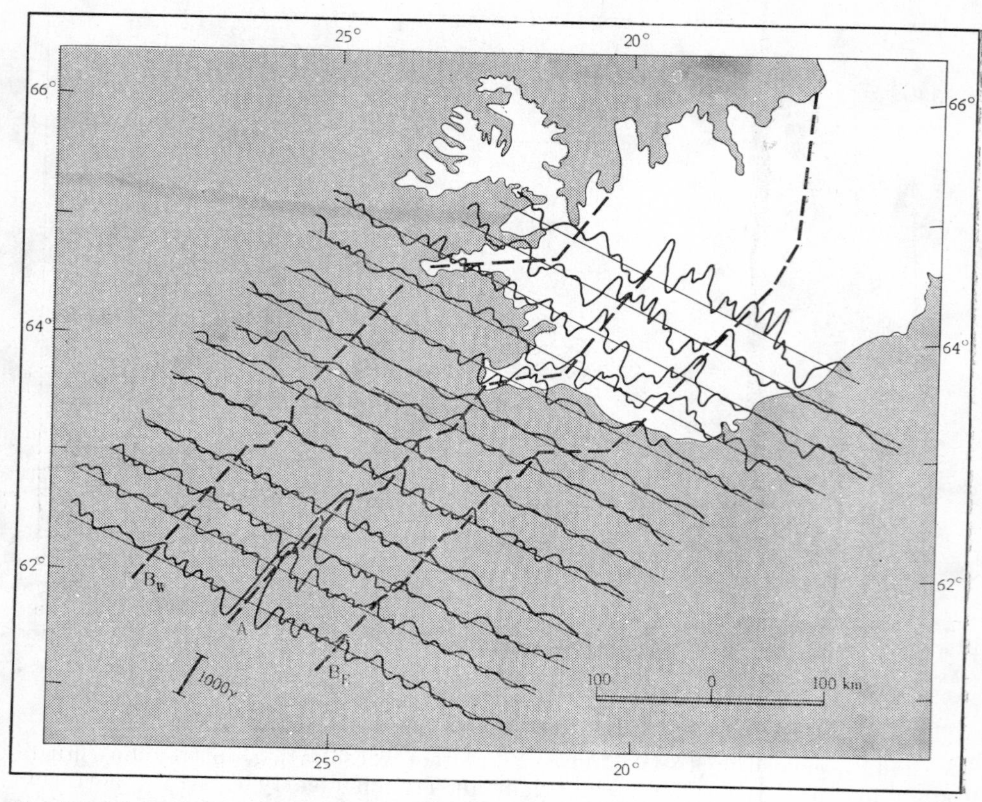

Fig. 129. Continuation of the magnetic anomalies of the Mid-Atlantic Ridge on the land in Iceland (after Einarsson, 1967). The dashed lines represent the most intensive anomalies. According to the scale of "ocean-floor spreading" anomalies B_W and B_E should be eight million years old; on land, however, they continue along a belt of much younger (Holocene and Pleistocene) volcanoes

When the submarine magnetic anomalies of the Mid-Atlantic Ridge are traced on land (in Iceland), incidentally, the mistaken character of the determinations of their age is immediately revealed (see Fig. 129). Apart from the axial anomaly of the Reykjanes Ridge, two other major anomalies are traceable on land at a distance of 80 km from the ridge's axis. One of them continues in the central graben of Iceland, the other on the Snaefellsnes Peninsula. According to the "fossilised geomagnetic time scale" the age of these two anomalies is eight million years. On land the same anomalies correspond in fact to manifestations of Pleistocene and Holocene volcanism of a considerably younger age.

Magnetic measurements by instruments towed near the bottom have indicated that the bands apparent from the surface of the ocean consist in fact of many isolated ovals that may be joined together in a very different way. The regular occurrence of bands of anomalies may therefore be doubted. The symmetry of the magnetic anomalies along both slopes of a ridge away from its axis, as required by the conception, is also not always observed.

It is hardly necessary to dwell on the other objections. What has been said is sufficient, it seems to me, to take a negative view of the whole system of conceptions of the "new global tectonics". This system is a premature generalization of still very inadequate data on the structure of the ocean floor, based on a schematizing that leads not to simplification of the phenomena but to their distortion.

It is therefore quite understandable that attempts to employ this conception to explain concrete structural situations on a local rather than a global scale lead to increasingly complicated schemes in which it is suggested that local axes of spreading develop here and there, that they shift their position, die out, and reappear, that the rate of spreading alters repeatedly and often ceases altogether, and that lithospheric plates are broken up into an even greater number of secondary and tertiary plates. All these schemes are characterised by a complete absence of logic, and of patterns of any kind. The impression is given that certain rules of the game have been invented, and that the aim is to fit reality into these rules somehow or other.

But the ideas of the "new global tectonics" have become widely accepted in our day.

How has this come about?

The approach of the "new global tectonics" to problems of the Earth's structure and evolution, free of outmoded prejudices, seemingly should not mean that everything built up by the previous research of "continental" geologists ought to be ignored. The new workers coming into the Earth sciences from other fields, however, having created their own geology of the oceans and being unable to demonstrate its correctness, would like to see continental geology altered accordingly.

It is significant that "continental" geologists themselves are actively involved in this appraising of the values and discrediting of the results that generations of geologists have considered their greatest achievements. The question arises which is more important here—the well-founded conviction that such a "purge" of old concepts and conceptions is really necessary, or the hypnotic effect of the "newcomers'" mathematical symbols and ruthless machine logic based on the simple contrast of "yes" or "no". This writer considers the answer to be clear from the foregoing critical survey, but it must be said, unfortunately, that geologists trained in geological survey in continental conditions have proved poorly prepared for the attack of the new forces and unable to stand up for their views (which they had long considered indisputable) with proper conviction.

This whole situation, characteristic of the times we are living in, is often called a "revolution in the Earth sciences". If by "revolution" is meant the drawing into our science of a vast body of new data throwing light onto the structure of the Earth's interior and of the ocean floor, and the combining of the various Earth sciences to tackle common problems and to employ new, above all quantitative, techniques to solve them, one can agree with such a definition of the current position in geotectonics. But one can hardly call the persistent introduction of the ideas of the "new global tectonics" in their concrete present-day expression a revolution.

One must agree, of course, that the new conception's very direct logic has a certain, at first glance, fascinating beauty. One must also recognise that the "new global tectonics" has drawn our attention to a set of new problems, the study of which will continue irrespective of what theoretical notions prevail. The claims of this new system of views to generality are also impressive, a generality that has long been lacking in geological theory since the collapse of the contraction hypothesis.

CHAPTER 26

The Search for a Synthesis

What generalization might be proposed in place of the "new global tectonics", which does not, it seems to us, to stand up to criticism? We must admit that we have no ready-made generalizing geonomic theory at our disposal, and unlike the supporters of the "new global tectonics" we do not claim to be able to explain everything that is happening inside the Earth. We are convinced, moreover, that it is simply impossible now either to put forward an exhaustive theory or even to advance one that embraces endogenous processes at all broadly. The reason lies in the immense gaps in our knowledge of the Earth's interior. Several of the most important aspects of its structure and of the processes taking place there are quite unknown to us.

The position, moreover, has not improved in recent years but has worsened, strangely enough just because of the enormous progress that has been made in study of the structure both of deep-lying layers and of the ocean floor. We know so much that is new that we can no longer be satisfied by simple global schemes, either those of general contraction or pulsation of the Earth or those of the "new global tectonics". The advances in the study of the interior have revealed the extraordinary complexity of the deep processes and the great non-uniformity of the deep structures. In addition they have surprised us by the global generality of these processes. Separate sectors of the tectonosphere develop differently, yet the individual development of the separate zones is at the same time controlled by a rhythm common to the whole planet. We do not yet have the knowledge to resolve this contradiction between the individual and the general.

For all that, however, we can try, on the basis of what we already know, to explain the most probable links between the various endogenous processes, and so to outline a hypothetical deep mechanism giving rise to them and determining the main patterns of their development. Such a hypothetical scheme, even if it did not explain everything, could at least help us choose the direction for further research.

A General Scheme of Deep Processes

In summing up what we said in earlier chapters about the principal features of the structure of continents and oceans, we can, it seems to us, formulate the following propositions about the causes and conditions of the tectonosphere's evolution.

Underlying the Earth's evolution is the differentiation of matter. The initial material is to be found in the lower mantle. From it heavy components sink, forming the core, and light ones rise; and ultimately the upper mantle and crust are formed by them.

In the early stages of the Earth's evolution the distribution of matter and the thermodynamic conditions in the interior favoured the formation of a layer of partial melting, the asthenosphere, at a certain depth in the upper mantle. Partial melting and the differentiation of basalt were an important stage in the evolution of the material on its way to the surface. The composition of oceanic crust indicates that it may have been formed as a result of direct fusion of material from the asthenosphere. The formation of continental crust is linked with a more complicated physicochemical process that involves both the formation of compounds of various degrees of mobility and acid leaching.

Since the density of the upper layers of the asthenosphere is lower than the mean density of the lithosphere, density inversion develops between them, with its corresponding mechanical instability. The degree of this instability depends on the temperature in the asthenosphere and the scale of partial melting, which leads, together with the lowering of mean density, to a reduction in the viscosity of the asthenosphere's material.

The asthenosphere is the level at which matter rising from the lower mantle stops. The upward floating of this material is the main means of transporting heat from hotter deep regions to the upper mantle. Wherever an ordinary batch of deep matter rises into the asthenosphere, the latter's temperature is raised. That, too, is the mechanism of "excitation" of the asthenosphere, as a result of which melting is intensified within it, its viscosity is lowered, and its density declines, which intensifies the density inversion between it and the lithosphere and the mechanical instability at their interface, and creates a situation for asthenospheric diapirism.

The pulsed character of endogenous processes directly connected with the upper mantle has its own cause in the rise of matter from the lower mantle occurring discontinuously, in definite portions, rather than continuously. After excitation, quiescence sets in as a result of cooling of the asthenosphere. Each regular excitation creates the conditions for an intensification of asthenospheric diapirism and consequently for a raising of the rate, amplitude, and contrast of block-undulatory oscillatory movements, and also for an intensifying of the evolution of mantle magma. During cooling all these processes become less intense.

Thermal excitation in the crust, however, which is expressed in geosynclines in the form of regional metamorphism and granitization, occurs with a time lag in relation to a disturbance in the *asthenosphere* and coincides with the time when the latter has already begun to quieten down. Regional metamorphism, granitisation, and the strongest plastic deformations therefore occur in the crust on a background of a weakening of the contrasts of block-undulatory oscillations and an attenuation of mantle magmatism. The periodic excitation and damping of the asthenosphere is also responsible for the rhythm of general oscillations both in geosynclines and orogenic zones and on platforms.

Concrete endogenous regimes, however, are not only governed by the state of the asthenosphere, but are also controlled by its interaction with the lithosphere, and the regime that is established depends on the lithosphere's properties and reaction to effects from the asthenosphere. The conditions for the origination of regimes will be discussed below.

The transition from a permobile stage to an unstable protogeosynclinal stage, and then to a stable geosynclinal-platform one, means that excitation of the asthenosphere was initially distributed almost uniformly over the whole area of continents and then passed into a process of greater and greater spatial localisation, at first unsteadily and then stably. The reduction of areas occupied by active regimes, i.e., geosynclinal and orogenic and the increase in areas of quiet, platform regimes, is linked with a progressing localisation of excitation processes in the asthenosphere.

On this background of increasing localisation of zones of excitation of the asthenosphere, phenomena of tectonic and magmatic "activization" of platform regimes occur as local and temporary deviations from the general trend of the lithosphere's evolution.

The role of deep (slash) faults in the structure of the crust increases in the course of geological history.

We can only understand the origin and evolution of the *lithosphere of oceans and deep seas* in the light of the data now available by assuming the existence of the forming of oceanic lithosphere on the site of a previously existing continental lithosphere, and that without drift and moving apart of the latter. It is a matter of the process that has been called "basification" or "oceanisation" of the continental lithosphere. In any case it is impossible, without the idea of such a process, to explain the formation of such seas as the Mediterranean, the Gulf of Mexico, and the marginal seas of the Western Pacific. In all these seas a preceding continental stage is quite obvious. In rejecting the idea of continental drift, however, we must inevitably assume this process of basification of the lithosphere also in the oceans, since there are signs in them, too, of a past existence of sectors of land in places.

As the Early Mesozoic and all the preceding history of the oceans, however, is unknown, we cannot say when and how far these processes of basification occurred on the sites of today's oceans, whether they were global, or whether the division of the lithosphere into continental and oceanic represents some "original" or very ancient non-uniformity in the composition and development of the Earth's deep interior.

During the Mesocenozoic, however, there was a deepening of the oceans, a deepening that is impossible, while maintaining isostasy, without an increase in the weight of the lithosphere through its basification. One can link the main act of the basification and formation of oceanic basins with the vast outburst of volcanism at the end of the Palaeozoic and beginning of the Mesozoic, which led to the formation of the third and second layers of oceanic crust. Since volcanism initially embraced the whole area of the oceans and was subsequently concentrated in an ever narrower belt along the crust of median ridges, it is natural to suppose that these ridges are zones of the most protracted processes of basification, which may be continuing on them to this day.

Conditions for the Development of Continental Endogenous Regimes

The regimes linked with an excited state of the asthenosphere are the geosynclinal, orogenic, and rift regimes. Underlying all of them is asthenospheric diapirism caused by intensification of the density inversion at the interface of the asthenosphere and lithosphere. In the upper part of the asthenosphere there is a build-up of the basalt fused from it and of separate massifs of the initial ultrabasic rocks remaining in it. This material, namely basalt plus inclusions of peridotites, builds up into protuberances, or diapirs, that arise on the surface of the asthenosphere. In addition to diapirs, which retain their link with the asthenosphere, "rootless asthenoliths" are also formed that are completely separated from the asthenosphere and rise separately.

Inasmuch as the diapirs are formed beneath the lithosphere, their shape and location very much depend on the latter's mechanical properties and capacity to move up and down, subject to movements of the top of the asthenosphere. Where the lithosphere is divided into long blocks by parallel deep faults, the diapirs take on the shape of these same parallel ridges separated by troughs in the roof of the asthenosphere. Asthenoliths also acquire a lengthened form.

When the asthenosphere is in a strongly excited state, in which the mechanical instability of its roof is intensified, a situation is created that promotes considerable amplitudes, rates, and contrast of vertical movements in its material. These are the conditions in which *eugeosynclinal and orogenic regimes* develop.

What exactly results from them depends on the character of the lithosphere's penetrability. If it is dispersed or diffused the basalt rising in diapirs and asthenoliths is injected into it and penetrates it along a dense network of channels, and sometimes pours out onto the surface. Ultrabasic magma entrained in the flow of basalt is also injected into the lithosphere with the latter and, on reaching the crust, is serpentinised and forms layered protrusions.

The magma solidifies on penetrating the lithosphere and the latter is basified as the volume of solidified basic and ultrabasic magma increases, and becomes heavier and heavier. The degree of basification varies in different places and is highest above spots where diapirs and asthenoliths have risen from the roof of the asthenosphere. It is at these spots therefore that bands of the lithosphere become heaviest; zones of subsidence are formed there on the surface, while zones of a relatively elevated position of the lithosphere are formed above the areas separating diapirs. Because of this mechanism the lithosphere's vertical movements are a mirror image of the movement of material in the roof of the asthenosphere, and that is why subsidence prevails on the whole over uplift.

A eugeosynclinal regime thus develops on the surface in its ophiolitic stage, which is characterised by contrasted block-undulatory oscillatory movements, with a general preponderance of subsidence and vigorous disseminated effusive and intrusive basic and ultrabasic magmatism, which is especially intensive in zones of subsidence, i.e., in intrageosynclines.

Basification, expressed in the formation of numerous layered and cross-cutting basic and ultrabasic intrusions in the crust, leads to alteration of the crust's properties, bringing them close to those of oceanic crust. It is only in this purely physical, and not in the least geological, sense that eugeosynclinal crust can be compared with oceanic.

Our ideas on the ophiolitic stage of a eugeosynclinal regime are illustrated in Figure 130, *I* and *VI*.

The transition from the ophiolitic stage to the inversion stage coincides with an epoch of cooling of the asthenosphere and weakening of asthenospheric diapirism. The contrast of vertical movements in its roof diminishes and some of the previously molten basalt crystallizes within the asthenosphere and (according to Lutz's scheme outlined above) leads to the liberation and rise of solutions that penetrate the crust after a number of physicochemical reactions, transporting heat to it and operating as the cause of regional metamorphism and granitisation.

Recrystallisation of the crust's material leads to a closing of the channels previously existing in it and to the formation of a crystalline shell. The character of the crust's penetrability is strongly altered; from being diffused it becomes concentrated and confined to separate faults only, predominantly located on the boundaries between zones of uplift and subsidence. At the same time diffuse permeability is eliminated throughout the lithosphere and that, because of the suppression of through circulation of magma in it, leads to a rapid closing of the previously existing channels, thereby promoting a reduction in the volume of basalt fused from the asthenosphere.

The change in the character of the lithosphere's permeability alters the relationship between its movements and those of the roof of the asthenosphere. Since mantle magma does not penetrate the lithosphere, there are no longer the conditions for a mirror relation between these movements. The relationship becomes direct; the lithosphere rises above asthenospheric diapirs, and sinks in the areas between them.

Such is the mechanism of *partial inversion*. Whereas the penetration of large amounts of mantle magma into the lithosphere had caused it to subside, now that this is stopped the lithosphere rises above asthenospheric diapirs, i.e., central uplifts are formed in intrageosynclines. At the same time the conditions for a predominance of subsidence over uplifts disappear, and the two are temporarily balanced.

Regional metamorphism and granitization are linked with the formation of "deep diapirs" in the crust. The general heating of the crust by solutions rising from the mantle leads to a lowering of the viscosity of the rocks, which is a factor facilitating the development of strong folded deformations in the crust.

The inversion stage of the eugeosynclinal regime is depicted in Figure 130 (*II*, *III*, and *VII*).

Having surveyed the conditions for the development of the inversion stage of the eugeosynclinal regime, we have thereby also shed some light on the set-up of the origin of an *orogenic regime*. It also originates in conditions of concentrated permeability of the lithosphere. But in contrast to the conditions of the inversion stage, strong excitation of the asthenosphere is needed for an orogenic regime, as for the ophiolitic stage of the eugeosynclinal re-

Fig. 130. Hypothetical deep profile of the lithosphere for various endogeneous regimes and their stages of development (after Beloussov): ►

1—continental crust; *2*—ophiolitic formation; *3*—lower terrigenous formation; *4*—limestone formation; *5*—upper terrigenous formation; *6*—molasse formation; *7*—platform sedimentary cover; *8*—molten basalt; *9*—asthenosphere (the density of hatching corresponds to the rate of differentiation of basalt); *10*—substratum (uppermost solid layers of the upper mantle); *11*—Golitsyn layer; *12*—region of an increase in the volume of crust on a platform; *13*—eclogites; *14*—metamorphism of the granulitic facies; *15*—metamorphism of the amphibolite facies; *16*—granitization; *17*—magma of acid and intermediate composition; *18*—deep mantle asthenolith; *19*—residual foci of magma of acid and intermediate composition; *20*—residual foci of magma of basic composition; *21*—volcano; *22*—tectonic fault; *23*—magma of alkaline composition; *24*—deep diapirism in the crust

I—platform (Pl), parageosyncline (PGS), eugeosyncline in the ophiolitic stage (EGS), miogeosyncline (MGS); *II*—inversion stage (MT—marginal trough; CU—central uplift; IT—intramontane trough); *III*—continuation of the development of the inversion stage (build up of contrasts of vertical movements); *IV*—orogenic regime (Fd—foredeep); *V*—orogenic regime (continuation) and intensification of downwarping on the site of the foredeep (right); *VI*—orogenic regime and the origination of a eugeosynclinal regime on the site of a former foredeep (cf. *IV* and *V*); *VII*—build up of contrasts of vertical movements in zones of an orogenic regime and the inversion stage in a eugeosyncline (cf. *VI*); *VIII*—orogenic regime in all zones; *IX*—platform regime; *X*—(*a*) rift regime, and (*b*) epiplatformal orogenic regime

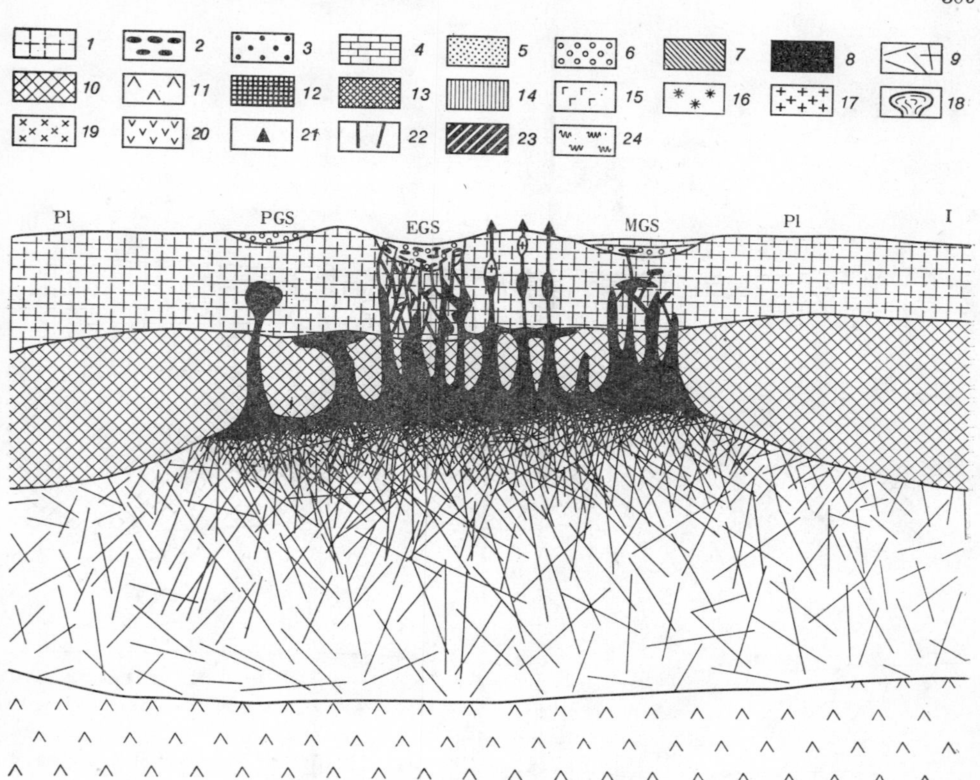
1
2
3
4
5
6
7
8
9
10
11
12
13
14
15
16
17
18
19
20
21
22
23
24
Pl
PGS
EGS
MGS
Pl
I

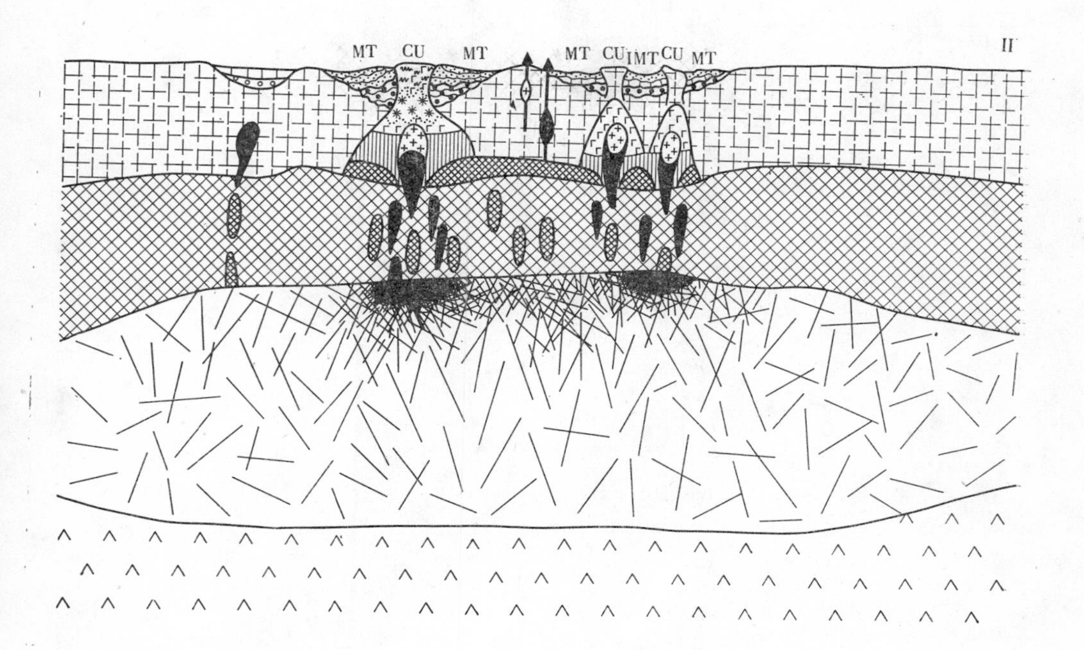
MT
CU
MT
MT
CU
IMT
CU
MT
II

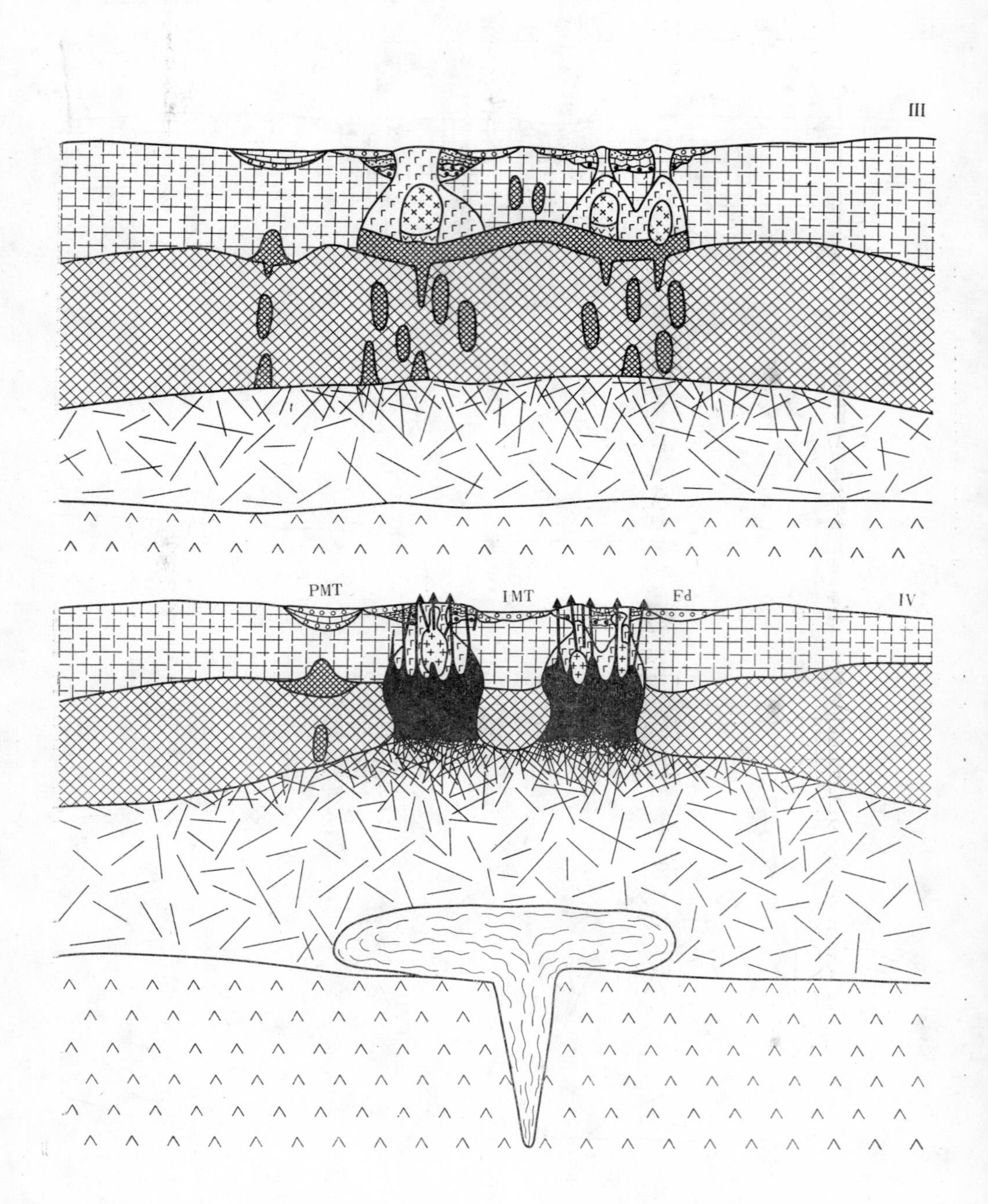
III
PMT
IMT
Fd
IV

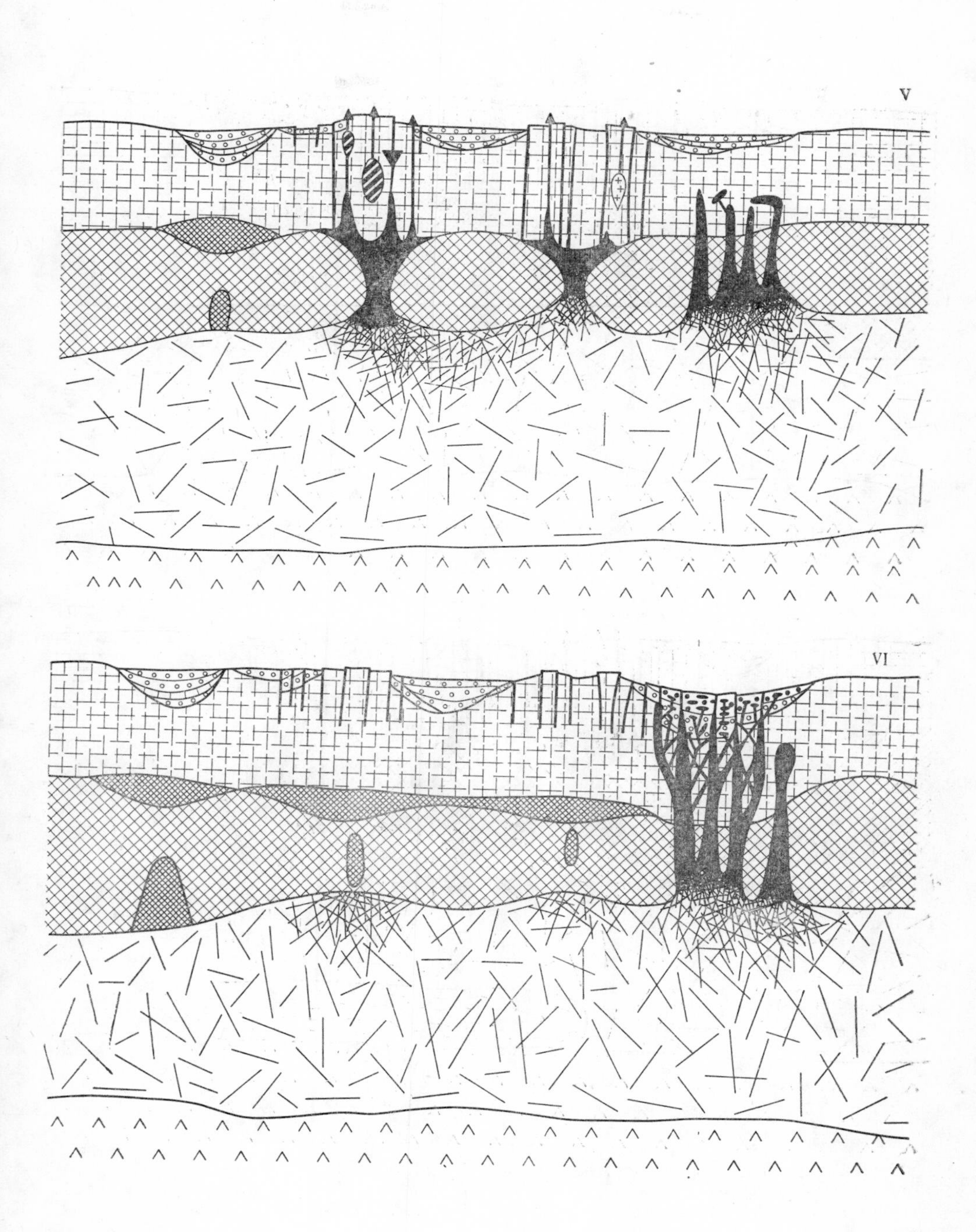
V
VI

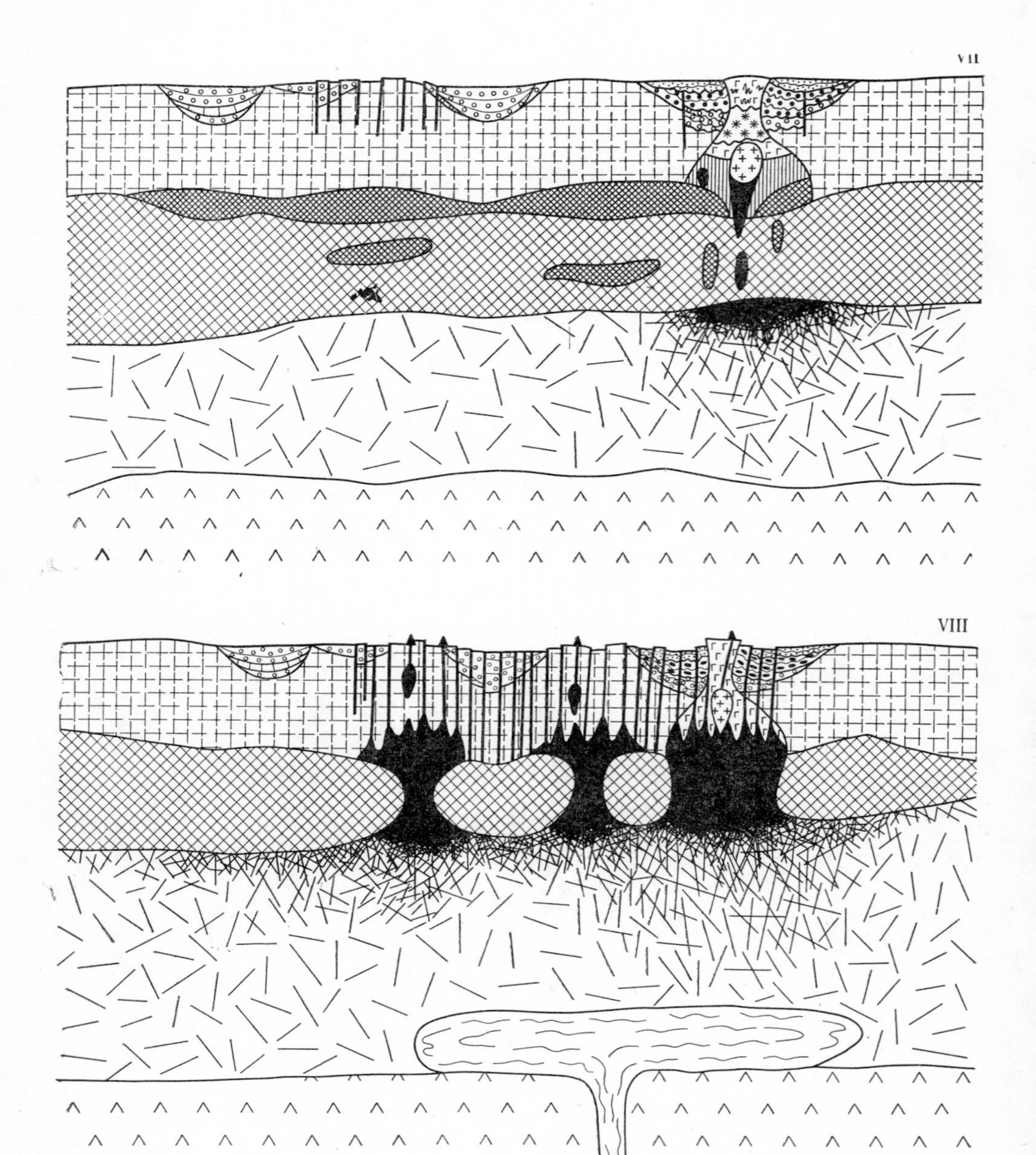
VII
VIII

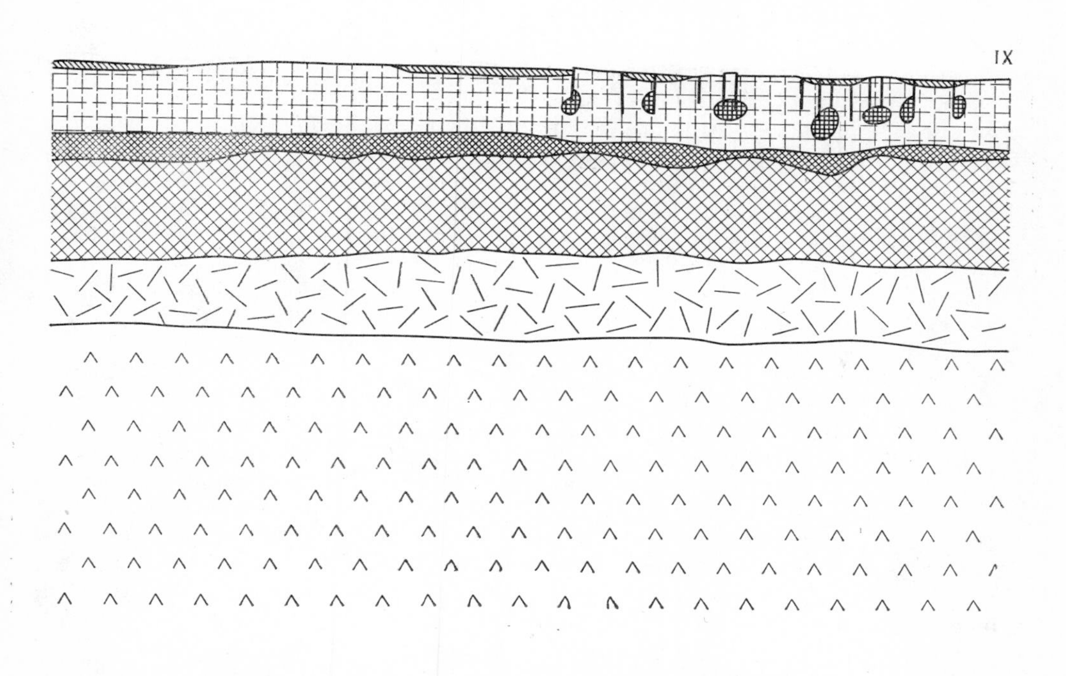
IX

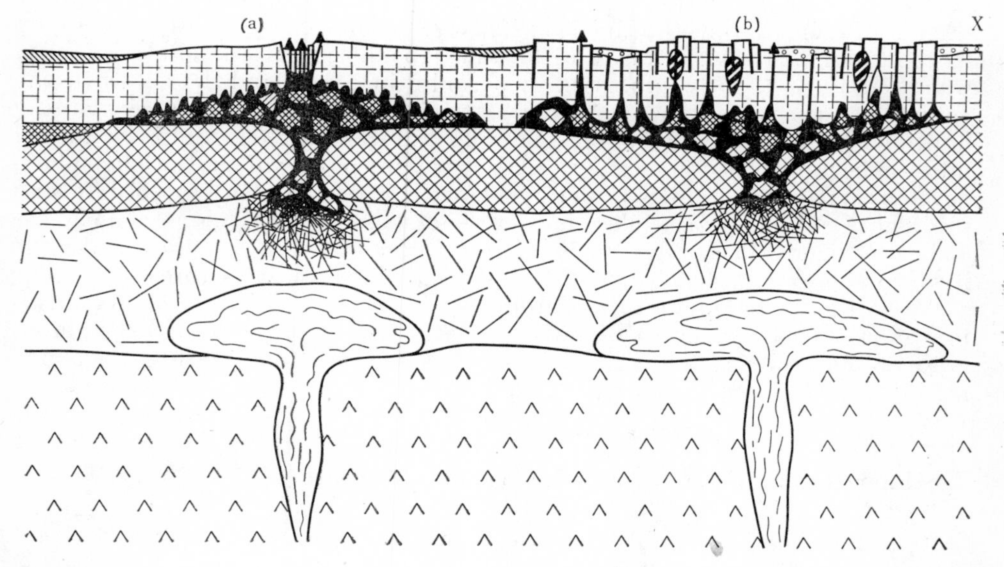
(a)
(b)
X

gime. The high diapirs forming on the surface of the asthenosphere encounter a lithosphere broken by deep faults into long blocks that are already crystallised and so lacking diffuse permeability, and push them upward, while blocks of the lithosphere sink into the spaces between the diapirs. This is again a direct relation between the movements of the lithosphere and asthenosphere.

As there are no conditions here for the lithosphere to become heavier, and the considerable heating of the asthenosphere and intensifying of partial melting in it lead to a general increase in its volume, uplifts predominate on the whole over subsidence (Figure 130, *IV*, *V*, *VIII*, *Xb*).

If the excited asthenosphere encounters a very strong and impenetrable lithosphere, such as exists in platforms, especially ancient ones, an external obstacle is created to the formation of many separate diapirs above the site of its intensified melting, since the crust is incapable of fractional vertical movements. In that case the basalt fused from the asthenosphere forms a broad, gently sloping accumulation above the site of excitation of the asthenosphere, and the lithosphere bends into a gently sloping arch, in which it is stretched as a result of the upwarping. The tension is increased by the effect of sideways gravitational flow of the protrusion of the asthenosphere beneath the lithosphere. The tension leads to cracking of the lithosphere either along the axis of the arching or along previously weakened zones.

Because very high pressure is required to overcome the great resistance of solid crust, the pressure creates special conditions for the fusing of basaltic magma in the asthenosphere; already, given high gas pressure at a not very great depth, alkaline basalts are fused rather than tholeiitic ones. This same high pressure causes the magma to attack the floor of the crust very aggressively, and a layer of a mixture of material from the mantle and fragments of crust is formed there.

All that leads to a *rift regime* at the surface (Fig. 130, *Xa*).

Platform regimes correspond to a quiescent state of the asthenosphere (Fig. 130, *IX*). Its pulses of heating, followed by epochs of cooling, show up in them only as alternating uplifts and subsidences of the platform as a whole, and are directly expressed in periodic increases and decreases in the volume of the asthenosphere. An increase in its volume and a corresponding uplift at the surface occurs at the end of an endogenous cycle and is sustained into the beginning of the next cycle, after which cooling of the asthenosphere develops until the middle of the cycle, with a decrease in its volume and subsidence at the surface. The density inversion between the asthenosphere and lithosphere does not play a marked dynamic role on platforms, the inversion being minimal, since the asthenosphere is weakly expressed.

We will not allude here to the problems of the inner division of platforms into subgeosynclines and subgeanticlines because our purpose at the moment is to try and outline the conditions for the origin of the main endogenous regimes in very general form.

The Interrelation of Continental Endogenous Regimes

It is customary to think that the ophiolitic stage of a eugeosynclinal regime belongs to the beginning of an endogenous cycle, while an orogenic regime is manifested at the end of a cycle. The latter regime, as we said

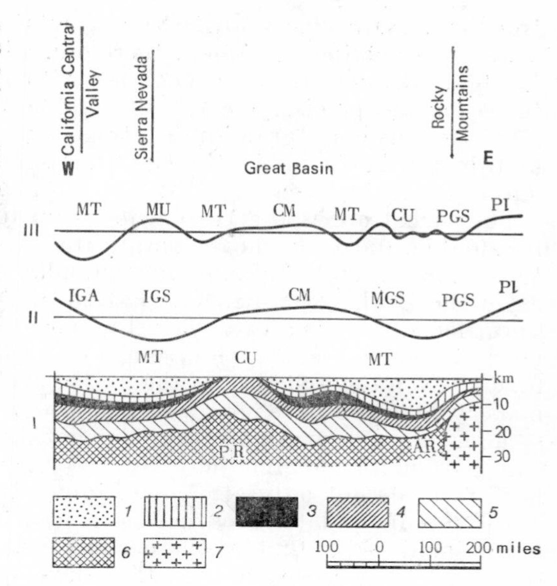

Fig. 131. Schematic diagram of the development of oscillatory movements in the Cordilleras of the USA in the Hercynian and Mesozoic cycles (after Beloussov):

I—palaeotectonic schematic profile through the Cordilleras for the end of the Permian period (end of the Hercynian cycle); *II*—troughs and uplifts at the beginning of the Mesozoic cycle; *III*—troughs and uplifts at the end of the Mesozoic cycle; *1*—Permian; *2*—Carboniferous; *3*—Devonian; *4*—Silurian; *5*—Cambrian; *6*—Proterozoic; *7*—Archaean; IGS—intrageosyncline; IGA—intrageanticline; CU—central uplift; MT—marginal trough; CM—central massif; MGS—miogeosyncline; PGS—parageosyncline; Pl—platform; MU—marginal uplift

above, is an independent state of the interior not necessarily connected with the preceding geosynclinal development. Furthermore, as we have already stressed, the orogenic regime and the ophiolitic stage of a eugeosynclinal regime do not develop at different times but simultaneously, only in different places.

The ophiolitic stage that belongs to the beginning of the Caledonian cycle usually developed in fact during the Cambrian period when the orogenic regime that completed the preceding Baikalian cycle was still continuing in other places. From this angle the Late Silurian and Early Devonian were times both of the ending of the Caledonian cycle and beginning of the Hercynian. Finally, the eugeosynclinal regime in many geosynclines of the Alpine cycle began in the Triassic, while the Hercynian was being completed by orogeny in other regions (see Fig. 106).

This mutual interpenetration of cycles is in fact only apparent, and the idea a simple conventionality. The states in the asthenosphere necessary for the ophiolitic stage of a eugeosynclinal regime and for an orogenic regime are in fact identical; both require its excitation and vigorous heating. This state is created with each pulse of excitation and is confined each time to the boundary between two cycles. If a pulse leads to a eugeosynclinal regime in one place and to orogeny in another, that depends rather on the state of the lithosphere and the character of its permeability. A rift regime, too, arises in similar deep conditions, but also requires a special state of the lithosphere.

From this standpoint it is quite natural for a eugeosynclinal regime to be followed by orogeny with the next pulse of excitation of the asthenosphere, since diffuse permeability of the crust and the whole lithosphere has given way to concentrated penetrability in the inversion stage of the eugeosynclinal regime as a result of regional metamorphism and recrystallization of the rocks.

A crystalline shell is not, however, formed everywhere in geosynclines but predominantly in central uplifts, while marginal and intramontane troughs remain uncrystallized and retain their diffused permeability. If a new geosyncline is formed, therefore, where one of a preceding cycle was located, the new intrageosynclines with eugeosynclinal regimes will have to be located precisely in the site of these troughs. Intramontane and marginal troughs may "grow" into intrageosynclines in the next cycle. Examples of this can be found in the Cordilleras of North America, the Tien Shan, and other regions (see Fig. 131).

For such troughs the two cycles seem to merge into one. The conditions for double and even more complicated cycles of geosynclinal development to originate thereby become understandable: for example, a Palaeozoic cycle as in the Urals, or a Riphean-Palaeozoic-Mesozoic one, as in the Canadian Cordilleras. This is a case when the mechanism giving rise to regional metamorphism and granitisation somehow or other does not "work" during the first, or even the first two, of the united cycles. Then, when the next endogenous cycle begins once more with a pulse of strong excitation of the asthenosphere, the geosyncline's development becomes "compounded".

A geosynclinal regime may also be renewed where a crystalline shell has been formed, but only after new, intensive breaking up of the lithosphere. The trend of the later geosyncline, and of its separate inner zones, may then not coincide with that of the earlier one. An example of nonconformity of the trend of two geosynclines that developed one after the other was cited above (see Fig. 103).

As for the possibility of a geosynclinal or orogenic regime's originating in sectors that have been in a platform condition for some time, the "regeneration" of active regimes is the less probable the longer the platform regime has existed and the better deep faults have been "healed" in connection with long immobility. "Regeneration" is more probable, therefore, on young platforms than on ancient ones, and an orogenic regime with its concentrated

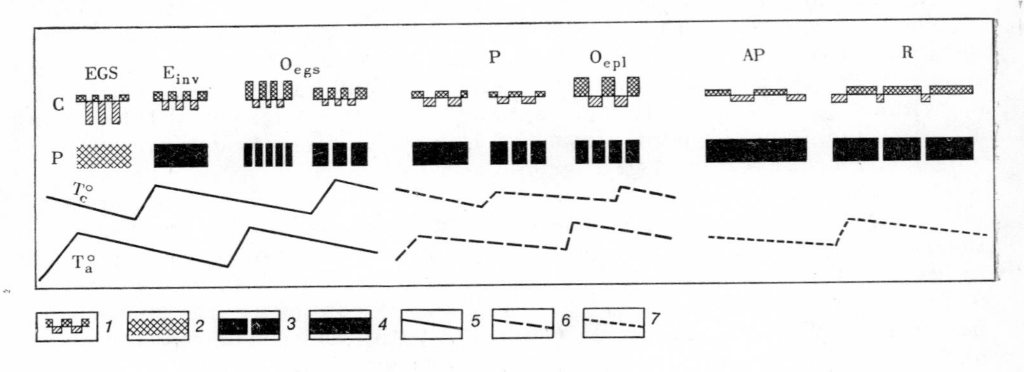

Fig. 132. Basic scheme of deep processes (after Beloussov):

C—contrast and predominance of rising and sinking block-undulatory movements; *P*—penetrability of the lithosphere; $T^0{}_c$—relative temperature change in the crust; T^0_a—relative temperature change in the asthenosphere;

1—ratio of rising and sinking block-undulatory oscillatory movements; *2*—diffuse permeability of the lithosphere; *3*—concentrated penetrability of the lithosphere; *4*—impenetrable lithosphere; *5*—curves of the rise and fall of temperature in the asthenosphere and crust for the ophiolitic and inversion stages of a eugeosynclinal regime and an epigeosynclinal orogenic regime; *6*—curve of the rise and fall of temperature in the asthenosphere and crust for the regimes of young platforms and epiplatformal orogeny; *7*—curve of t e rise and fall of temperature in the asthenosphere for the regimes of ancient platforms and rifts.

Regimes: EGS—eugeosyncline (ophiolite stage); E_{inv}—inversion stage of a eugeosynclinal regime; O_{egs}—epigeosynclinal orogenic regime; YP—young platform; O_{epl}—epiplatformal orogenic regime AP—ancient platform; R—rift regime penetrability is therefore more probable than a geosynclinal one, for which much more intensive fracturing of the lithosphere is needed.

It is the orogenic regime that is, in fact, the most common form of epiplatformal activization, which is mainly manifested on young platforms. On ancient platforms, and on those sectors of young ones that have proved to be strongest, excitation of the asthenosphere generally leads to a rift regime.

A basic scheme of the deep processes that lead, in combination with the lithosphere's reactions, to one endogenous regime or another on the surface is depicted in Figure 132.

Conditions for Differentiation of the Tectonosphere into Continental and Oceanic

The higher temperature of the oceanic tectonosphere compared with the continental may be the consequence of a lower differentiation of the material, since the degree of differentiation determines the greater or lesser dispersion of radioactive elements with depth. This higher temperature, however, may be regarded as the cause rather than the consequence of low differentiation of the material. If we take Lutz's view we can consider it necessary, that there should be partial cooling of basalts at a certain stage in the asthenosphere so that granitising and metamorphising solutions will be differentiated from it, i.e., for the asthenosphere's strongly excited state to give way to a quieter one; the high temperature can make such a cooling impossible, and we can consider that crust of continental type cannot emanate from the upper oceanic mantle.

In considering this matter in its general aspect we may note that the material's capacity to differentiate to one degree or another depends on temperature; with a given composition and pressure there will always be a temperature that is most favourable for maximum fractional differentiation of its components by density in a gravitational field. This temperature will mainly be that at which a multiphase system is created comprising solid, liquid, and gaseous components. In a gravitational field such a system readily differentiates into phases of various density. A lower temperature hampers differentiation by reason of the solidifying of matter and rise in its viscosity, which makes relative migration of particles either too slow or quite impossible. Too high a temperature, however, at which all the components prove to be equally molten and mixed, is also unsuitable. Within this interval there are intermediate cases for which, at each temperature, there is equilibrium of a quite definite degree of differentiation.

The following point arises from these arguments: could the continental tectonosphere be transformed into oceanic as a result of a rise in its temperature? And could continental crust thereby be transformed into oceanic as a partial manifestation of this same change in the whole tectonosphere?

Let us imagine the following chain of events.

As the result of a specially high rise in temperature, a rather larger volume of the material of the upper mantle than usual is involved in partial melting. An especially large amount of basalt is thereby formed, which, seeping upward, accumulates in the upper layers of the asthenosphere, forming an emulsion with a liquid content up to 50% rather than the normal 10% or

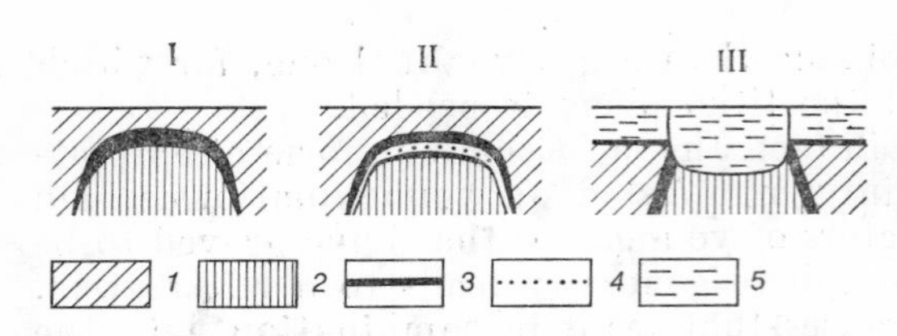

Fig. 133. The formation of a collapse caldera (after Richey, 1948):

I and *II*—underground collapse; *III*—surface caldera; *1*—country rock; *2*—sinking block; *3* and *4*—magma rising from a deep-lying magma chamber; *5*—lava erupted to the surface

15%. To some extent, too, moreover, there is full melting of the ultrabasic material of the mantle.

All that would (1) greatly intensify the density inversion between the asthenosphere and the lithosphere, and (2) lead to a density lower than that of the lithosphere becoming distributed to a depth of several hundred kilometres in the upper mantle. The mean density of the lithosphere (i.e., of the crust and substratum together) is probably close to 3.0 g/cm^3. It may even be a little higher if it is supposed that the increased heating has caused metamorphism not only of the granulite facies in lower layers of the crust but also of the eclogite facies, with complete loss of water from the minerals. The molten matter of the upper mantle, however, would have a density of 2.7 to 2.8 g/cm^3. These are the figures for a pressure of one atmosphere. The point here, however, is not the absolute value of the densities but the difference between them, which should be maintained to a considerable depth irrespective of the pressure.

In these conditions, the lithosphere, if it is broken up by faults, should consequently sink very deep into the asthenosphere in separate blocks, while molten matter from the asthenosphere would rise to the surface. The penetration of basaltic mantle melts into the lithosphere, including the crust, by horizontal channels as well as by vertical ones should thereby be very important, and as a result blocks of lithosphere completely surrounded by magma would be isolated and sink readily into the asthenosphere.

This process is indistinguishable on the whole from that of "roof foundering", described by Daly or from the mechanism of the formation of the ring intrusions of Scotland and the Northern Appalachians (see Fig. 133).

The blocks of the substratum and crust that have sunk into the asthenosphere might remain solid for a certain time in the material of the latter, forming a "crust-mantle mix" with it. These same blocks might undergo melting, wholly or partially, and the product of this melting might be ejected to the surface in the form of a magma of predominantly andesite or basalt-andesite composition. Finally, the blocks might become completely liquefied in the asthenosphere and so mixed with its material as to wholly lose their compositional and structural features. In that case their material would become part of the asthenosphere and be more or less uniformly distributed through it. This type of evolution would have to be considered most common in the area of the oceans.

This is undoubtedly the most difficult and least clear part of the whole conception. We can so far only appeal to the general principle mentioned above that for every temperature there must be a corresponding degree of differentiation of the upper mantle's material in the Earth's gravitational field. At a certain temperature, significant differentiation would occur and continental crust be formed at the surface. When the temperature in the mantle rose steeply, the previous distribution of matter would become unbalanced and a new equilibrium would have to be achieved through a certain homogenization of the tectonosphere, and crust again dissolved into the mantle.

This process of the dissolving of continental crust continues at present in mid-ocean ridges. In that connection it is there that we should look for remnants of continental lithosphere preserved in the oceans.

The American geologist A. Meyerhoff has pointed out that a considerable area of the Mid-Atlantic Ridge at 45° NL is taken up by rocks of the continental crust like gneiss, granite gneiss, granite, etc. Their age, by radioactive dating, is 1550 to 1690 million years, i.e., Proterozoic. The view has been expressed that these are boulders carried there by icebergs, but the accumulation of them is so large, and the individual blocks are so big (as underwater photographs have shown), not to mention that they are too far south of the latitude of icebergs, as to make this interpretation doubtful. It is more probable that they are the remains of continental crust broken up by processes of basification.

Not far from this point on the crest of the Mid-Atlantic Ridge ultrabasic rocks have been found with a composition of continental Alpine-type ultrabasalts. The view has been expressed *a propos* of them that they are remnants of continental mantle that have remained since the separation of the African and American continents at the place where the original joint between the continents was located, and that they have been preserved because they proved to be in the "immobile zone" between the two diverging convection currents transporting the continents. The artificiality of this explanation speaks for itself.

The crust beneath Iceland has a continental thickness up to 50 km. And although the mean seismic velocities in it are "basaltic", the great volume of acid lavas, which constitute up to 12% of the total in certain areas of Iceland, allows us to suggest that there could be blocks of ancient continental crust in its composition still unassimilated by the mantle.

There should be an active asthenospheric diapir beneath a median ridge. Its top is the lens of material of lower density that has been discovered by gravimetry (see Fig. 122). Sideways flow of the top of the diapir under the force of gravity should lead to stretching of the lithosphere. The graben on the arch of a median ridge is one of the results of this stretching. Another result is the formation of numerous fissures on the slopes of the ridges and running parallel to them. The fissures are filled with dykes of various age, and therefore belonging to different geomagnetic epochs. The various dykes correspondingly have a residual magnetisation of different polarity, which is the cause of the banded magnetic anomalies on the slopes of median ridges. The view that a "fossilised" time scale is concealed in these magnetic anomalies seems to us illusory.

Differentiation and the General Evolution of the Earth

Because of the low heat conductivity of the Earth's material, the heating of its inner regions caused by radioactive decay can lead to the accumulation of heat. If differentiation occurred and relatively light material separated out in deep regions and rose to the outer geosphere, it would carry some of the heat accumulating in the deep interior with it and heat the upper layers of the Earth. It is in this phenomenon that we must seek the cause of the periodic heating ("excitation") of the asthenosphere.

The mechanisms have been suggested for this periodic transport of heated material from the interior to the surface. One of them, described by E.V. Artyushkov (1968), assumes that differentiation begins in the lower mantle, in the shell whose density is closest of all to the Earth's mean density. The periodicity of the upward transport of heated material is explained as due to the fact that great masses of light material must accumulate deep down before they can migrate upward, and that takes time. According to Stokes' law the velocity of movement of less dense matter through denser matter under the force of gravity is directly proportional to the square of the radius of the rising body, given the same difference in density and unaltered viscosity of the medium. When the rising bodies are small in size, the velocity of their movement may consequently be too low, considering the high viscosity of the Earth's material, even on the geological time scale.

The second mechanism, proposed by A.N. Tikhonov and his associates (1969) is based on calculations that indicate that radioactive heating would lead in certain conditions to the generation of a zone of melting in the upper mantle at a depth of several hundred kilometres (probably around 500 km). This molten layer is more efficient, owing to convection, than solid matter of the same composition. The increase in heat conductivity may be as much as one order of magnitude. In that connection heat would be transferred through the molten layer with an adiabatic gradient lower than the geothermal gradient in solid rocks and the gradient of their temperature of melting connected with the change in pressure with depth. As a result heat transfer through the liquid would lead to melting of the solid rocks lying on top of the molten layer and to a moving upward of this layer. At the same time crystalline material would settle to the bottom of it with the result that the whole molten layer would move upward.

Because the most easily fusible material, which is also usually the lightest material, melts out of the rocks of the roof first, the upward movement of the molten layer proves to be a kind of "zone refining" by which not only is the upward transport of heat intensified but the medium through which the layer passes is also differentiated. It entrains the relatively lighter constituents, and as it moves upward it becomes thinner because the temperature of the external medium becomes lower, and dies out altogether at a depth of a score or more of kilometres below the surface. Then after a certain time, heat would be transferred from the interior solely by normal thermal conductivity, i.e., its flow to upper layers would weaken. But heat would accumulate once more in the deeper layers, which would again lead to formation of a molten layer, which in turn would migrate upward, and so on. With thermal parameters of the medium probably close to nature the calculated periodicity of the intensification and attenuation of heat transfer in the upper mantle is close to that of endogenous cycles (see Fig. 134).

The second mechanism is very interesting because it may explain the regular periodicity of the rise of deep-seated material, but the proposed calculations cannot be considered wholly convincing, since they do not take account of the fact that radioactive elements are also carried to the surface in each act of "zonal refining", which would essentially alter the conditions for the next act of melting. It is very probable that the whole mechanism suggested by this conception would very soon grind to a halt because of the migration of radioactive elements. The mechanism suggested by Artyushkov therefore seems more probable, though the periodicity of excitation of the asthenosphere connected with it would not be so regular.

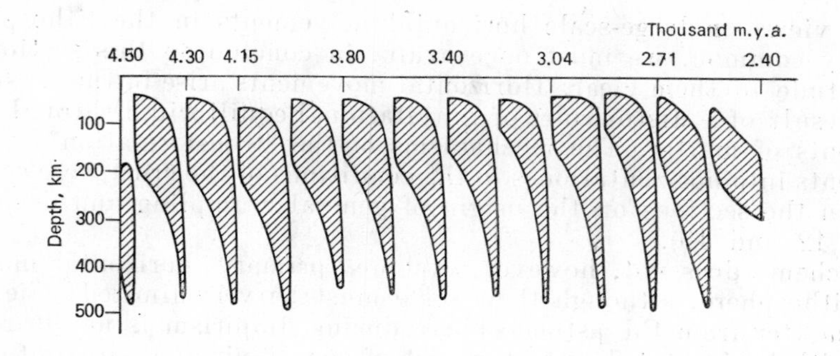

Fig. 134. The process of periodic melting in the Earth's mantle (after Tikhnov *et al.*, 1969):

Time is calculated from the moment of the Earth's formation. Hatched shapes illustrate the history of partial melting of layers of the mantle. The average length of a cycle is 170 million years, with an efficiency of thermal conductivity of the molten layer ten times higher than the thermal conductivity of the solid matter of the mantle

The fusing of light material does not occur everywhere in a continuous medium of high viscosity, but along certain channels within which the material is much less viscous. This idea of channels of lower viscosity is very useful for understanding the localisation of active regimes and the general evolution of the tectonosphere.

One can imagine that there was a dense network of channels in the mantle in the permobile stage such that material could rise almost everywhere to the asthenosphere from the deep interior.

Channels became rarer in the transition stage but were not yet stably localised, and light matter accumulating in certain regions of the lower mantle could continue its upward course first in one place and then in another.

The development of stable ancient platforms and their later expansion that characterises the stable geosynclinal-platform stage, obviously reflects a corresponding stable localisation of channels, the width of whose zones of manifestation become narrower from one cycle to another.

We can hardly attempt to resolve the problem of the causes of the origin and subsequent evolution of channels of reduced viscosity here; our propositions about it would be too indeterminate. An essential step would be made in this direction if the patterns of distribution and history of deep faults on the surface of the Earth, which still remain unknown, could be studied.

We would like, however, to draw attention to one circumstance. A reduction in the number of operative channels of lower viscosity would have to lead in the course of geological time to this, that an ever greater volume of the relatively light material accumulating on the lower mantle through its differentiation would not find an outlet upward. Such a situation could ultimately lead to an "explosion", which would be expressed in the breaking of new channels and extremely strong excitation of the asthenosphere above them through the rapid transport of a large volume of highly heated material to it.

Could such an "explosion" have been the cause of the forming of oceans, with their highly heated mantle? We have perforce to leave that question still unanswered.

Since views on large-scale horizontal movements in the lithosphere are now very common, we must once again, in concluding this section, make our attitude to them clear. Horizontal movements arise in the crust mainly as the result of a disturbance of gravitational equilibrium through vertical movements of matter. The mechanisms of the "transformation" of vertical movements into horizontal ones in the conditions of the crust were considered earlier in the sections on the origin of general crumpling and deep folding (Chaps. 12 and 13).

Our scheme does not, however, exclude "primary" horizontal movements of the lithosphere, although their scale must be very limited. The vertical rise of matter from the asthenosphere during diapirism is necessarily associated with horizontal flow to the root of the diapir and with outward flow from the top of the diapir. We associate the tension in rift belts with such horizontal flow, as should be clear from the foregoing. In this case and others the movement of the lithosphere is caused by friction of the flowing material of the asthenosphere against the floor of the lithosphere. Tension in one place supposes compression in another, so that it is also possible for "primary" compression to arise in the lithosphere.

The horizontal movements then also prove to be connected with vertical migration of matter, but only at a deeper level. These movements do not go beyond the limits of separate, local tectonic zones such as, for example, intrageosyncline or an intrageanticline, a subgeosyncline or a subgeanticline, and occur within the lithosphere, retaining their own continuity, and on the whole remaining in one spot. They have nothing in common with the horizontal shifting of isolated blocks of the lithosphere for thousands of kilometres suggested by the "new global tectonics".

The hypothetical scheme of deep processes set out above is very simple in principle.

At the bottom of the Earth's evolution is differentiation of its material, which begins in the lower mantle. From it heavy and light fractions respectively sink into the core and rise to the upper mantle. It is this light fraction that mainly interests us.

The Earth is, however, at the same time, a kind of heat engine with valves. The closed valve is the low thermal conductivity of its material, which fosters accumulation of heat in the interior before it can find release. The open valves are channels of reduced viscosity, by which strongly heated material rises from the lower mantle to the asthenosphere, causing pulsed excitation of the matter. The concrete endogenous regimes are determined by the reciprocal relationship of the asthenosphere and lithosphere.

We consider the tectonosphere of the oceans and deep seas to be mainly secondary, formed by basification of the continental tectonosphere. Basification is essentially homogenization of tectonosphere, earlier more differentiated, and is associated in our scheme with strong excitation of the asthenosphere through the same transport of deep material and heat, but which takes on an "explosive" character in these cases.

All the other features and history of endogenous processes are a consequence of the effect of the concrete conditions in which these deep processes develop.

Our conception calls for further study of the relationship between endogenous regimes and both the deep structure and the geophysical fields of the tectonosphere. It calls for study of the section of the crust below oceans, without which it is impossible to speak about their origin. Much, in the history of continents, however, is also unclear, especially in pre-Phanerozoic times. The problem of the interrelation of tectonic, magmatic, and metamorphic processes is becoming very important. All ideas about endogenous processes and regimes would acquire proper accuracy and be clarified if they could be described quantitatively.

There is an immense field of work in studying the physicochemical processes taking place in the tectonosphere. The problem of the conditions and mechanism of the differentiation and homogenization of the Earth's material is also one of prime importance.

All the matters listed above are complex problems and call for a geonomic approach that would include combined use of geological, geophysical, and geochemical techniques and of the methods of the latest advances in mechanics, physics, and physical chemistry.

The need for a comprehensive, operational approach to the study of the problems facing the Earth sciences is a basic and major difficulty, but for all that it has begun in recent years to be surmounted. Further efforts are needed so that the errors that stem from a too one-sided and simplified conception of what is happening within our planet will be overcome at the same time.

Literature

MAIN LITERATURE

Beloussov, V.V. Osnovy geotektoniki (Backgrounds of Geotectonics), Moscow, Nedra, 1975, 262 pp.

Gilbert, G.K. Lake Bonneville. *U.S. Geol. Survey, Monogr.* 1, Wash., 1890, 438 pp.

Hills, E.Sh. Elements of Structural Geology, 2nd ed. Chapman and Hall, London, 1972, 502 pp.

Khain, V.E. Obshchaya tektonika (General Tectonics), 2nd ed., Moscow, Nedra, 1973, 511 pp.

Schmidt-Thomé, P. Tektonik-Lehrbuch der Allgemeinen Geologie, Bd. II. Ferd-Enke, Stuttgart, 1972, 579 pp.

Stille, H. Grundfragen der vergleichenden Tektonik. Borntraeger, Berlin, 1924, 443 pp.

Tetyayev, M.M. Osnovy geotektoniki (Backgrounds of Geotectonics), 2nd ed. Leningrad, Geolizdat, 1941, 288 pp.

Verhoogen, J., Turner, F.Y, Fyfe, W.S. The Earth. An Introduction to Physical Geology. Holt, Rinehart, Winston, N.Y. 1970, 748 pp.

PRESENT-DAY AND RECENT OSCILLATORY MOVEMENTS

Meshcheryakov, Yu.A. Sovremennye dvizheniya zemnoi kory (Present-day Movements of the Earth's Crust), *Priroda*, No. 9, 1958, p. 15-24.

Meshcheryakov, Yu.A. Strukturnaya geomorfologiya ravninnykh stran (Structural Geomorphology of the Flat Countries). Moscow, Nauka, 1965, 340 pp.

Milanovsky, E.E. Noveyshaya tektonika Kavkaza (Recent Tectonics of the Caucasus). Moscow, Nedra, 1968, 482 pp.

Nikolaev, N.I. Neotektonika i eyo vyrazhenie v strukture i reliefe territorii SSSR (Neotectonics and its Reflection in Structure and Relief of the Territory of the USSR). Moscow, Gosgeoltekhizdat, 1962, 392 pp.

Schultz, S.S. Analiz noveishey tektoniki i relief Tyan-Shanya (Analysis of Recent Tectonics and Relief of the Tien-Shan). Moscow, Geografgiz, 1949, 223 pp.

Tsuboi, Chuji. Deformations of the Earth's Crust as Disclosed by Geodetic Measurements, In: *Ergebnisse der Kosmischen Physik*, IV, 1939, p. 106-156.

ANCIENT OSCILLATORY MOVEMENTS

Beloussov, V. V. Bolshoi Kavkaz. Opyt geotektonicheskogo analiza (The Great Caucasus. An Attempt at a Geotectonic Analysis), I-III, Moscow, 1938-1940.

Ronov, A. B. Istoriya osadkonakopleniya i kolebatelnykh dvizheniy evropeiskoi chasti SSSR (History of Sedimentation and Oscillatory Movements of the European Part of the USSR). *Proc. of the Geophys. Inst. Acad. Sci., USSR*, Moscow, 1949, 341 pp.

DEEP FAULTS

Burtman, V.S. Talaso-Ferganskiy sdvig i sdvig San-Andreas (Talas-Ferghana Transcurrent Fault and Transcurrent Fault of San-Andreas). In: *Razlomy i gorizontalnye dvizheniya zemnoi kory* (Faults and Horizontal Movements of the Earth's Crust), Moscow, Nauka, 1963, p. 128-151.

Cloos, M. The Ancient European Basement Blocks. *Trans. Am. Geophys. Union*, vol. 28, No. 1, 1948, p. 99-103.

Glubinnye razlomy (Deep Faults). Collection of papers, ed. by Y.K. Dzhevanovsky, Moscow, Nedra, 1964, 214 pp.
Glubinnye razlomy Yuzhnogo Tyan-Shanya (Deep Faults of the Southern Tien-Shan). Lvov, Lvov State University, 1973, 163 pp.
Hobbs, W.H. Repeating Patterns in the Relief and the Structure of the Land. *Geol. Soc. Am. Bull.*, vol. 22, No. 2, 1911, p. 123-176.
Peive, A. V. Glubinnye razlomy v geosinklinalnykh oblastyakh (Deep Faults in Geosynclinal Areas), *Izv. Acad. Sci. USSR, geol. ser.*, 1945, No. 5, p. 23-46.
Sonder, M. Mechanik der Erde, Schweizbart, Stuttgart, 1956, 291 pp.
Tarakanov, R.Z. Stroenie fokalnoi zony Kurilo-Kamchatskoi ostrovnoi dugi (Structure of Focal Zone of Kuril-Kamchatka Island Arc). In: *Zemnaya kora ostrovnykh dug i dalnevostochnykh morei* (Earth's Crust of Island Arcs and Far-East Seas), Moscow, Nauka, 1972, p. 215-234.

FOLDING

Ampferer, O. Über die Bewegungsbildung von Faltengebirgen. *J. d-d. K. K. Geol. Reichsanstalt*, Bd. 56, Wien, 1906, p. 539-622.
Beloussov, V.V. La Géologie Structurale, Moscow, Mir, 1978, 295 pp.
Beloussov, V.V. Skladchatost i osnovnye tipy tektonicheskikh deformatsiy (Folding and Main Types of Tectonic Deformations). *Bull. MOIP, otd. geol.*, v. 44, No. 4, 1969, p. 5-23.
Diapirism and Diapirs. Symposium. J. Braunstein and G. O'Brien, ed., *Am. Ass. Petrol. Geol. Mem.*, No. 8, Tulsa, 1968.
Heim, A. Geologie der Schweiz, Bd. II., H. 1, 1921; H. 2, 1922, 1019 pp.
Lebedeva, N. B. A Model of a Folded Zone. *Tectonophysics*, vol. 7, No. 4, 1969, p. 339-351.
Ocherki strukturnoi geologii slojno-dislotsirovannykh tolshch (Essays on Structural Geology of Highly Deformed Rocks). *Coll. of papers*, V. Beloussov, V. Ez., ed., Moscow, Nauka, 1970, 304 pp.
Ramberg, H. Gravity, Deformation and the Earth's Crust, as Studied by Centrifuged Models. Acad. Press, London, 1967.
Skladchatye deformatsii zemnoi kory, ikh tipy i mekhanizm obrazovaniya (Folded Deformations of the Earth's Crust, Their Types and Mechanism of Formation). *Coll. of papers*, V. Beloussov a. I. Kirillova, ed., Moscow, Ac. Sci. USSR, 1962, 335 pp.
Sorsky, A.A. Tipy geosinklinalnoi skladchatosti Kavkaza i usloviya ikh formirovaniya (Types of Geosynclinal Folding of the Caucasus and Conditions of Their Origin). In: *Skladchatye oblasti Evrazii* (Folded Zones of Eurasia). Moscow, Nauka, 1964, p. 303-317.

ENDOGENOUS CONTINENTAL REGIMES

Beloussov, V.V. Zemnaya kora materikov (Earth's Crust of Continents). Moscow, Nauka, 1966, 123 pp.
Beloussov, V.V. Endogennye rezhimy materikov (Endogenous Regimes of Continents). Moscow, Nedra, 1978, 232 pp.
Beloussov, V.V. Basic Trends in the Evolution of Continents. *Tectonophysics*, vol. 13, No. 1-4, 1972, p. 95-117.
Cloos, H. Hebung-Spaltung-Volkanismus. *Geol. Rundschau*, Bd. 30, H. 4A, 1939, p. 405-527.
Garetzky, R.G. Tektonika molodykh platform Evrazii (Tectonics of Young Platforms of Eurasia). Moscow, Nauka, 1972, 299 pp.
Illies, Y.H. The Rhine Graben Rift System. *Geophys. Surveys*, Vol. 1, No. 1, 1972, p. 27-60.
Kay, M. North American Geosynclines. *Geol. Soc. Am. Mem.*, 48, No. 9, N.Y. 1951, 143 pp.
Khain, V.E., Muratov, M.V. Crustal Movements and Tectonic Structure of Continents, *Geophys. Mon.* 13, *Am. Geoph. Un.*, 1969, p. 523-538.
Ronov, A.B. Nekotorye obshchie zakonomernosti razvitiya kolebatelnykh dvijenii materikov (po dannym obyomnogo metoda) [Some General Trends in Development of the Oscillatory Movements of Continents (On 'Volume' Methods)]. *Problemy tektoniki*, Moscow, Gosgeoltekhizdat, 1961, p. 118-164.
Ronov, A.B., Migdissov A.A., Barskaya N.B. Zakonomernosti razvitiya osadochnykh porod i paleogeograficheskykh usloviy sedimentatsii na Russkoi platforme (opyt ka-

lichestvennogo issledovania) [Regularity in Development of Sedimentary Rocks and of Palaeogeographic Conditions of Sedimentation on Russian Platform (An Attempt at Quantitative Study)]. *Litologiya i poleznye iskopaemye*, 1969, No. 6, p. 3-36.
Steinmann, G. Die Ophiolitischen Zonen in den Mediterranean Kettengebirgen. *Congrès Géol. Intern., XIV Sess.* Espagne, 1926, fasc, II, Madrid, 1927, p. 637-668.
Zorin, Y.A. Noveishaya structura i izostazia Baikalskoi riftovoi zony i sopredelnykh territorii (Recent Structure and Isostasy of the Baikal Rift Zone and Adjacent Territories). Moscow, Nauka, 1971, 168 pp.

TECTONICS OF THE OCEANS

Beloussov, V.V. Zemnaya kora i verkhnyaya mantiya okeanov (Earth's Crust and Upper Mantle of Oceans). Moscow, Nauka, 1968, 255 pp.
Drake, Ch.L. Continental Margins. *Geophys. Mon*, 13, P. Y. Hart, ed., Am. Geophys. Un. Wash., 1969, p. 549-556.
Heezen, B.C. Inland and Marginal Seas. The Upper Mantle. Developments in Geotectonics, 4. Ellsevier Co., 1972, p. 293-308.
Maxwell, Y.C. The Mediterranean Ophiolites and Continental Drift. The Mega-tectonics of Continents and Oceans. H. Yohnson, ed., Rutgers Univ. Press., 1969, p. 167-193.
Vogt, P.P., Schneider, E.D., Yohnson, G.L. The Crust and Upper Mantle Beneath the Sea. *The Earth's Crust and Upper Mantle, Geophys. Monograph*, 13, P. Y. Hart, ed., Am. Geophys. Union, Wash., 1969, p. 556-617.

THE EARTH'S DEEP STRUCTURE

Artemiev, M.E. Izostaticheskie anomalii sily tyazhesti i nekotorye voprosy ikh geologicheskogo istolkovaniya (Isostatic Anomalies of Gravity Forces and Some Questions of Their Geologic Interpretation). Moscow, Nauka, 1966, 138 pp.
Artyushkov, E.V. Gravitatsionnaya konvektsiya v nedrakh zemli (Gravitational Convection in the Interior of the Earth). *Fizika Zemli*, 1968, No. 9, p. 3-17.
Cook, K.L. The Problem of the Mantle-crust Mix: Lateral Inhomogeneities in the Earth's Mantle. *Advances in Geophysics*, vol. 9, 1962, p. 245-360.
Heezen, B.C., Gray, C., Segre, A.G., Zarudskii, E.F. Evidence of Foundered Continental Crust Beneath the Central Tyrrhenian Sea. *Nature*, vol. 229, No. 5, 283, 1971, p. 327-329.
Lee, W.H.K. On the Global Variations of Theoretical Heat-flow. *Phys. of the Earth and Planetary Interiors*, vol. 2, No. 5, 1970, p. 332-341.
Lyubimova, E.A. Termics of the Earth and the Moon. Moscow, Nauka, 1968, 278 pp.
Magnitsky, V.A. Vnutrennee stroenie i fizika Zemli (Internal Structure and Physics of the Earth). Moscow, Nedra, 1965, 379 pp.
Ringwood, A.E. Composition and Evolution of the Upper Mantle. The Earth's Crust and Upper Mantle, *Geophys. Mon.* 13, P. Hart, ed., Am. Geophys. Un., 1969, p. 1-17.
Sobolev, V.S. Stroenie verkhnei mantii i sposoby obrazovaniya magmy (Structure of the Upper Mantle and Ways of Formation of Magmas). In: *V.I. Vernadsky's Readings*, XIII, Moscow, Nauka, 1973, 34 pp.
Vinogradov, A.P. O proiskhozhdenii veshchestva zemnoi kory (On Origin of the Substance of the Earth's Crust). Geonomia, 1961, No. 1, p. 3-29.
Upper Mantle, A.R. Ritsema, ed., *Developments in Geotectonics*, 4, Elsevier, 1972, 637 pp.

GENERAL VIEWS

Beloussov, V.V. Against the Hypothesis of Ocean-floor Spreading. *Tectonophysics*, 9 (1970), p. 489-511.
Beloussov, V.V. Die Tektonosphäre der Erde, Idee und Wirklichkeit, *Zeitsch. f. geol. Wissensch äften*, H. 11, 1974, Berlin, s. 1251-1285.
Beloussov, V.V. Gravitational Instability and the Development of the Structure of Continents (Attempt at a Synthesis). *Energetics of Geological Processes.* S. Safena and S. Bhattacharji, ed., Springer-Verlag, N.Y., 1977, p. 3-18.
Coleman, R.G. Plate Tectonic Emplacement of Upper Mantle Peridotites Along Continental Edges. *Journ. Geophys. Res.*, 76, No. 5, 1971, p. 1212-1222.
Dewey, Y.F., Bird, I. Plate Tectonics and Geosynclines. *Tectonophysics*, 10, No. 5/6, 1970, p. 625-638.

Dewey, Y.F., Bird, Y.M. Mountain Belts and the New Global Tectonics. *Journ. Geophys. Res.*, **75**, No. 14, 1970.

Dietz, R.S. Continent and Ocean Basin Evolution by Spreading of the Sea Floor. *Nature*, **190**, No. 4779, 1961, p. 854-857.

Isacks, B., Oliver, J., Sykes, L.R. Seismology and the New Global Tectonics. *Journ. Geophys. Res.*, **73**, No. 18, 1968.

Karig, D.E. Origin and Development of Marginal Basins in the Western Pacific. *Journ. Geophys. Res.*, **76**, No. 11, 1971, p. 2542-2561.

Le Pichon, X. Marine Magnetic Anomalies, Geomagnetic Field Reversals and Motions of the Ocean Floor and Continents. *Journ. Geophys. Res.*, **73**, No. 6, 1968, p. 2119-2136.

Vine, F.Y., Matthews, D.H. Magnetic Anomalies Over Oceanic Ridges. *Nature*, **199**, No. 4897, 1963, p. 947-949.

Subject Index*

Ancient oscillatory movement(s) 13, 28, 66
Anticlinoria 155
Arch 157
Asthenosphere 270

Block-undulatory oscillatory movement(s) 62, 63, 66, 103
Boudinage 115
Brachyanticlinal structure(s) 90

Central uplift 155
Continental rise 210, 211
Continental shelf 210, 211
Continental slope 210, 211
Crust, types of 259
— continental 259
— oceanic 262
— subcontinental 263
— suboceanic 262
Crustal tectonic movement(s) 8
Cycle(s)
— Alpine 147
— Baikalian 147

Deep diapir 114, 119
Deep oceanic basin(s) 225
Deep trench 212
Density inversion 121
Denudation surface(s) 22
Diaphthoresis 89
Diapir
— granite 119
— granite-gneiss 119
— metamorphic 119
— surficial 121
Diapiric fold(s) 90
Disjunctive tectonic movement(s) 8, *69*
Dome-shaped structure(s) 90
Dyke
— Breven 81
— Dolerite 81
— Great Rodesian 81

Earth, age of 258
Earth, composition of 254
Earth's heat regime 247
Earth's magnetic field 250
Earth's rotation 241
Earth's seismicity 243
East European platform 65
Epeirogenesis 5
Epeirogenic tectonic movement(s) 5
Epigeosynclinal orogenesis 201
Epiplatform orogenesis *154*
Eustatic fluctuation(s) 18

Facies, metamorphic
— amphibolite 117
— eclogite 117
— granulite 117
— greenschist 117
Fault(s) classification of 69, 70
— deep 69, 70
— intracrustal 124
— normal 70, 80
— open (*see* Separations) 129
— reverse 70
— shear 74, 132
— slash 69, 70, 71, 79, 131
— thrust 129
— *transcurrent 73*, 74
Fault plane 128

* Key words and page numbers set in *italics* refer to the headings.

Folding
— *block* 8, *84*, 101; 119
— deep 84
— *injection* 8, 84, *90*, 101, 119
Folding of general crumpling 84, *101*, *103*, 120
— *origin of crumpling folding 103*
Foliation 115
Foredeep(s) 203
Formation of flysch 141
Formation of sedimentary facies 46
Formation of thicknesses 43

General oscillation(s) 62
Geosynclinal belt 186
Geosynclinal region 145
Geosynclinal system 145
Geosyncline
— Alpine 70
— Hercynian 70
Geothermal gradient 117, 249
Graben(s) 157
— Red Sea 158
Gradients of recent oscillatory movements 26
Granitization 117, 120, *136*, 262

Heat regime of tectonosphere 272
Horst-anticlinorium 201

Impermeability of the crust *136*
Intrageanticline(s) 138, 141, 145
Intrageosyncline(s) 138, 141, 145
Inverted synclinoria 155
Island arc
— type-I 168
— type-II 170

Joint system(s) 135

Layer(s) of oceanic crust
— first *218*
— Golitsyn 256, 265
— granitic-gneissic 261
— granulitic-basic 262
Leptogeosyncline(s) 138
Lithofacies 46

Macro-oscillatory movement(s) 8
Magmatic activization 157
Magmatic movement(s) 8
Magmatism *136*
Marginal trough(s) 141
Metamorphism 119, 120, 121
Method of facies 57
Method of thicknesses 54
Method of volume 60
Mid-ocecnic ridge(s) 225
Molasses 151

Nappe(s) 95, 109
— Helvetian 111, 114
— Pennine 112, 114, 122
New global tectonics 291

Orogenic tectonic movement(s) *5*
Oscillatory tectonic movement(s) 8
Overthrust (*see* Faults)

Parageosyncline(s) 55
Permeability of the crust *136*
— concentrated 136
— diffused 136
Planation surface(s) 22
Posthumous trough(s) 150
Present-day oscillatory movement(s) *13*, *14*
Protogeosyncline 184
Protoplatform 184

Rapakivi granite(s) 184
Recent oscillatory movement(s) 13, *20*, 24
Regime(s) 136
— continental margin 137, 165
— — Atlantic type 137, 165, 168, 171
— — Pacific type 137, 165, 168, 171
— epigeosynclinal 137, 151, 154
— epiplatformal 137, 151
— eugeosynclinal 137, 138, 145
— *geosynclinal* 136, 137, *138*
— marginal 137
— miogeosynclinal 137, *142*, 145
— of ancient platforms 137, *147*
— of central intrusions 137, 147
— of magmatic platforms activization 137, *163*
— of median masses 137, *143*
— of young platforms 137, *147*
— orogenic 137, 146, *150*, 308

— orthogeosynclinal 137
— parageosynclinal 137, *142*, 145
— plateau-basalt 137, *163*
— *platform* 137, *146*
— *rift* 137, *157*
Regional metamorphism 117, *136*, 141
Reverse block fold 119
Ridge(s)
— aseismic 213
— mid-oceanic 213
Riecke's principle 118
Rift valley(s) 81
Rock(s)
— granitized 115, 119
— igneous 118
— metamorphosed 115
— migmatised 115
— sedimentary 118

Seamount 215
Second layer of the oceanic crust *222*
Sedimentary formation(s) 67
Seismic waveguide 265
Separation(s) 129
Serpentization 89
Shear 129
— lateral *131*
— vertical *131*
Slash 89
Stage
— ophiolitic *138*
— orogenic 150
— permobile 183
— protogeosynclinal 184
Subduction 134
Subgeanticline 146, 147
Subgeosyncline 146, 147
Surface tectonic movement(s) 8
Synclinoria 155
Syneclise 155

Tectonic deformation 115
Tectonic nappes of deep or Pennine type 111
Tectonosphere 257
Trough(s)
— foredeep 153
— intermontane 153
— marginal 153
Types of tectonic movement 8

Ultrametamorphism 118, 119
Undulatory oscillatory movement 8, 62, 66

Waveguide(s) 265

Zone(s)
— Benioff 79, 80
— crush 71
— folded 142, 154
— rift 128, 133
Zoning of continents 201
— *by types of endogenous regime*
— *tectonic 202*

Suggested Further Readings

A. Shcheglov, D.Sc.

Fundamentals of Metallogenic Analysis

19 figures, 12 tables, 335 pages. 1979
Cloth, 13 × 20 cm.

The book deals with general problems of the metallogenic analysis. It is intended for broad sections of specialists and students of higher educational institutions studying deposits of essential minerals, conditions of their formation and regularity of distribution in the earth's crust.
Contents. **Metallogeny—an Independent Branch of Geological Science.** Methodological and Methodical Fundamentals of Regional Metallogenic Analysis. Basic Regularities Governing Distribution of Ore Deposits in Space and Time. Regional Metallogenic Zonality. Study of Ore Deposits by Using Regional Metallogenic Analysis. Regional Geochemistry and Metallogenic Analysis. Regional Geophysics and Metallogenic Analysis. Erosional Section and Long-Range Valuation of Ore-Bearing Territories.
Major Structural Elements of the Earth's Crust (Metallogeny of Regions). On Principal Structural Elements of the Earth's Crust. Metallogeny of Geosynclinal-Folded Regions. The Metallogeny of Platforms. Metal-

Mir
Publishers
Moscow

logeny of Regions of Autonomous Tectonic-Magmatic Activization. Metallogeny of Median Masses. Metallogeny of the Seas and Oceans Floor. New Global Tectonics and Metallogeny. Further Trends in the Development of Metallogenic Analysis and Tasks Facing Regional Metallogeny. Research Conclusion. Bibliography. Subject Index.

Yu. M. Vasiliev, V. S. Milnichuk,
M. S. Arabaji

General and Historical Geology

280 figures, 35 tables, 470 pages. 1981
Cloth, 13 × 20 cm.

A basic introduction to general and historical geology, the scientific and practical significance of the geological sciences, and the techniques of studying geological processes and phenomena. Discusses various hypotheses of the origin of the solar system and the Earth, of the internal structure and geotectonics of Earth, describes the techniques of reconstructing the past of the Earth's crust, provides a background on palaeontology and its techniques, and gives a detailed description of the development of the crust in the Precambrian, Early Palaeozoic, Late Palaeozoic, Mesozoic, and Cainozoic eras.
Is intended as a textbook for students of the non-geological departments of oil, mining, and civil engineering faculties, and the geological prospecting departments of technical colleges. The authors lecture at the Gubkin Institute of Petrochemistry and Gas Industry in Moscow.
Contents. The Earth as a Planet. The Processes of the Earth's External Dynamics. Processes of the Earth's Internal Dynamics. Reconstruction of the Geological Past of the Crust. Principles of Palaeontology. Main Eras in the Evolution of the Crust.

Mir
Publishers
Moscow